D1643912

30107 004 458 417

Alan K. Mitchell
Pasi Puttonen
Michael Stoehr
Barbara J. Hawkins
Editors

Frontiers of Forest Biology: Proceedings of the 1998 Joint Meeting of the North American Forest Biology Workshop and the Western Forest Genetics Association

Frontiers of Forest Biology: Proceedings of the 1998 Joint Meeting of the North American Forest Biology Workshop and the Western Forest Genetics Association has been co-published simultaneously as *Journal of Sustainable Forestry*, Volume 10, Numbers 1/2 and 3/4 2000.

Pre-publication REVIEWS, COMMENTARIES, EVALUATIONS . . .

"In the preface to this book the editors suggest that 'the emphasis in forest research is shifting from productivity-based goals to goals related to sustainable development.' It is clear, from the contributions contained within, that this is beginning to happen, but it is also clear that there is a long way to go. A very wide range of research topics are approached in the various articles that

More pre-publication
REVIEWS, COMMENTARIES, EVALUATIONS . . .

comprise these proceedings, but there is clear emphasis on the genetics, genecology, and physiology of trees, particularly temperate trees. To be fair, there is also quite a noticeable effort to fill the obvious information gaps that exist for non-timber tree species, such as paper birch. . . .

On the other hand, these proceedings are also testimony to what does or should distinguish forest biology from other sciences: a focus on intra- and inter-specific interactions between forest organisms and their environment, over scales of both time and space. This evolution or maturation of the discipline (or 'inter-discipline') is evident in some . . . repeating themes that weave through many of the contributions. Included here is an emphasis on genotype by environment interactions, phenotypic plasticity, spatial and seasonal aspects of physiology, biotic influences on nutrient cycling, system response to disturbance, and variability in species- and population-level responses to environmental stress (e.g., aluminum, root disease, frost, UV-B, and drought).

Most of the individual contributions have been kept concise and to the point, and they all meet a uniformly high editorial and scientific standard. These are very focussed bite-sized morsels of valuable information. The longer articles, still very efficiently packaged, provide stimulating overviews of contemporary issues in, for example, conservation genetics and below ground processes. I found these to be the most enlightening. As a physiologist, however, I was also keenly interested in those more specific papers that involved new techniques or dealt with topics that have received only passing attention in the past (e.g., stem respiration, respiratory efficiency, and winter photosynthesis)."

Robert D. Guy, PhD
Associate Professor
Department of Forest Sciences
University of British Columbia

More pre-publication
REVIEWS, COMMENTARIES, EVALUATIONS . . .

"As an outgrowth of the June 1998 joint meeting of the North American Forest Biology Workshop and the Western Forest Genetics Association, an edited volume containing many of the papers given at the conference has been edited by four scientists under the leadership of Alan K. Mitchell. . . . Forty-two papers have been divided into four sections: Genetics, Physiology, Silviculture, and Conservation. Most papers are quite short: however, two excellent reviews are found in the book. Topics range from a review of *in situ* gene conservation of forest trees to management schemes to reduce the impacts of the exotic weed, Scotch broom . . .

In the review on gene conservation, considerations of adaptability, the levels of variation associated with trees, the relative long life cycle of trees, climate change and the fragmented nature of the landscape in which *in situ* reserves would be placed are all discussed. . . .

The next section on physiology contains 13 chapters: this section . . . links with the previous section and is diverse with papers spanning from the effects of root disease on the water relations of *Pinus strobus* to winter-time whole-tree and stand transpiration in *P. taeda* and *P. elliottii*. . . .

There are a number of papers which deal with belowground biology and function. . . . in the book, a number of papers integrate various aspects of physiology, belowground biology and silviculture. . . .

One of the topical issues that emerged from the workshop dealt with the sustainability of forest practices and the importance of not affecting site productivity during silvicultural manipulations. . . .

A series of papers follows which are largely silvicultural in nature. . . .

The last section of the book deals with issues of conservation and restoration biology. . . . Each of these chapters highlights critical features associated with the canopies of old-growth trees. Each of these chapters highlights critical features associated with understanding and maintaining diverse and healthy ecosystems. . . . The book ends with a paper emphasizing biodiversity in canopies and, by implication, the important of having such canopies. In a sense the book has come full circle since the first paper emphasized the important of having *in situ* reserves for maintaining gene diversity."

Thomas M. Hinckley, PhD
Professor/Chair
College of Forest Resources
University of Washington
Seattle, WA

Frontiers of Forest Biology: Proceedings of the 1998 Joint Meeting of the North American Forest Biology Workshop and the Western Forest Genetics Association

Frontiers of Forest Biology: Proceedings of the 1998 Joint Meeting of the North American Forest Biology Workshop and the Western Forest Genetics Association has been co-published simultaneously as *Journal of Sustainable Forestry*, Volume 10, Numbers 1/2 and 3/4 2000.

The *Journal of Sustainable Forestry* Monographic "Separates"

Below is a list of "separates," which in serials librarianship means a special issue simultaneously published as a special journal issue or double-issue *and* as a "separate" hardbound monograph. (This is a format which we also call a "DocuSerial.")

"Separates" are published because specialized libraries or professionals may wish to purchase a specific thematic issue by itself in a format which can be separately cataloged and shelved, as opposed to purchasing the journal on an on-going basis. Faculty members may also more easily consider a "separate" for classroom adoption.

"Separates" are carefully classified separately with the major book jobbers so that the journal tie-in can be noted on new book order slips to avoid duplicate purchasing.

You may wish to visit Haworth's website at . . .

http://www.haworthpressinc.com

. . . to search our online catalog for complete tables of contents of these separates and related publications.

You may also call 1-800-HAWORTH (outside US/Canada: 607-722-5857), or Fax 1-800-895-0582 (outside US/Canada: 607-771-0012), or e-mail at:

getinfo@haworthpressinc.com

Frontiers of Forest Biology: Proceedings of the 1998 Joint Meeting of the North American Forest Biology Workshop and the Western Forest Genetics Association, edited by Alan K. Mitchell, Pasi Puttonen, Michael Stoehr, and Barbara J. Hawkins (Vol. 10, No. 1/2 & 3/4, 2000). *Based on the 1998 Joint Meeting of the North American Forest Biology Workshop and the Western Forest Genetics Association, Frontiers of Forest Biology addresses changing priorities in forest resource management. You will explore how the emphasis of forest research has shifted from productivity-based goals to goals related to sustainable development of forest resources. This important book contains fascinating research studies, complete with tables and diagrams, on topics such as biodiversity research and the productivity of commercial species that seek criteria and indicators of ecological integrity.*

"There is clear emphasis on the genetics, genecology, and physiology of trees, particularly temperate trees. . . . These proceedings are also testimony to what does or should distinguish forest biology from other sciences: a focus on intra- and inter-specific interactions between forest organisms and their environment, over scales of both time and place." (Robert D. Guy, PhD, Associate Professor, Department of Forest Sciences, University of British Columbia, Vancouver, Canada)

Contested Issues of Ecosystem Management, edited by Piermaria Corona and Boris Zeide (Vol. 9, No. 1/2, 1999). *Provides park rangers, forestry students and personnel with a unique discussion of the premise, goals, and concepts of ecosystem management. You will discover the need for you to maintain and enhance the quality of the environment on a global scale while meeting the current and future needs of an increasing human population. This unique book includes ways to tackle the fundamental causes of environmental degradation so you will be able to respond to the problem and not merely the symptoms.*

Protecting Watershed Areas: Case of the Panama Canal, edited by Mark S. Ashton, Jennifer L. O'Hara, and Robert D. Hauff (Vol. 8, No. 3/4, 1999). *"This book makes a valuable contribution to the literature on conservation and development in the neo-tropics. . . . These writings provide a fresh yet realistic account of the Panama landscape." (Raymond P. Guries, Professor of Forestry, Department of Forestry, University of Wisconsin at Madison, Wisconsin)*

Sustainable Forests: Global Challenges and Local Solutions, edited by O. Thomas Bouman and David G. Brand (Vol. 4, No. 3/4 & Vol. 5, No. 1/2, 1997). *"Presents visions and hopes and the challenges and frustrations in utilization of our forests to meet the economical and social needs of communities, without irreversibly damaging the renewal capacities of the world's forests." (Dvoralai Wulfsohn, PhD, PEng, Associate Professor, Department of Agricultural and Bioresource Engineering, University of Saskatchewan)*

Assessing Forest Ecosystem Health in the Inland West, edited by R. Neil Sampson and David L. Adams (Vol. 2, No. 1/2 & 3/4, 1994). *"A compendium of research findings on a variety of forest issues. Useful for both scientists and policymakers since it represents the combined knowledge of both." (Abstracts of Public Administration, Development, and Environment)*

Frontiers of Forest Biology: Proceedings of the 1998 Joint Meeting of the North American Forest Biology Workshop and the Western Forest Genetics Association

Alan K. Mitchell
Pasi Puttonen
Michael Stoehr
Barbara J. Hawkins
Editors

Frontiers of Forest Biology: Proceedings of the 1998 Joint Meeting of the North American Forest Biology Workshop and the Western Forest Genetics Association has been co-published simultaneously as *Journal of Sustainable Forestry*, Volume 10, Numbers 1/2 and 3/4 2000.

Food Products Press
An Imprint of
The Haworth Press, Inc.
New York • London • Oxford

Published by

Food Products Press, 10 Alice Street, Binghamton, NY 13904-1580 USA

Food Products Press is an imprint of The Haworth Press, Inc., 10 Alice Street, Binghamton, NY 13904-1580 USA.

Frontiers of Forest Biology: Proceedings of the 1998 Joint Meeting of the North American Forest Biology Workshop and the Western Forest Genetics Association has been co-published simultaneously as *Journal of Sustainable Forestry*, Volume 10, Numbers 1/2 and 3/4 2000.

Cover design by Thomas J. Mayshock Jr.

Library of Congress Cataloging-in-Publication Data

Joint Meeting of the North American Forest Biology Workshop and the Western Forest Genetics Association (1998: University of Victoria)

Frontiers of forest biology: proceedings of the 1998 Joint Meeting of the North American Forest Biology Workshop and the Western Forest Genetics Association / Alan K. Mitchell . . . [et al.].

p. cm.

Held June 21 to 26, 1998 at the University of Victoria.

"Frontiers in forest biology . . . has been co-published simultaneously as Journal of sustainable forestry, volume 10, numbers 1/2 and 3/4 2000."

Includes bibliographical references.

ISBN 1-56022-070-8 (alk. paper)–ISBN 1-56022-079-1 (alk. paper)

1. Forests and forestry–North America–Congresses. 2. Forest management–North America–Congresses. 3. Forest genetics–North America–Congresses. 4. Forest ecology–North America–Congresses. 5. Forest conservation–North America–Congresses. I. Mitchell, Alan Kenneth, 1950-II. Western Forest Genetics Association. Meeting (1998: University of Victoria) III. North American Forest Biology Workshop (1998: University of Victoria) IV. Title.

SD140.J65 1998
634.9–dc21 00-027916

INDEXING & ABSTRACTING

Contributions to this publication are selectively indexed or abstracted in print, electronic, online, or CD-ROM version(s) of the reference tools and information services listed below. This list is current as of the copyright date of this publication. See the end of this section for additional notes.

- *Abstract Bulletin of the Institute of Paper Science and Technology*
- *Abstracts in Anthropology*
- *Abstracts on Rural Development in the Tropics (RURAL)*
- *AGRICOLA Database*
- *Biostatistica*
- *BUBL Information Service, an Internet-based Information Service for the UK higher education community, <URL:http://bubl.ac.uk/>*
- *CNPIEC Reference Guide: Chinese National Directory of Foreign Periodicals*
- *Engineering Information (PAGE ONE)*
- *Environment Abstracts. Available in print-CD-ROM–on Magnetic Tape. For more information check: www.cispubs.com*
- *Environmental Periodicals Bibliography (EPB)*
- *FINDEX, www.puplist.com*
- *Forestry Abstracts; Forest Products Abstracts (CAB Abstracts), www.cabi.org/*
- *GEO Abstracts (GEO Abstracts/GEOBASE)*
- *Human Resources Abstracts (HRA)*
- *Journal of Planning Literature/Incorporating the CPL Bibliographies*
- *Referativnyi Zhurnal (Abstracts Journal of the All-Russian Institute of Scientific and Technical Information)*
- *Sage Public Administration Abstracts (SPAA)*
- *Sage Urban Studies Abstracts (SUSA)*
- *Wildlife Review*

(continued)

Special Bibliographic Notes related to special journal issues (separates) and indexing/abstracting:

- indexing/abstracting services in this list will also cover material in any "separate" that is co-published simultaneously with Haworth's special thematic journal issue or DocuSerial. Indexing/abstracting usually covers material at the article/chapter level.
- monographic co-editions are intended for either non-subscribers or libraries which intend to purchase a second copy for their circulating collections.
- monographic co-editions are reported to all jobbers/wholesalers/approval plans. The source journal is listed as the "series" to assist the prevention of duplicate purchasing in the same manner utilized for books-in-series.
- to facilitate user/access services all indexing/abstracting services are encouraged to utilize the co-indexing entry note indicated at the bottom of the first page of each article/chapter/contribution.
- this is intended to assist a library user of any reference tool (whether print, electronic, online, or CD-ROM) to locate the monographic version if the library has purchased this version but not a subscription to the source journal.
- individual articles/chapters in any Haworth publication are also available through the Haworth Document Delivery Service (HDDS).

Frontiers of Forest Biology: Proceedings of the 1998 Joint Meeting of the North American Forest Biology Workshop and the Western Forest Genetics Association

CONTENTS

ABOUT THE EDITORS

Alan K. Mitchell, PhD, is Research Physiologist with the Canadian Forest Service based at the Pacific Forestry Centre in Victoria, British Columbia, Canada. He holds Bachelors and Masters degrees in Biology from the University of Victoria, Victoria, BC and a PhD in Forest Resources from the University of Washington, Seattle, Washington. He conducts a program of physiological research into the effects of forestry practices on ecosystem processes and is engaged in the development of physiological and morphological indicators of stress in shade-tolerant conifers. These indicators are being applied to comparisons of alternative silvicultural systems for sustainable productivity in coastal montane forests in British Columbia.

Pasi Puttonen, PhD, is Professor of Silviculture at the University of Helsinki, Helsinki, Finland. He holds a Bachelor of Forestry from the University of Helsinki, Masters degrees in Forestry from the University of Helsinki and from Oregon State University, Corvallis, Oregon, USA, and a Doctorate in Agriculture and Forestry from University of Helsinki. He is Registered Professional Forester in Finland and in British Columbia. His research interests are in physiological mechanisms underlying tree growth and development under alternative silviculture systems and young stand development. He has held the position of Research Leader, Forest Production Processes and Technical Advisor in the Silvicultural Systems Research and Development Program, BC Ministry of Forests, Research Branch, Victoria, British Columbia. He is currently developing a program of research into development and silviculture of young, heterogenous stands in Finland at the University of Helsinki.

Michael Stoehr, PhD, is Research Geneticist with the Research Branch of British Columbia Ministry of Forests based in Victoria, British Columbia, Canada. He holds Bachelor and Masters degrees in Forestry from Lakehead University, Thunder Bay, Ontario and a PhD from University of Toronto in Forest Genetics in 1989 and has completed two post-doctoral research terms in molecular genetics at McMaster University, Hamilton, Ontario and University of Victoria, Victoria, BC. His work involves assessment of genetic diversity in seed orchard crops, effects of seed orchard management on genetic diversity and genetic worth of orchard seedlots. The development and application of molecular markers for paternity analysis of orchard crops is currently the main emphasis of the work.

Barbara J. Hawkins, PhD, is Conifer Seedling Physiologist with the Centre for Forest Biology at the University of Victoria, Victoria, British Columbia, Canada. She holds a Bachelor of Science in Forestry degree from the University of British Columbia, Vancouver, British Columbia; and a PhD in Forestry from the University of Canterbury, Christchurch, New Zealand. Her research covers two major areas: conifer nutrition and cold hardiness. Currently, she is investigating intra- and interspecific differences in nutrient productivity in several conifer species to gain an understanding of the relative variability at these two levels, and to partition the variability into genetic and environmental components.

Preface

The theme for this meeting, Frontiers of Forest Biology, grew out of the recognition that new approaches in Forest Science will be needed to address changing priorities in forest resource management. The emphasis in forest research is shifting from productivity-based goals to goals related to sustainable development of forest resources. For example, biodiversity research has emerged from past demands for estimates of damage and loss of commercial timber to address questions about introduced and indigenous species and the loss of organisms that underpin ecological processes. Research on the productivity of commercial species has grown into studies that seek criteria and indicators of ecological integrity. We are challenged to promote a viable forest industry while assuring that non-timber values are protected.

The Keynote speakers at the meeting set the social, biological and industrial context of the Forest Biology Frontier. Plenary speakers suggested approaches by which Genetics, Physiology, Silviculture and Conservation can be moved beyond the present frontiers of forest research. Papers in this volume indicate the depth and breadth of current initiatives in forest research.

A quick analysis of strengths, weaknesses, opportunities, and threats in forest biology research reveals interesting issues. A clear strength is the large number of researchers working in this field. Without doubt, the knowledge generated by forest biology research over the last 100 or so years has contributed significantly in achieving economical timber production. There is a strong tradition in forest biology on which to build.

[Haworth co-indexing entry note]: "Preface." Mitchell, Alan K. et al. Co-published simultaneously in *Journal of Sustainable Forestry* (Food Products Press, an imprint of The Haworth Press, Inc.) Vol. 10, No. 1/2, 2000, pp. xvii-xix; and: *Frontiers of Forest Biology: Proceedings of the 1998 Joint Meeting of the North American Forest Biology Workshop and the Western Forest Genetics Association* (ed: Alan K. Mitchell et al.) Food Products Press, an imprint of The Haworth Press, Inc., 2000, pp. xvii-xix. Single or multiple copies of this article are available for a fee from The Haworth Document Delivery Service [1-800-342-9678, 9:00 a.m. - 5:00 p.m. (EST). E-mail address: getinfo@haworthpressinc.com].

The next challenge in forest biology is to maintain a relevant contribution to sustainable development. A requirement for this is the continuing development of new theories and hypotheses, their critical examination, in laboratory and field tests. Forest biology has a good tradition but opportunities for more quantitative approaches are available. A clear strength of research in forest biology is an international critical mass, well exemplified by the various nationalities attending this meeting.

Unfortunately, forestry is plagued by some weaknesses that seem to be global. Forestry and forest sciences have a poor public perception. Forest research is still regarded as a closed circle. It is not the fastest sector of the society to change when the society changes. Especially, our communication to the public leaves much to be desired. In most cases, forest research is done with public funding and thus it is only fair that we justify our work in terms of how it serves the needs of society. For example, in British Columbia, clear-sighted, well-designed research programs are essential as an underpinning for sustainable forest management in the province. But what are we going to do? Obviously, we need good extension, technology transfer and communication as a part of those programs, but we have not yet found a balance between extension and research.

Forest biology is an elemental and integral science that deals with natural resources and environment and has a purpose in providing knowledge for forest management. At the same time, we can advocate the stewardship of the forests. The role of science in forest management in the 21st century will be much greater than it is today. In forest biology, we are fortunate to have several opportunities. For example, opportunities are available for a better integration of ongoing research with graduate, post-graduate and post-doctorate studies. This integration would enlarge the critical mass needed for new findings. Another challenge and opportunity is how to integrate forest biology into the whole forest sector. To achieve this, perhaps we should think of products before forests. Future challenges and opportunities may lie in a better integration of forest biology with the whole forest sector, including forestry practices, the mechanical and chemical forest industry, machinery and equipment, and technology and information systems. This integration would increase the relevance of forest biology by broadening the knowledge base serving the forest sector. At the same time, the science of forest biology should also address social issues. This would serve to develop public confidence that forest research can provide answers to questions concerning the sustainability of our forest resources.

Alan K. Mitchell
Canadian Forest Service
Pacific Forestry Centre
Victoria, B.C.

Pasi Puttonen
University of Helsinki
Helsinki, Finland

Michael Stoehr
British Columbia Ministry of Forests
Research Branch
Victoria, B.C.

Barbara J. Hawkins
University of Victoria
Centre for Forest Biology
Victoria, B.C.

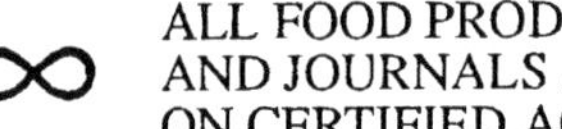

ALL FOOD PRODUCTS PRESS BOOKS
AND JOURNALS ARE PRINTED
ON CERTIFIED ACID-FREE PAPER

Acknowledgment

The 1998 Joint Meeting of the North American Forest Biology Workshop and the Western Forest Genetics Association was realized with the support of the Canadian Forest Service, the British Columbia Ministry of Forests, the University of Victoria, Centre for Forest Biology, and the Society of American Foresters. This meeting could not have been conducted without the commitment and energy of the local organizing committee, Al Mitchell and Pasi Puttonen (Co-Chairs), Michael Stoehr, Barry Jaquish, Barbara Hawkins, Jack Woods, John King, and Jill Peterson. The committee also thanks Pat McGuire (Conference Management, University of Victoria) for her organizational skills and foresight. Special recognition is also due the session moderators and volunteers who contributed significantly to the success of the conference.

Thanks are extended to the authors and reviewers whose contributions made this volume possible and to Tina Kerr who assisted in the compilation of this Proceedings.

Alan K. Mitchell
Canadian Forest Service
Pacific Forestry Centre
Victoria, B. C.

Pasi Puttonen
University of Helsinki
Helsinki, Finland

Michael Stoehr
British Columbia Ministry of Forests
Research Branch
Victoria, B. C.

Barbara J. Hawkins
University of Victoria
Centre for Forest Biology
Victoria, B. C.

PART ONE: FRONTIERS OF FOREST GENETICS

Plenary Address: Conserving Adaptive Variation in Forest Ecosystems

S. N. Aitken

INTRODUCTION

The primary goal of *in situ* gene conservation is to maintain or enhance genetic potential for adaptation to future conditions. Despite increased understanding of theoretical issues surrounding gene conservation, some very basic questions persist about *in situ* conservation. How should natural populations be characterized genetically for conservation purposes? What hierarchical levels of genetic organization

---

S. N. Aitken is affiliated with the Department of Forest Sciences, University of British Columbia, 3041-2424 Main Mall, Vancouver, BC, Canada V6T 1Z4 (E-mail: aitken@unixg.ubc.ca).

[Haworth co-indexing entry note]: "Plenary Address: Conserving Adaptive Variation in Forest Ecosystems." Aitken, S. N. Co-published simultaneously in *Journal of Sustainable Forestry* (Food Products Press, an imprint of The Haworth Press, Inc.) Vol. 10, No. 1/2, 2000, pp. 1-12; and: *Frontiers of Forest Biology: Proceedings of the 1998 Joint Meeting of the North American Forest Biology Workshop and the Western Forest Genetics Association* (ed: Alan K. Mitchell et al.) Food Products Press, an imprint of The Haworth Press, Inc., 2000, pp. 1-12. Single or multiple copies of this article are available for a fee from The Haworth Document Delivery Service [1-800-342-9678, 9:00 a.m. - 5:00 p.m. (EST). E-mail address: getinfo@haworthpressinc.com].

should be targeted for conservation and which populations selected to represent genetically similar groups? How large do populations need to be to conserve evolutionary potential? And finally, how can *in situ* reserves be managed in the face of unprecedented rates of climatic change? I will address these questions in the context of conserving genetic diversity in natural populations of widespread temperate and boreal forest trees.

UNDERSTANDING CURRENT LEVELS OF WITHIN-POPULATION VARIATION

The current amount and distribution of genetic variation provides a snapshot in evolutionary time of one stage in a dynamic process. Within-population variation for selectively neutral traits reflects the current balance between increases in genetic variance due to mutation and losses due to genetic drift. The expected equilibrium level of genetic variance is the product of the input of variance due to mutation per generation (V_m) and the effective population size, N_e (Lande, 1995). A high level of variation can result from a large N_e (and thus little genetic drift), or from high mutation rates. A rather different situation exists for adaptive traits, where the equilibrium level of genetic variance largely reflects a balance between mutation and selection, except when populations are small. For adaptive traits, the equilibrium level of genetic variance varies with the width of the fitness function (i.e., the variance equivalent of a plot of phenotype versus fitness) as well as N_e and V_m (Turelli, 1984; Lande, 1995). For traits under selection, high mutation rates, weak selection, or disruptive selection for multiple niches within a mating neighborhood (e.g., Campbell, 1979) can be invoked to explain the maintenance of high levels of genetic variation. Unfortunately, little is known about either mutation rates or the strength and nature of natural selection in tree populations.

Estimates of the per locus mutation rates for short-lived experimental organisms are typically 10^{-5} to 10^{-8} per generation, while the corresponding rates for mutations affecting quantitative (polygenic) traits in the same species are 10^{-2} to 10^{-3} per generation (Lynch, 1996). Quantitative traits have much higher effective mutation rates than single gene traits due to the large number of genetic loci affecting them, any one of which may mutate. Estimates of mutation rates for short-lived laboratory organisms may not be appropriate for extrapolation to the genomes of trees and other woody plants. The high genetic load of some forest tree species (summarized in Lynch and Walsh, 1998) may be evidence of either high mutation rates or of large historic N_e. It has long been speculated that trees may also accumulate substantial genetic variation from somatic mutations (Geber and Dawson, 1993).

Methods are needed to estimate mutation rates in trees. For example, a 'somatic mutation accumulation' study is underway at UBC to assess the contribution of somatic mutations to quantitative trait variation (S.N. Aitken, in progress). Some hypervariable genetic markers such as microsatellites, found in non-coding regions of DNA with relatively high mutation rates, may be useful for comparing mutation rates between trees and short-lived experimental organisms.

Mutation rates for quantitative traits have been used to estimate the minimum N_e needed to both conserve adaptive variation under mutation/selection balance and to avoid the accumulation of mildly deleterious mutations that collectively can reduce the fitness of small populations to the point of extinction (Lynch et al., 1995). Franklin (1980) and Soule (1980) concluded that if $N_e \approx 500$, over 90 percent of the current genetic variation for quantitative traits would be maintained under mutation/drift balance. More recently, Lande (1995) suggested that an N_e one order of magnitude larger ($\approx$5,000) is necessary to maintain variation, as only those mutations with relatively weak effects, comprising around ten percent of all mutations, will persist and contribute to genetic variance for quantitative traits. Since above $N_e \approx$ 1000, further increases in population size will not enhance the equilibrium level of genetic variation maintained, Lynch (1996) recommended conservation programs strive to maintain population sizes above this threshold. It appears that as long as N_e is in the thousands rather than the hundreds, and if trees have mutation rates similar to other organisms, genetic diversity will be conserved adequately. It should be noted that a given N_e, which assumes panmictic mating and equal contributions from all individuals to the next generation, typically requires a census number N three to ten times greater than N_e to achieve (Lynch, 1996).

There are additional considerations that may result in larger reserve sizes than that necessary for maintaining adaptive variation. Genotypes may be distributed patchily within populations, with groups of similar genotypes clustered spatially. While these patches may have similar mean phenotypes for any given adaptive trait, different genotypes can produce similar phenotypes with different allelic compositions across the loci affecting the trait. Holsinger (1993) estimated that patch sizes should be approximately 60 times the dispersal distance (estimated as the square root of the sum of dispersal distances squared), and that reserves should contain around ten patches. For example, he estimated that this would require *in situ* reserves of 3,500 to 7,000 ha containing hundreds of thousands of individuals. Large reserves are also desirable in that they will experience less pollen contamination from sexually mature plantations of genetically improved trees than smaller reserves.

CHARACTERIZING PATTERNS OF AMONG-POPULATION VARIATION

The standard tools currently used for characterizing genetic diversity and identifying 'Evolutionarily Significant Units' (ESUs) for gene conservation are molecular markers (e.g., Moritz 1994). The recent explosion in available molecular markers has provided more informative tools for characterizing intraspecific phylogenies and within-population levels of genetic variation for selectively neutral or near-neutral traits. Markers are particularly useful for identifying phylogenies at the species, subspecies or racial levels, revealing genetic differences resulting from population divergence hundreds of thousands, or even millions of years ago. However, marker variation has often been used inappropriately to make inferences about amounts and patterns of adaptive variation (Lynch, 1996; Lewontin, 1984). Rules of thumb have been proposed for determining if populations are sufficiently differentiated to justify conservation based on molecular markers (e.g., Moritz, 1994). However, these guidelines seem inappropriate for trees, as large adaptive differences can be found among populations that are very similar in terms of allele frequencies for some types of widely used markers (e.g., Aitken and Libby, 1994).

If N_e has been small historically (a few hundred or less), and populations have been relatively isolated, the amount of within-population variation and genetic differentiation among populations for both neutral and adaptive traits will be largely due to genetic drift (Lynch, 1996). In this case, assessing neutral and adaptive traits may well lead to similar conclusions regarding the genetic similarity of populations.

However, if historical effective population sizes have been large, as is the case for most temperate conifers, selectively neutral markers and adaptive traits may have both substantially different patterns of variation among populations, and different levels of within-population variation. Thus, conclusions with respect to conservation strategies will depend greatly on methods used to quantify diversity.

The evolutionary footprint of events such as bottlenecks on levels of genetic diversity will persist longer for neutral traits than for quantitative traits as the latter accumulate mutational variation much more quickly than the former (Lynch, 1996). This may be useful for assessing the rate at which mutational variance for quantitative traits accumulates. For example, western red cedar (*Thuja plicata*) has very little detectable variation for genetic markers, but significant variation both within and among populations for adaptive traits (Copes, 1981; Rehfeldt, 1993). If the timing of the Pleistocene bottleneck can be determined, populations derived from the ancestral bottlenecked population identified, and historic population sizes and growth rate estimated using genetic markers combined with paleoecological evidence, it may be possible to estimate rates of accumulation of mutational variance for adaptive traits.

How should populations be selected for conservation when marker-based genetic distances are small yet populations vary considerably for adaptive traits? Genecological studies have repeatedly found clinal variation in adaptive traits for widespread conifers that correspond to temperature and moisture regimes (e.g., Rehfeldt, 1993, 1995; Campbell and Sugano, 1979; Campbell, 1987), particularly for annual developmental cycle traits. Shoot growth phenology and cold hardiness are key traits that exhibit strong clinal patterns of variation. Traits relating to water use and drought response, including biomass allocation, vulnerability to cavitation and water use efficiency, also vary geographically in some species (e.g., White, 1987; Kavanagh et al., 1998; and Zhang et al., 1994). Selection varies among environments for these traits, and thus populations have diverged considerably for them.

For conservation purposes populations can be compared using the concept of population replaceability (S.N. Aitken, in prep.) Under this model, a population can be identified that it is capable of replacing (phenotypically) a set of similar current populations through selection, while maintaining both a large N_e (>1,000) and most (i.e., at least

95 percent) of the original within-population genetic variation. The ability to replace one population with another, phenotypically, will depend on the amount of variation for the traits of interest, the heritability of those traits, and the genetic correlations among traits. Particular interest should be paid to peripheral populations as they are more likely to contain novel genotypes and be the most difficult to replace with other populations (Lesica and Allendorf, 1995). The appropriate quantitative traits for such an analysis would be those exhibiting the strongest clinal variation as they have likely experienced the strongest differences in selection intensity and direction among environments. Bud phenology can be quickly and easily scored in existing provenance trials or seedling common-garden tests, and is a likely candidate trait for this purpose.

ADAPTATIONAL LAG

Local populations of temperate and boreal trees often do not appear to be optimally adapted to current local environments. In provenance trials, population samples originating from milder climates (e.g., one or two degrees of latitude south) often out-grow local sources without experiencing greater injury or mortality. This trend may reflect adaptation of populations to extreme, infrequent climatic events not experienced during relatively short-term provenance trials. It is more likely, however, that the trend indicates that populations are tracking but lagging behind optimum phenotypes as environments change. It is unlikely that any population will ever be optimally adapted to its current climate (Lynch and Lande, 1993), but the long generation time of many forest trees likely exacerbates 'adaptational lag' in a changing environment (Stettler and Bradshaw, 1994).

Mature forests exhibit considerable homeostasis, as illustrated by the ability of mature trees to withstand considerable environmental change over centuries or even millennia (Stettler and Bradshaw, 1994). Most mortality occurs at the seedling establishment stage, with additional competition-induced mortality after crown closure (e.g., Campbell, 1979). While this homeostasis suggests that mature forests may persist for long periods of time despite rapid environmental change, it also means that mature forests may be well enough adapted to persist, yet be unable to replace themselves, as seedlings become maladapted to new conditions. As the rate of climatic change accelerates, or generations lengthen, the magnitude of adaptational lag will increase. If a population becomes too out-of-phase with its environment, reproductive success will drop below that necessary for population replacement, and the population will decline (Lynch and Lande, 1993).

Population responses to rapid climatic change must be considered when planning genetic conservation strategies. During periods of rapid environmental change, populations have three possible fates: migration, adaptation or extinction. Migration appears to have been the most common response of plant species to climate change in the past and the strategy most likely to be successful in the future (Geber and Dawson, 1993). However, the potential for migration across managed, fragmented landscapes will be limited, particularly for woody plants that are not a high priority in management objectives, or for those with low rates of natural reproduction. Predicted rates of environmental change would require migration rates higher than the post-glacial maxima estimated from the palynological record for populations to track environments across the landscape (Huntley, 1991).

It has been suggested that large changes over time in species distributions with little associated change in morphology is evidence of a lack of adaptive response to changing environments (Huntley, 1991). However, the ability of populations to survive by adapting is evidenced by the considerable adaptive differences among populations descended from a single glacial refugium in the last ten or so millennia. For example, there is considerable variation among coastal populations from southern Oregon to British Columbia in adaptive traits such as phenology (e.g., Campbell and Sugano, 1979). Variation in physiological traits such as phenology cannot be observed in the fossil record. Whether populations can adapt rapidly enough to track environmental change depends on the rate of environmental change per generation, the amount of genetic variation for adaptive traits, and the strength of selection for these traits (Lynch and Lande, 1993). For woody plants, extensive generation lengths are likely to be a limiting factor for adaptation to a rapidly changing climate, resulting in a greater rate of change per generation. It should be recognized, however, that the average generation length would vary greatly among tree species depending on species demographics and life history characteristics such as age of sexual reproduction and frequency of disturbances.

EFFECTS OF FRAGMENTATION

The genetic structure of widespread, common species is more likely to be strongly affected by fragmentation than that of historically rare species with isolated populations (Holsinger, 1993). Species that have had relatively high rates of gene flow historically may experience reduced levels of genetic diversity as a result of reductions in effective population size, which may in turn slow responses to climate change to the extent that they are limited by genetic diversity. Paradoxically, fragmentation may also increase the opportunity for populations to adapt to environmental conditions previously outside the ecological amplitude of the species by reducing gene flow carrying maladapted alleles from interior to marginal populations, and thus allowing marginal populations to diverge (Holsinger, 1993). Under some conditions, reductions in gene flow to *in situ* reserves may allow for adaptation of these populations to conditions outside the current range (Namkoong, 1989). However, a larger reserve size becomes necessary in the absence of gene flow as gene flow increases N_e.

Fragmentation can also result in changes in mating systems that affect population fitness. For example, in a relatively isolated population of ponderosa pine in the Willamette Valley fragmented by agriculture, urbanization and forest species conversions in the last century, the outcrossing rate in six stands averaged just 60 percent, compared to 90 percent elsewhere in the range of this species (Gooding, 1998). As ponderosa pine exhibits considerable inbreeding depression (Sorensen and Miles, 1974), this change in mating system could considerably reduce the mean fitness of populations and further reduce N_e.

GENETIC ASPECTS OF IN SITU RESERVE MANAGEMENT

The long-term management of *in situ* reserves for genetic conservation purposes will involve more than simply placing reserve boundaries on maps and eliminating resource extraction from these areas. There will be little room for migratory responses to changing environments in many reserves, except possibly for those spanning considerable elevational ranges. Migration between reserves will be hindered by land use and management objectives. How, then, can *in situ* reserves be managed to best promote population persistence, conservation of genetic diversity and adaptation? One possible strategy would be to promote relatively short generation lengths in reserves, for example, through frequently opening gaps for natural regeneration. In this way, the success of natural regeneration over time could be monitored, as natural regeneration failure is likely to be the first warning sign of mal-adaptation to changing climates. This would also facilitate maintenance of the target species composition and successional stages. However, this management strategy will likely be in direct conflict with other objectives, as development and maintenance of old-growth stand structures is often a primary goal for multipurpose reserves. Reserves with the primary purpose of gene conservation such as Gene Resource Management Units (Riggs, 1982; Millar and Libby, 1991) are likely to be less common than multipurpose reserves such as parks, in areas like British Columbia with considerable existing reserves (e.g., Yanchuk and Lester, 1996).

Another strategy, one that recognizes that as climates continue to change, aging tree populations will be increasingly out-of-phase, adaptationally with environmental conditions, is to manage reserves so that trees achieve relatively long life spans. Under this strategy, *in situ* reserve populations will become increasingly '*ex situ*,' but the homeostasis associated with mature forest stands would allow populations to persist. Over time, the reserve population would become a genetic resource for a range of temperature and moisture found elsewhere in the species' range, rather than for the geographic area it currently occupies.

It is important that *in situ* reserves are naturally regenerated whenever possible. Artificial regeneration removes or alters selection pressures at the life history stage in which they are likely strongest, during seedling establishment. When selection on a quantitative trait is relaxed, the result is usually a decline in fitness of a few percent per generation (Falconer and Mackay, 1996). Artificial regeneration may result in selection for different traits than natural regeneration. For example, part of the demise of Pacific Northwest salmon stocks has been attributed to changes in selection for reproductive traits and early survival in hatcheries compared to natural environments, yielding hatchery stocks with lower fitness in the wild, on average, than wild populations (Hindar et al., 1991). Some artificial regeneration may be needed in reserves if natural regeneration fails or if artificial migration between *in situ* reserves becomes necessary. In this event, planting densities should be high to allow for selection among planted individuals (Campbell, 1987; Ledig and Kitzmiller; 1992).

The paleoecological history and current genetic structure of most temperate and boreal trees indicates a great capacity of these species to persist, migrate and adapt to new conditions. Reserves should be managed to maintain and even enhance evolutionary processes, rather than simply as static collections of current genetic diversity. Large current population sizes, high levels of genetic variation, homeostasis of mature trees and other demographic factors indicate a low probability of extinction for most of these species. However, the maintenance of adapted populations of these ecosystem dominants over highly variable and rapidly changing environments will require *in situ* reserves that are sufficiently large to maintain genetic variation for adaptive traits. Populations should be identified for conservation based on genetic variation for adaptive traits such as growth phenology. Management strategies aimed at maintaining old growth stand structure will need to recognize that reserve populations may become less fit in reserve environments over time, although they remain an important genetic resource for other geographic areas.

REFERENCES

Aitken, S.N. and W.J. Libby. 1994. Evolution of the pygmy-forest edaphic subspecies of *Pinus contorta* across an ecological staircase. Evolution 48:1009-1019.

Campbell, R.K. 1987. Biogeographical limits of in Southwest Oregon. For. Ecol. And Mgmt. 18:1-34.

Campbell, R.K. 1979. Genecology of in a watershed in the Oregon Cascades. Ecology 60:1036-1050.

Campbell, R.K. and A.I. Sugano. 1979. Genecology of bud-burst phenology in: Response to flushing temperature and chilling. Bot. Gaz. 140(2):223-231.

Copes, D.L. 1991. Isoenzyme uniformity in western red cedar seedlings from Oregon and Washington. Can. J. For. Res. 11:451-453.

Falconer, D.S. and T.F.C. Mackay. 1996. Introduction to quantitative genetics. 4th Ed. Longman Sci. and Tech., Harlow, U.K.

Franklin, I.R. 1980. Evolutionary changes in small populations. In M.E. Soule and B.A. Wilcox (Eds.), Conservation Biology: An Evolutionary-Ecological Perspective. Sinauer Associates, Sunderland, MA, pp. 135-149.

Geber, M.A. and T.E. Dawson. 1993. Evolutionary responses of plants to global change. In P.M. Kareiva, J.G. Kingsolver, and R.B. Huey (Eds.), Biotic Interactions and Global Change. Sinauer Associates, Inc., pp. 179-197.

Gooding, G. 1998. Genetic variation and mating system of Ponderosa pine in the Willamette Valley. M.Sc. thesis, Department of Forest Sciences, Oregon State University.

Hindar, K., N. Ryman and F. Utter. 1991. Genetic effects of cultured fish on natural fish populations. Can. J. Fish. Aquat. Sci. 48:945-957.

Holsinger, K.S. 1993. The evolutionary dynamics of fragmented plant populations. In P. M. Kareiva, J.G. Kingsolver, and R.B. Huey (Eds.), Biotic Interactions and Global Change. Sinauer Associates, Inc., pp. 198-216.

Huntley, B. 1991. How plants respond to climate change: Migration rates, individualism and the consequences for plant communities. Annals of Botany 67:15-22.

Kavanagh, K.L., B.J. Yoder, S.N. Aitken, B.L. Gartner and S. Knowe. 1998. Root and shoot vulnerability to xylem cavitation in four populations of seedlings. Tree Physiology. In press.

Lande, R. 1995. Mutation and conservation. Conservation Biology 9(4): 782-791.

Ledig, F.T. and J.H. Kitzmiller. 1992. Genetic strategies for reforestation in the face of global climate change. For. Ecol. & Mgmt. 50:153-169.

Lesica, P. and F.W. Allendorf. 1995. When are peripheral populations valuable for conservation? Cons. Biol. 9(4):753-760.

Lewontin, R.C. 1984. Detecting population differences in quantitative characters as opposed to gene frequencies. Am. Nat. 123(1):115-124.

Millar, C.I. and W.J. Libby. 1991. Strategies for conserving clinal, ecotypic and disjunct population diversity in widespread species. In D.A. Falk and K.E. Holsinger (Eds.), Genetics and Conservation of Rare Plants. Oxford University Press, New York, New York, pp. 149-170.

Lynch, M. 1996. A quantitative-genetic perspective on conservation issues. In J.C Avise and J.L. Hamrick (Eds.), Conservation Genetics: Case Studies from Nature. Chapman & Hall, New York, pp. 471-501

Lynch, M., J. Conery and R. Burger. 1995. Mutation accumulation and the extinction of small populations. Am. Nat. 146(4):489-518.

Lynch, M. and Walsh, B. 1998. Genetics and analysis of quantitative traits. Sinauer Associates, Inc. Sunderland, MA, 980 pp.

Lynch, M. and R. Lande. 1993. Evolution and extinction in response to environmental change. In P.M. Kareiva, J.G. Kingsolver, and R.B. Huey (Eds.), Biotic Interactions and Global Change. Sinauer Associates, Inc. pp. 234-250.

Moritz, C. 1994. Defining 'Evolutionarily Significant Units' for conservation. TREE 9(10):373-375.

Namkoong, G. 1989. Population genetics and the dynamics of conservation. In L. Knutson and A.K. Stoner (Eds.), Biotic Diversity and Germplasm Preservation, Global Imperatives. Kluwer Academic Publishers, Dordrecht, Netherlands.

Rehfeldt, G.E. 1995. Genetic variation, climate models and the ecological genetics of Larix occidentalis. For. Ecol. Mgmt. 78:21-37.

Rehfeldt, G.E. 1993. Genetic structure of western red cedar populations in the Interior West. Can. J. For. Res. 24:670-680.

Riggs, L.R. 1982. Genetic resources: An assessment and plan for California. California gene resource program, Sacramento.

Sorensen, F.C. 1969. Embryonic genetic load in coastal Douglas fir, *Pseudotsuga menziesii* var. *menziesii*. American Naturalist 103: 389-398.

Sorensen, F.C. and R.S. Miles. 1974. Self-pollination effects on Douglas fir and ponderosa pine seeds and seedlings. Silvae Genet. 23:135-138.

Soule, M.E. 1980. Thresholds for survival: Maintaining fitness and evolutionary potential. In M.E. Soule and B.A. Wilcox (Eds.), Conservation Biology: An Evolutionary-Ecological Perspective. Sinauer Associates, Sunderland, MA, pp. 151-170.

Stettler, R.F. and H.D. Bradshaw. 1994. The choice of genetic material for mechanistic studies of adaptation in forest trees. Tree Physiology 14: 781-796.

Turelli, M. 1984. Heritable genetic variation via mutation-selection balance: Lerch's zeta meets the abdominal bristle. Theor. Pop. Biol. 25:138-193.

White, T.L. 1987. Drought tolerance of southwestern Oregon Douglas fir. Forest Science 33(2): 283-293.

Yanchuk, A.D. and D.T. Lester. 1996. Setting priorities for conservation of the conifer genetic resources of British Columbia. The Forestry Chronicle 72(4):406-415.

Zhang, J.W., L. Fins and J.D. Marshall. 1994. Stable carbon isotope discrimination, photosynthetic gas exchange, and growth differences among western larch families. Tree Physiol. 14:531-539.

Tree Spacing Affects Clonal Ranking in *Eucalyptus grandis* × *E. urophylla* Hybrids

J. S. Brouard
S. E. T. John

INTRODUCTION

The Problem

As wood fibre becomes scarcer, there is increasing global interest in high-yield, clonal forestry. Clonal selection and identification of optimal cultural conditions need to be carried out together as it is possible that clones will respond differently to changes in environment, including spacing. The highest yielding clone in a clonal block planting at a given spacing may not outperform other clones if the spacing is changed. Yield-density relationships are important when applying inferences from genetic tests to production plantations.

It is commonly believed that individual tree height is relatively insensitive to planting density. Although individual tree diameter is known to be strongly influenced by stand density, stand basal area and stand volumes are sometimes assumed to be unaffected. In the absence

J. S. Brouard and S. E. T. John are affiliated with Isabella Point Forestry Ltd., 670 Isabella Point Road, Salt Spring Island, BC, Canada V8K 1V2 (E-mail: johnbro@saltspring.com).

[Haworth co-indexing entry note]: "Three Spacing Affects Clonal Ranking in *Eucalyptus grandis* × *E. urophylla* Hybrids." Brouard, J. S., and S. E. T. John. Co-published simultaneously in *Journal of Sustainable Forestry* (Food Products Press, an imprint of The Haworth Press, Inc.) Vol. 10, No. 1/2, 2000, pp. 13-23; and: *Frontiers of Forest Biology: Proceedings of the 1998 Joint Meeting of the North American Forest Biology Workshop and the Western Forest Genetics Association* (ed: Alan K. Mitchell et al.) Food Products Press, an imprint of The Haworth Press, Inc., 2000, pp. 13-23. Single or multiple copies of this article are available for a fee from The Haworth Document Delivery Service [1-800-342-9678, 9:00 a.m. - 5:00 p.m. (EST). E-mail address: getinfo@haworthpressinc.com].

of detailed knowledge, relative ranking of genetic entries is also assumed to be unaffected by planting density, at least within the range used for normal, operational plantings.

In practice, genetic selections are commonly based on relative ranking for juvenile height even though mature stand volume is the ultimate trait of interest. It has also been suggested that foresters might plant at close spacings and measure early in the hope that relative rankings will be similar to those for the same materials planted at wider spacings and measured at later ages (Franklin, 1979). Vegetative multiplication by rooted cuttings in clonal forestry systems allows the separation of genetic and environmental effects. Thus clonal spacing trials can be useful in understanding density and competition effects.

The Experimental Approach

In order to test clonal response to planting density, a systematic plaid factorial spacing (Marynen) design was used, with tested densities ranging from 1111 to 2500 stems per hectare (3×3 m^2 to 2×2 m^2 square spacing). Nine clones of hybrid *Eucalyptus* (*Eucalyptus grandis* Hill ex. Maiden × *E. urophylla* Blake) were tested in monoclonal plots at a research site near São Mateus in Espirito Santo, Brazil (18° 43′S and 39° 52′W). The usual pulpwood rotation in the region for clonal eucalyptus is seven years. Detailed measurements of the trial were taken at ages one and two years from planting.

METHODS

Marynen (1963) developed these designs which systematically apply a range of spacings, and used them to study oil palms in Congo. The design was also applied to poplars by Delvaux (1967) and Eucalyptus by Delwaulle (1979). Marynen designs are similar to the plaid factorial design described by Lin and Morse (1975). Salient features include efficient use of space and material, and the possibility of valid statistical inferences while testing a range of densities.

Each of the nine clones in this study was represented by 21 blocks of 25 trees (Figure 1), 525 measured trees per clone, and a total of 4725 trees. Three blocks were felled in each of the first and second years for detailed measurements. The 75 trees felled per clone per year

FIGURE 1. Details of plot design for Marynen spacing trial.

x	Γ	Γ	Γ	Γ	Γ	x	Γ	Γ	...
x	Γ	Γ	Γ	Γ	Γ	x	Γ	Γ	...
x	x	x	x	x	x	x	x	x	...
x	d	d	e	f	f	x	Γ	Γ	...
x	d	d	e	f	f	x	Γ	Γ	...
x	b	b	c	e	e	x	Γ	Γ	...
x	a	a	b	d	d	x	Γ	Γ	...
x	a	a	b	d	d	x	Γ	Γ	...
x	x	x	x	x	x	x	x	x	...

KEY

Spacing treatment	#/block & type	Stocking
a 3.0 m × 3.0 m	4 centered	1111 stems/ha
b 3.0 m × 2.5 m	4 eccentric	1333 stems/ha
c 2.5 m × 2.5 m	1 eccentric	1600 stems/ha
d 3.0 m × 2.0 m	8 centered	1667 stems/ha
e 2.0 m × 2.5 m	4 eccentric	2000 stems/ha
f 2.0 m × 2.0 m	4 centered	2500 stems/ha

Measured trees *25/block*
x Surrounds *24/block (note internal surrounds are shared)*
Total trees *49/block*
Γ *Treatment trees from adjacent blocks*

were used to construct individual clone volume equations at both ages. The remaining (standing) trees were measured for diameter at breast height and total height. Underbark volume was estimated using the individual clone volume equations.

An analysis of variance was performed on plot means for the densities tested. The treatment means were based on averages of four measurements for treatments a, b, e and f, and eight measurements for treatment d. There was only one plot tree per block for the highly asymmetrical treatment c. Because of this, treatment c was deleted from analyses of variance, but not from assessment of treatment means.

Model: $Y_{ijk} = [\mu + r_i + c_j + \varepsilon_{,ij}$ [Error (a)] $+ t_k + ct_{ik} + \varepsilon_{,ijk}$ [Error (b)] where:

Y_{ijk} represents the average response of the k'th spacing treatment in the j'th clone in the i'th replication;

i = 1, 2, . . . r blocks, (r = 21 at age one-year and r' = 18 at age two-years);
j = 1, 2, . . . c clones, (c = A, B, . . . I);
k = 1, 2, . . . t spacing treatments, (t = a, b, . . . f);
N = 21 blocks × 9 clones × 25 trees/block = 4725 trees in the first year; and
N' = 18 × 9 × 25 = 4050 trees in the second year.

Clones were considered random and treatments fixed. Analysis of variance was performed based on a split-plot design, with spacing treatments as subplots, and clones as whole plots (Table 1).

Variance components were calculated for replications (blocks), clones, block by clone interactions, clone by treatment interactions and residual error by equating observed mean squares from the analyses of variance with their expectations in Table 1. The mean squared treatment effects were calculated in a similar manner. The ANOVA procedure of SAS (1985) was applied to plot means in order to estimate the sums of squares.

Coppice regrowth from felled blocks was deleted for the second-year analyses, giving a total of 18 blocks for undisturbed treatments. Dependent variables were overbark diameter at breast height (DBH), total height (HT), and underbark volume (VOL).

Broad-sense heritabilities were calculated for diameter, height and volume in order to examine the different responses of these growth traits. As this experiment was not specifically designed to show progress under selection, the broad-sense heritability was interpreted as defining the proportion of variation among clones within a single unspecified spacing treatment unit. Thus:

$$h^2_b = (\sigma^2_C + \sigma^2_{CT})/\sigma^2_P$$

TABLE 1. Expected mean squares for ANOVA

Source	d.f.	Expected Mean Squares
R	$r-1$	$\sigma^2_e + t\sigma^2_{RC} + ct\sigma^2_R$
C	$c-1$	$\sigma^2_e + t\sigma^2_{RC} + rt\sigma^2_C$
Error (a), RC	$(r-1)(c-1)$	$\sigma^2_e + t\sigma^2_{RC}$
T	$t-1$	$\sigma^2_e + t\sigma^2_{RC} + rt/(t-1)\ \sigma^2_{CT} + rc/(t-1)\ \Sigma T^2$
CT	$(c-1)(t-1)$	$\sigma^2_e + t\sigma^2_{RC} + rt/(t-1)\ \sigma^2_{CT}$
Error (b), RT + RTC	$c(t-1)(r-1)$	σ^2_e

where the phenotypic variance, $\sigma^2_P = \sigma^2_C + \sigma^2_{CT} + \sigma^2_{RC} + \sigma^2_e$ with variance components as defined in Table 1. Approximate standard errors were calculated for the estimated heritabilities using the method of Dickerson (1969) described in Hallauer and Miranda (1981).

RESULTS

Analyses of variance for diameter, height and volume were performed on data from ages one and two years. The mean squares for clones, treatments and clone × treatment interactions are presented in Table 2 along with appropriate errors for F-tests, degrees of freedom, F values and significance levels. Variance components were calculated from the same ANOVA's in order to allow computation of broad sense heritabilities. These are summarized in Table 3.

TABLE 2. Summary of mean squares, F values and significance from ANOVA's

A. Source: Clone

Trait	MS(C)	d.f.	Error(a)	d.f.	F value	Significance
D1	68.4493	8	0.8166	160	84.3	**
H1	78.0869	8	0.5880	160	133.9	**
V1	150.6085	8	2.5856	160	58.2	**
D2	7.0037	8	0.6011	136	11.6	**
H2	67.0184	8	1.5717	136	42.6	**
V2	857.4380	8	42.2207	136	20.3	**

B. Source: Treatment

Trait	MS(C)	d.f.	MS(CT)	d.f.	F value	Significance
D1	0.8372	4	0.3375	32	2.5	n.s.
H1	0.0588	4	0.1421	32	0.4	n.s.
V1	1.6381	4	0.7916	32	2.4	n.s.
D2	23.9203	4	1.4330	32	16.7	**
H2	2.9239	4	0.5503	32	5.3	**
V2	1245.8592	4	98.9094	32	12.6	**

C. Source: Clone × Treatment Interaction

Trait	MS(C)	d.f.	Error	d.f.	F value	Significance
D1	0.3375	32	0.1395	720	2.4	**
H1	0.1421	32	0.0813	720	1.7	**
V1	0.7916	32	0.4342	720	1.8	**
D2	1.4330	32	0.2584	612	5.5	**
H2	0.5503	32	0.1478	612	3.7	**
V2	98.9094	32	13.8182	612	7.2	**

where: D1 = DBH at 1Y, H1 = HT at 1Y, V1 = VOL at 1Y, D2 = DBH at 2Y . . .
*n.s. = not significant, and ** = significant at the 1% level*

The proportions of total variance attributable to the various effects and their interactions are summarized in Table 4. Broad-sense heritabilities and their standard errors are presented in Table 5.

Mean diameter, height, individual tree volume and stand volume productivity responses to spacing at age two years are shown in Figures 2 to 5.

DISCUSSION

Analysis of Variance

Clone effects were highly significant for all three growth traits and at both ages, suggesting strong and sustained genetic control on growth. Spacing treatment effects were non-significant at age one year

TABLE 3. Variance components used in calculation of broad-sense heritability

Trait	Clone	Clone × Treat.	Rep × Clone	Error	Total phenotypic
D1	0.6442	−0.0181	0.1344	0.1395	0.9000
H1	0.7381	−0.0170	0.1013	0.0813	0.9038
V1	1.4097	−0.0683	0.4303	0.4342	2.2059
D2	0.0711	0.0370	0.0685	0.2584	0.4351
H2	0.7272	0.0454	0.2848	0.1478	1.2052
V2	9.0580	2.5195	5.6805	13.8182	31.0762

TABLE 4. Mean squares expressed as a percentage from the ANOVA

Trait	Age	Rep	Clone	Error(a)	Treat	C × T	Error(b)
DBH	1Y	2	89	4	0	1	3
HT	1Y	2	91	6	0	0	1
VOL	1Y	4	91	0	0	1	4
DBH	2Y	1	17	13	39	18	12
HT	2Y	0	90	2	2	3	3
VOL	2Y	3	40	−2	30	20	10

TABLE 5. Broad-sense heritabilities (and standard errors)

Trait	Age 1 Year	Age 2 Years
DBH	0.70 (0.63)	0.27 (0.17)
HT	0.80 (0.72)	0.63 (0.56)
VOL	0.63 (0.57)	0.40 (0.27)

FIGURE 2. Mean diameter response to density treatment at two years.

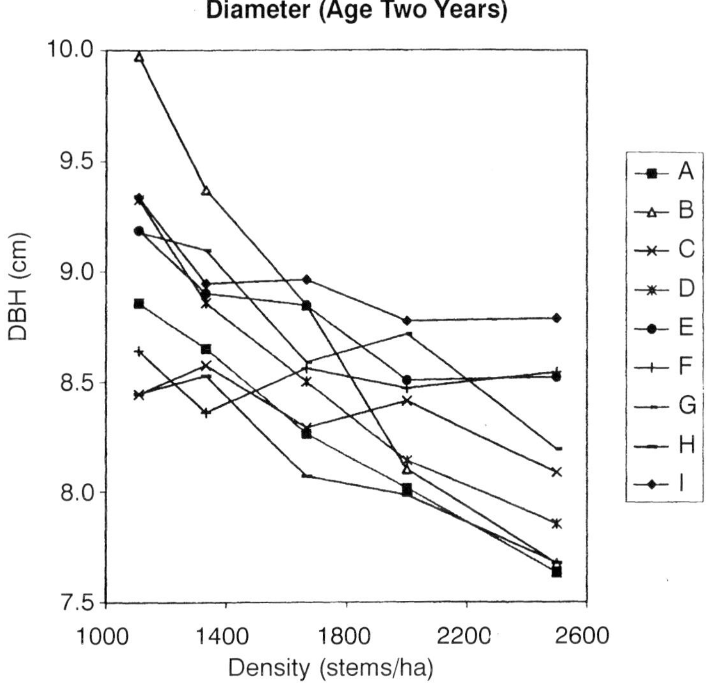

but highly significant at age two. This suggests that competition effects intensified after age one year. Clone by spacing interactions were highly significant at both ages for all three growth traits. This could be due to heterogeneity of variance in response to spacing, rank-order changes, or both. Inspection of plots of growth by spacing treatment reveals no rank-order changes at age one year, but large rank-order changes are apparent for diameter and volume at age two (Figures 2 and 4).

At age one, clones explain 90% of the total variance for all traits, and treatments explain none (Table 4). By age two, when competition has begun, density treatments explain approximately 40% of the variance in diameter and 30% of the variance in volume, but none of the variation in height, which is still mostly attributable to clone identity. Clone by treatment interaction variance explains 18 to 20% of the

FIGURE 3. Mean height response to density treatment at two years.

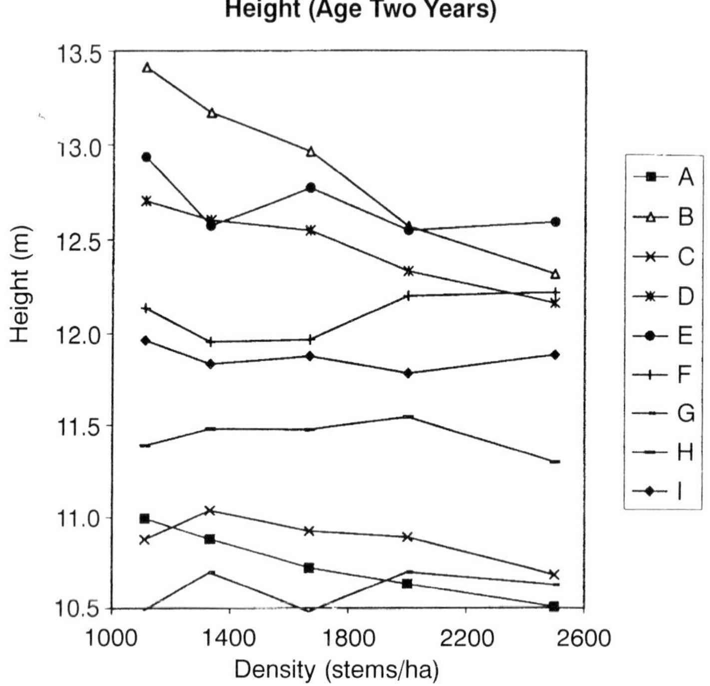

variation for diameter and volume, but only 3% of the variation for height, which was not sensitive to plant density at this age.

Heritability

Broad-sense heritability estimates for the three traits and two ages also show the effects of the onset of density-induced competition (Table 5). The broad-sense heritability of diameter and volume both decrease substantially from age one to age two as density effects become stronger, while the drop in heritability for height was smaller.

Yield-Density Response

At age one year there was no significant treatment effect for any of the response variables and few rank-order changes. The most striking responses occur at age two for diameter. Clones B and D exhibit very

FIGURE 4. Mean volume response to density treatment at two years.

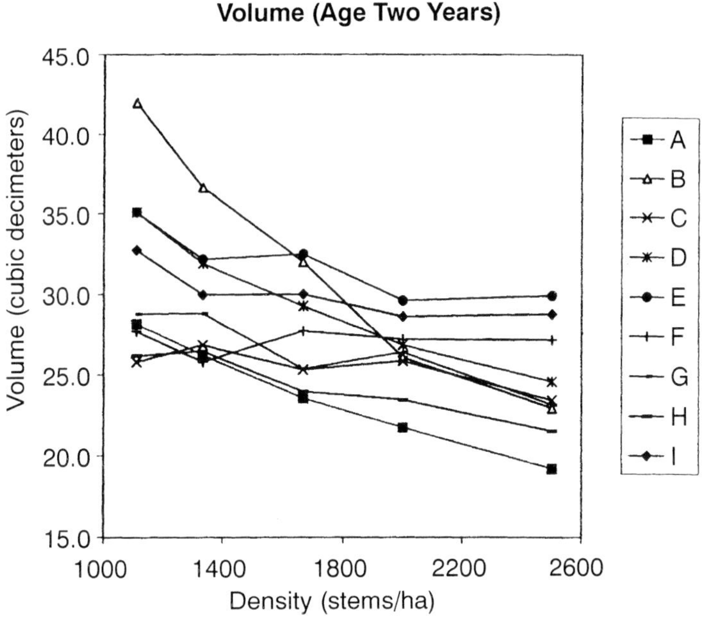

large rank changes between close and wide spacings. Clone B goes from first to last rank (Figure 2). These markedly varied responses to density are particularly striking since they occur at such a young age, and over a small range of densities, which are well within the limits of operational silviculture.

Selection

Selection on heritable traits in variable populations forms the genetic basis for most tree improvement programs. Selection of the top-ranked genetic entries is usually based on trials where an array of genotypes is tested under controlled conditions. Juvenile height has been found empirically to be a reasonably good predictor of rotation-age volume productivity. Thus, many tree improvement programs practice some form of indirect selection based on juvenile height at a fixed reference age, with the assumption that gains will be obtained in rotation age volume productivity.

In this experiment, the three tallest clones at close spacing at age

FIGURE 5. Mean volume productivity in response to density treatment at two years.

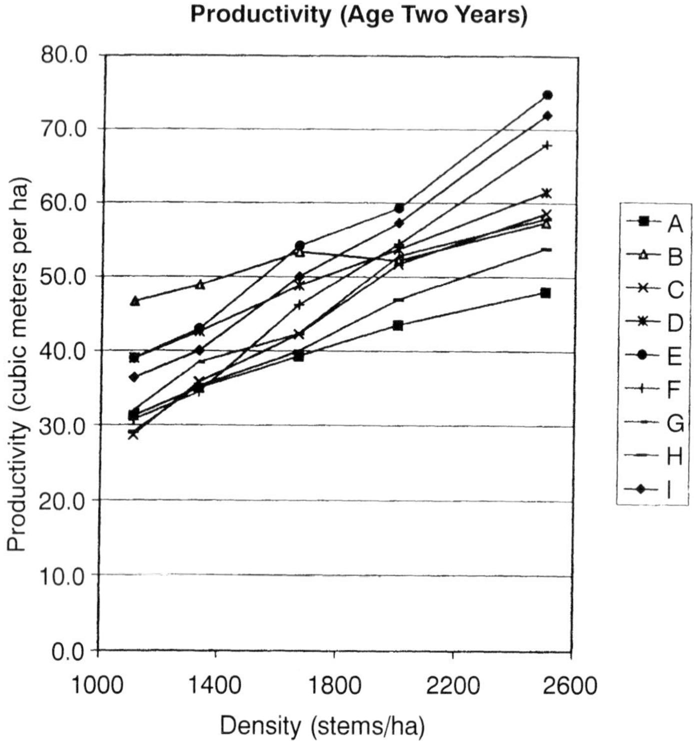

one were D, I, and B, with heights all close to 5 m. At age two, the largest average individual tree volumes were found in clones B, E, and C (ranging from 35 to 43 dm^3 per tree) at the widest spacing (Figure 3). Low correlations between one-year height at close spacing and two-year individual tree volume at wide spacing would suggest that selection may not be effective if we plant close and measure early.

Furthermore, the objective of selection is usually to maximize productivity per unit area rather than individual tree volume. The highest productivity at age two is achieved with clones E, I, and F planted at close spacing, with values ranging from 34 to 37 m^3 per ha per year (Figure 5). This suggests that early height growth is a poor predictor of later stand volume productivity.

CONCLUSIONS

- Individual clones responded very differently to changes in stem density; there were large and significant rank-order changes. Optimal spacing is clone-dependent.
- Affect on height by spacing was relatively little, but volume and diameter were strongly influenced.
- At age one year, density effects were weak, but by age two (30% of rotation age), all clones showed a decrease in size with increasing stem density.
- Broad-sense heritability estimates provide a good indication of the responses of the three growth traits to density.
- Variation in available growing space in genetic trials, particularly clonal screening trials, could confound plant density effects with genetic growth potential.

REFERENCES

Delvaux, J. 1967. Contribution a l'Etude de l'Education des Peuplements. V. Competition inter-clones chez le peuplier, and VII. Competition inter-individuelle chez le peuplier. Ministère de l'agriculture, Administration des eaux-et-forêts. Station de recherche des eaux-et-forêts, Groenendall-Hoeilaart, Belgique, Travaux–Série B, B, No 33.

Delwaulle, J.-C. 1979. Plantations forestières en Afrique tropicale séche. Techniques et especes à utiliser. Revue Bois et Forêts des Tropiques 184: 47-61.

Dickerson, G.E. 1969. Techniques for research in quantitative animal genetics. In: Techniques and Procedures in Animal Science Research. Am. Soc. Anim. Sci., Albany, New York.

Franklin, E.C. 1979. Model relating levels of genetic variance to stand development of four North American conifers. Silvae Genetica 28(5-6): 207-212.

Hallauer, A.R. and J.B. Miranda, Fo. 1981. Quantitative Genetics in Maize Breeding. Iowa State University Press, Ames, Iowa.

Lin, C.-S. and P.M. Morse. 1975. A compact design for spacing experiments. Biometrics 31:661-671.

Marynen, T. 1963. Contribution a l'étude de la densité de plantation chez les vegetaux. Pub. De l'I.N.E.A.C., Série Scientifique, No 102. 79 pp.

SAS Institute Inc. 1985. SAS User's Guide: Statistics, Version 5 Edition, Cary, NC, SAS Institute Inc. 956 pp.

Seed Source Testing of Paper Birch (*Betula papyrifera*) in the Interior of British Columbia

M. R. Carlson
V. G. Berger
C. D. B. Hawkins

INTRODUCTION

Paper birch (*Betula papyrifera*) is an ecologically early-successional species found at low to mid elevations in the more productive biogeoclimatic zones and sub zones of the British Columbia interior. Recent silvicultural and ecological research efforts in British Columbia have drawn attention to it's roles in nutrient cycling and disease and insect pest interactions (Simard, 1997). Retention of some birch in young conifer plantations to improve soil nutrient status and ameliorate root disease problems is becoming increasingly common. Birch may soon be considered an acceptable species for plantation silviculture, via both natural regeneration and planting. We have a rapidly

M. R. Carlson and V. G. Berger are affiliated with B.C. Ministry of Forests, Research Branch, Kalamalka Forestry Centre, 3401 Reservoir Road, Vernon, B.C., V1B 2C7.

C. D. W. Hawkins is affiliated with the University of Northern British Columbia, Forestry, 3333 University Way, Prince George, B.C., V2N 4Z9.

Address correspondence to: M. R. Carlson, Ministry of Forests, Research Branch, Kalmalka Forestry Centre, 3401 Reservoir Road, Vernon, B.C., Canada V1B 2C7 (E-mail: Mike.Carlson@gems3.gov.bc.ca).

[Haworth co-indexing entry note]: "Seed Source Testing of Paper Birch (*Betula papyrifera*) in the Interior of British Columbia." Carlson, M. R., V. G. Berger, and C. D. B. Hawkins. Co-published simultaneously in *Journal of Sustainable Forestry* (Food Products Press, an imprint of The Haworth Press, Inc.) Vol. 10, No. 1/2, 2000, pp. 25-34; and: *Frontiers of Forest Biology: Proceedings of the 1998 Joint Meeting of the North American Forest Biology Workshop and the Western Forest Genetics Association* (ed: Alan K. Mitchell et al.) Food Products Press, an imprint of The Haworth Press, Inc., 2000, pp. 25-34. Single or multiple copies of this article are available for a fee from The Haworth Document Delivery Service [1-800-342-9678, 9:00 a.m. - 5:00 p.m. (EST). E-mail address: getinfo@haworthpressinc.com].

evolving ecological and silvicultural knowledge base for the species but very little is known about patterns of genetic variation, i.e., relative magnitudes of genetic differences between trees within stands, between stands within geographic area and between broad geographic areas.

Evolutionary "strategies" for coping with heterogeneous environments (in time and space) differ among our western North American tree species (Rehfeldt, 1994). Some species are considered generalists, some specialists and some intermediate in their sensitivity to movement along such environmental gradients as elevation, latitude, longitude, temperature, precipitation, etc. Generalists (i.e., western red cedar and western white pine) tend to exhibit only modest phenotypic (and thus genotypic) differences among geographically and/or ecologically disparate seed sources when grown in common garden trials and are said to be "loosely" adapted to physical and biotic features of their environmental origins. Specialists, on the other hand (i.e., Douglas fir and lodgepole pine), when common garden grown, exhibit greater phenotypic (genotypic) differences among seed sources collected along environment gradients and are said to be more "tightly" adapted to local environments. The practical implications of these differing strategies to reforestation work is that seed sources of generalist species can be moved throughout a wider range of environments (away from origin) without risking maladaption than would be the case for specialist species. Seed source testing of many commercially important conifers has been underway in British Columbia and the U.S. Pacific Northwest for many decades resulting in operational guidelines based on biological and ecological rationale. However, very little is known about the evolutionary strategies or patterns of genetic variation of our broad-leaved tree species such as paper birch. The objective of this paper birch seed source trial is to provide information about patterns of variation in growth and adaptive traits across a broad geographic area in the B.C. interior.

MATERIALS AND METHODS

During late summer of 1994, seed from 18 stands of paper birch were collected across five forest regions ranging from Nelson, B.C. to Prince Rupert, B.C. (Figure 1). All stands were on productive low to mid elevation sites. Ten trees were selected in each stand, felled and

FIGURE 1. Seed collection and planting site locations.

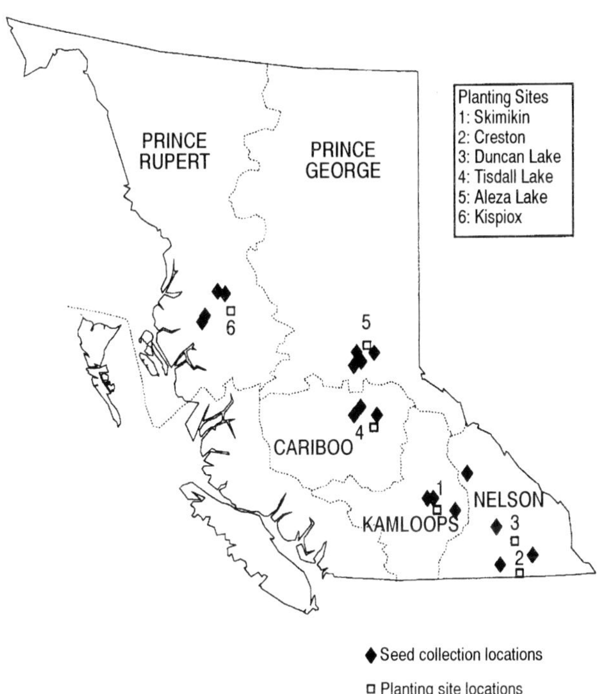

catkins collected. Seed was bulked by stand and sown at both the Red Rock Research Station near Prince George and the Kalamalka Forestry Centre near Vernon in spring 1995 in PSB 415D and 412D styroblocks (Beaver Plastics, Edmonton AB). In addition, two silver birch (*Betula pendula*) wind pollinated families from the Finnish national birch breeding program were sown and grown for comparison with our native birch. Six field sites were selected (Figure 1) and site prepared in the summer of 1995. One site was located within each the Prince Rupert, the Prince George, the Caribou and the Kamloops forest regions. Two sites were chosen in the Nelson forest region. Two of the six sites (Skimikin and Creston) are southern interior agricultural field sites. The remaining four sites are wild forested locations with Duncan Lake in the south and Tisdall Lake, Aleza Lake and Kispiox in the central interior.

In 1996, all sites were spring planted with cold stored seedlings grown at both the nurseries. The randomized complete block field design (16 trees per seed source planted as 4 tree row plots in 4 blocks for each of the 2 nurseries) allows for comparisons among forest regions (5), among seed sources within forest regions (4,3,3,4,4), among planting sites (6) and between nurseries (2). Total tree heights were measured in fall 1996 and fall 1997. A mixed-model ANOVA (SAS Proc GLM) was used to analyze height growth data with regions, nurseries and planting sites considered as fixed effects and seed sources as random. Cold damage was assessed in spring of 1997. Cold damage classes assessed were (a) dead, (b) no damage, (c) dieback and resprout from along the stem above ground level and (d) dieback and resprout from the ground.

RESULTS

Two year field data from the 18 seed sources were grouped by forest regions into 5 regional seed sources plus the bulked genetically improved silver birch families (2). Regional mean tree heights and frost damage assessments by planting site are presented in Figures 2 and 3, respectively.

Large site-to-site differences exist, with trees on the two agricultural sites averaging over 120 cm in height at 2 years versus an average of less than 60 cm across the four wild sites. Regional seed source height means were little differentiated except on the Tisdall Lake and Aleza Lake sites. The Tisdall Lake site is the least productive of all six with cold, wet soils, low organic fraction, short growing season (characteristic of the sub-boreal spruce zone) and high frost hazard. The Aleza Lake site is potentially much more productive than Tisdall Lake but experienced a severe frost event in the fall of 1996. The frost contributed to the observed seed source height differences due to greater top dieback among southern sources. The Prince George and Caribou regional sources had greater average tree heights than the other four regions on these two difficult sites but not significantly greater ($p = 0.10$). The silver birch seed source had less than average two year heights on the three southern sites but better than average height at Kispiox and superior heights on the two difficult Tisdall Lake and Aleza Lake sites.

Figure 3 shows marked differences between the (3) southern and the (3) northern sites in average severity of frost damages. Dead, top

FIGURE 2. Mean 2 year tree heights for regional seed sources by site.

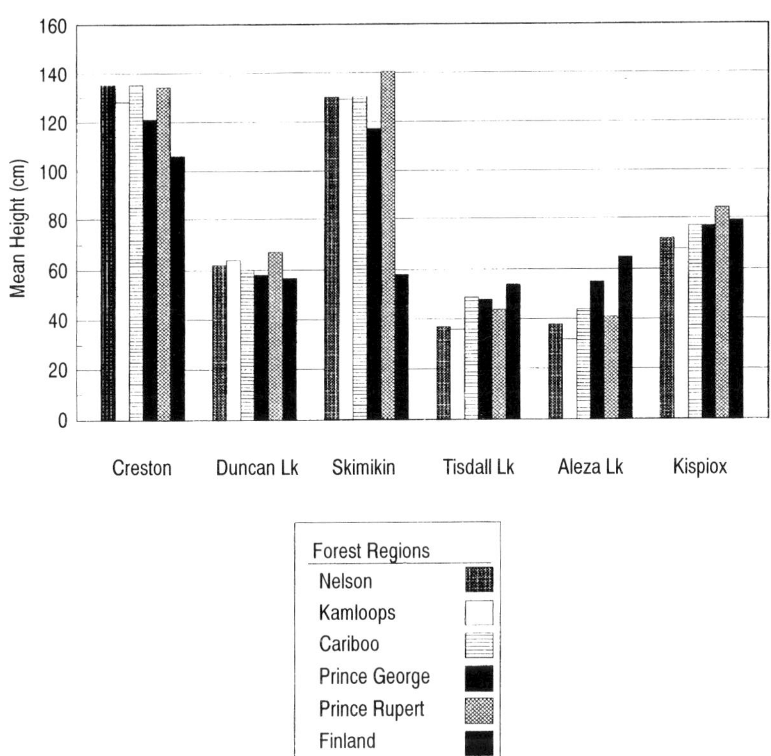

dieback and resprout damaged trees combined are in frequencies of about 30% and 70% for southern and northern planted trees, respectively. Regional differentiation in frost damage is shown in Figures 4 and 5. At the relatively safe southern Skimikin site (overall top dieback = 40%), the Prince George and Prince Rupert sources experienced about 25% dead and damaged trees compared with about 50% for the three more southerly regional sources (Figure 4). These regional differences are more extreme at the harsh Tisdall Lake site (overall dead, dieback plus resprout = 85%), whereas Nelson and Kamloops regional sources have close to 100% damaged trees compared with 70-75% damaged from the three more northerly regions. From analysis of variance, regions accounted for 25% of the total non-site related variation in frost damage while seed sources contributed only 4%.

FIGURE 3. Frost damage (%) during winter of 1996/97 by site.

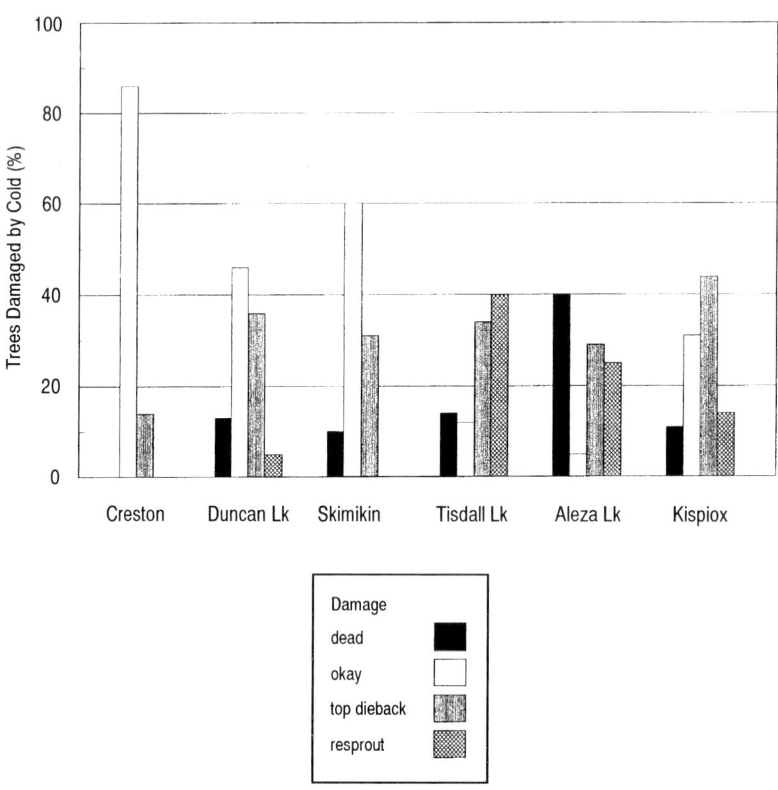

Silver birch had less frost damage than any of the paper birch seed sources at each of the six sites

No consistent differences between nurseries for tree heights or frost damage frequencies were found when comparisons were made for the Skimikin and Tisdall Lake sites (most productive and least productive site, respectively). Figures 6 and 7 show mean two year tree heights for individual seed sources at the Skimikin and Tisdall Lake sites, respectively. Differences among individual seed sources within regions are greater on average than differences among regional sources. From analysis of variance seed sources accounted for 29% of the total non-site related variation while regions contributed 9%. The silver birch source ranks second at Tisdall Lake and last at Skimikin.

FIGURE 4. Frost damage (%) during winter of 1996/97 for regional seed sources at Skimikin.

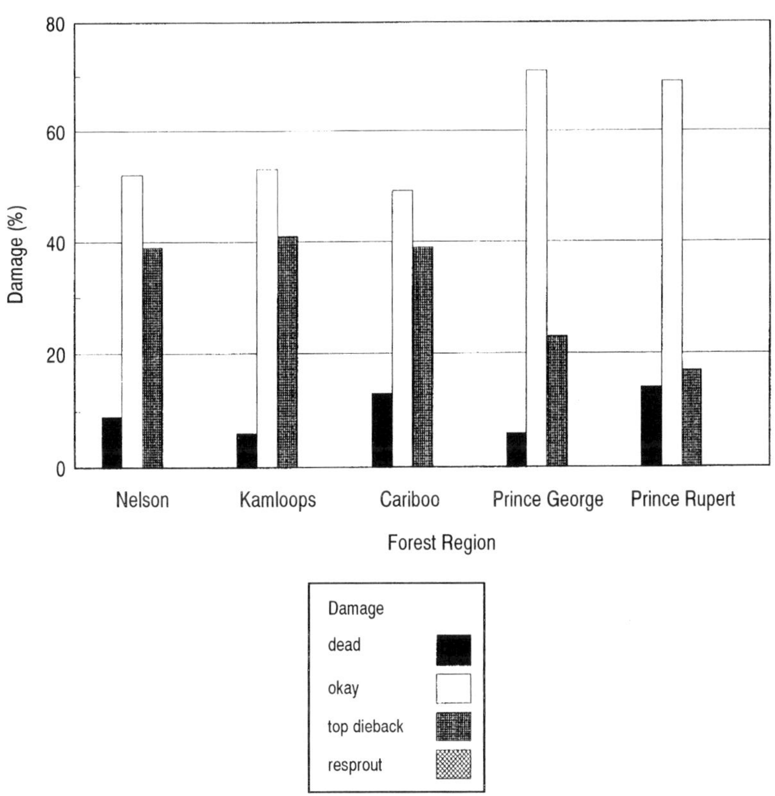

DISCUSSION

When individual stand seed sources are bulked into regional groups, height growth potentials are poorly differentiated except on the more difficult northern sites where first year fall frost events have resulted in excessive stem dieback of southern sources. It is somewhat surprising that northern regional sources, when grown on southern sites, are growing at about the same rates as southern sources. Most conifer species tested in this way show a clinal pattern of growth with height growth being inversely proportional to latitude of seed source.

Absolute height differences among individual stand seed sources (within regions) are sizable, particularly on the productive Skimikin

FIGURE 5. Frost damage (%) during winter 1996/97 for regional seed sources at Tisdall Lake.

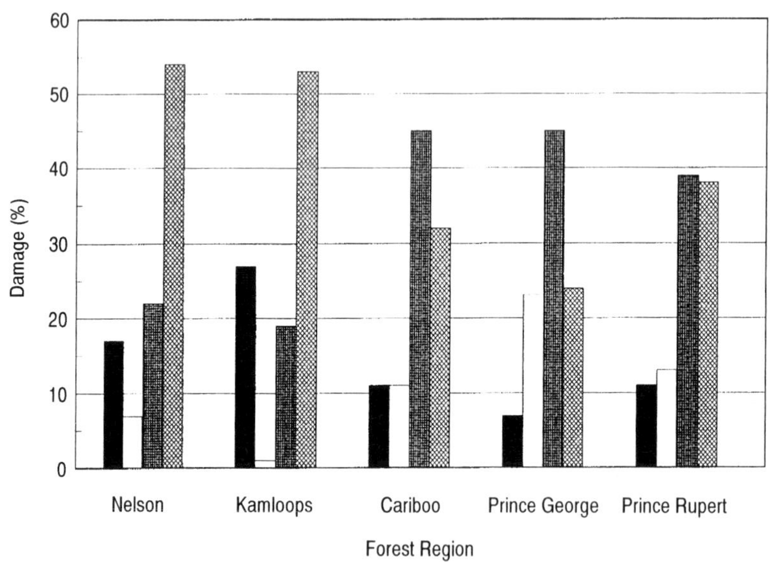

site. Individual stand differences appear to be greater than regional differences on average and suggest that the stand level component will be important in genetic improvement work (i.e., many stands within a region must be sampled).

Frost tolerance seems well differentiated at the regional level with more northerly region materials suffering less frost damage across both northern and southern sites. This suggests adaptive differences at least on a large geographic scale and less differentiation at the stand-to-stand level. Paper birch appears to be more of a generalist with regards to latitudinal movement and frost tolerance. Winter 1997/98

FIGURE 6. Two year tree heights at Skimikin for 18 paper birch and 1 silver birch seed source.

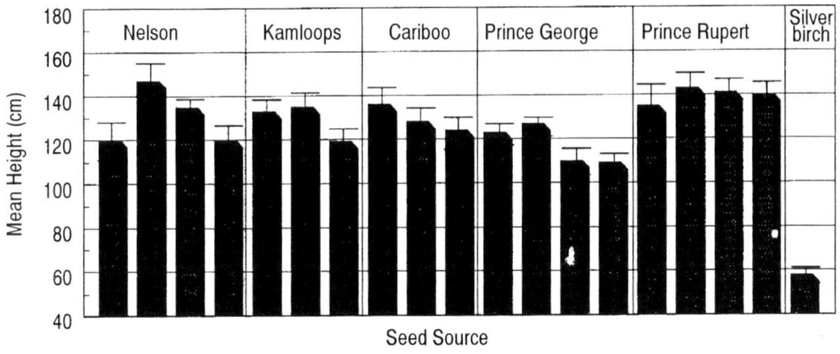

FIGURE 7. Two year tree heights at Tisdall Lake for 18 paper birch and 1 silver birch seed source.

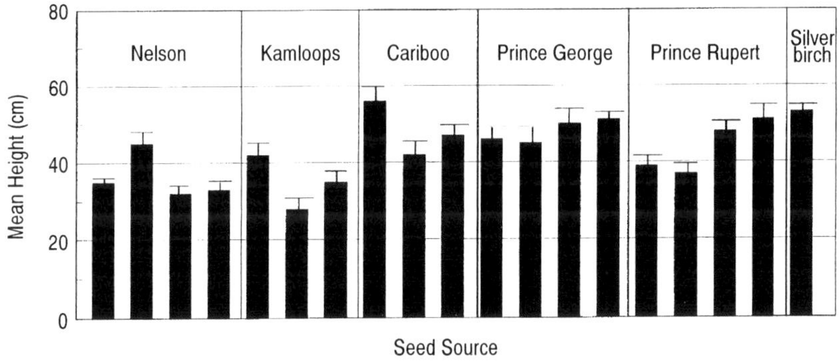

will be the first winter in which most sites will have trees above snowline allowing assessment of mid winter cold tolerances.

Laboratory frost hardiness assessments were done in late summer and fall of 1996. Levels of stem tissue damage measured in these tests for 16 of the 18 seed sources had high positive correlations (r = 0.71) with frost damage assessments across the five wild field sites (Simpson et al., 1998).

The first paper birch population improvement work in British Columbia began in 1998 with the outplanting of 195 wind pollinated families representing 195 phenotypically selected trees from 19 stands across the Kamloops and Nelson forest regions. Periodic measurements and analyses of this progeny test and the seed source trial reported here will improve our understanding of the genecology and production potential of paper birch in the B.C. interior.

REFERENCES

Rehfeldt, G.E. 1994. Evolutionary Genetics, The Biological Species, and the Ecology of the Interior Cedar-Hemlock Forests. In Proceedings of Interior Cedar-Hemlock-White Pine Forests: Ecology and Management. March 2-3, 1993, Washington State University, Pullman, Washington.

Simard, S. 1997. Ecological and Silvicultural Characteristics of Paper Birch in the Southern Interior of British Columbia. In Ecology and Management of B.C. Hardwoods FRDA Rpt. 225, October 1996, pp. 157-167.

Simpson, D., W. Binder and S. L'Hirondelle. 1998. Presented at the 1998 Joint Meeting of the North American Forest Biology Workshop and Western Forest Genetics Association. University of Victoria, Victoria, B.C. June 21-26, 1998.

Genetics of Elevational Adaptations of Lodgepole Pine in the Interior

M. R. Carlson
J. C. Murphy
V. G. Berger
L. F. Ryrie

INTRODUCTION

Lodgepole pine (*Pinus contorta* Dougl. ex Loud.) is one of the most wide ranging conifer species in western North America. Four geographic races are recognized with two found in British Columbia, *P. contorta* var. *contorta* on the coast and var. *latifolia* in the interior. Interior lodgepole pine currently contributes more to the annual allowable wood volume harvested in British Columbia (~35%) than any other single species. It is also the most frequently planted tree (105 million seedlings in 1997).

Ecologically, the *latifolia* variety is an early successional or pioneer species occupying a broad range of soil types, montane climates and elevations in the B.C. interior. Genecologically, it is considered a 'specialist,' displaying a steep cline in growth potential along elevational gradients and to a lesser extent, along geographic gradients of latitude and longitude in the Rocky Mountains of the U.S. (Rehfeldt

M. R. Carlson, J. C. Murphy, V. G. Berger, and L. F. Ryrie are affiliated with B.C. Ministry of Forests, Research Branch, Kalamalka Forestry Centre, 3401 Reservoir Road, Vernon, B.C., V1B 2C7.

[Haworth co-indexing entry note]: "Genetics of Elevational Adaptations of Lodgepole Pine in the Interior." Carlson, M. R. et al. Co-published simultaneously in *Journal of Sustainable Forestry* (Food Products Press, an imprint of The Haworth Press, Inc.) Vol. 10, No. 1/2, 2000, pp. 35-44; and: *Frontiers of Forest Biology: Proceedings of the 1998 Joint Meeting of the North American Forest Biology Workshop and the Western Forest Genetics Association* (ed: Alan K. Mitchell et al.) Food Products Press, an imprint of The Haworth Press, Inc., 2000, pp. 35-44. Single or multiple copies of this article are available for a fee from The Haworth Document Delivery Service [1-800-342-9678, 9:00 a.m. - 5:00 p.m. (EST). E-mail address: getinfo@haworthpressinc.com].

1988, 1994). Generally, low to mid elevation sources grow faster than high elevation sources on all but the very highest of sites. In long term B.C. provenance trials, lodgepole pine exhibits strong geographic and population differentiation for growth and survival, with less variation at the within population (tree-to-tree) level (Ying et al., 1985; Xie and Ying, 1995). Geographic differences in susceptibility to needle cast, stem gall and stem rust diseases have also been demonstrated (Wu and Ying, 1997; Wu et al., 1996).

A selective breeding program for lodgepole pine began in 1976. Since that time more than 1,200 parent trees have been wind-pollinated progeny tested in five geographically and ecologically defined seed planning zones (Figure 1). Rogued first generation grafted clonal seed orchards have been produced for all five of these zones. In one of these zones, the Thompson-Okanagan (TO), two seed orchards exist. One is to service low to mid elevation seed needs (700-1,200 meters) and one for higher elevation needs (1,200-1,600 meters). Despite this division of tested parents into two elevationally distinct seed production populations, the low orchard has parents ranging in elevation of origin from 700 meters to 1,300 meters. Despite their relatively high

FIGURE 1. Seed Planning Zones (5) with Lodgepole pine breeding programs and seed orchards.

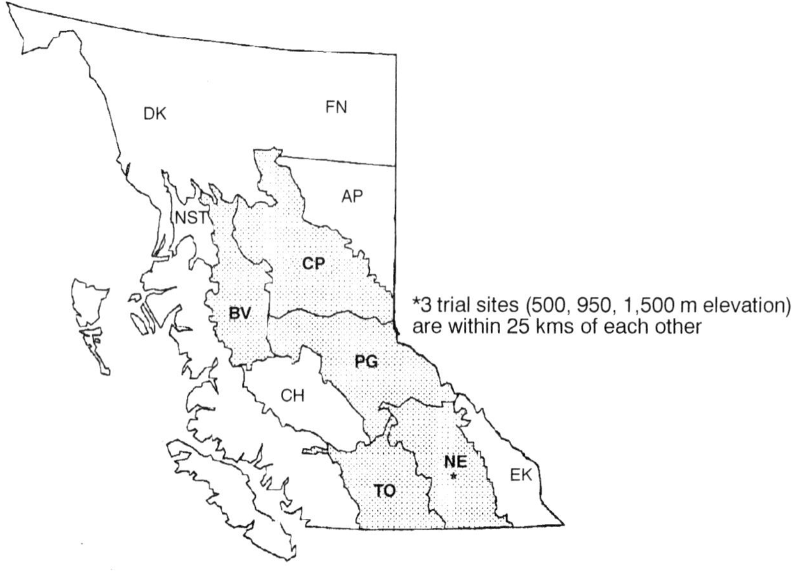

elevation origins, certain parent trees ranked high enough in progeny tests to be included in the low elevation orchard. The parent origin average for this orchard is 953 meters, but 15 of 65 parents are from above 1,100 meters and 6 are from above 1,200 meters. Seventeen parents are from below 800 meters elevation. Inclusion of trees from such an elevation range introduces the potential for departures from panmixia due to slight differences in timing of pollen shed and female strobilus receptivity between parents from extremes of elevation. More important are questions about the growth and adaptational character of seed resulting from the intercrossing of high and low elevation parents. Growth potential is negatively correlated with elevation of origin in lodgepole pine, as is tolerance of an important needle cast (rust) disease, *Lophodermella concolor*. Seed from low elevation parents has both greater growth potential and tolerance of needle cast when planted on low to mid elevation sites. Seed from high elevation parents has less growth potential at all planting site elevations and less needle cast tolerance when planted at low to mid elevation sites. Separate orchard seed lots from low and high elevation seed parents can be constructed by segregating cones at harvest but no control of pollen parents is possible so portions of both seed lots (high seed parent and low seed parent) will be of hybrid (high × low) origin. Decisions about how to deploy such seed lots will depend somewhat on the modes of inheritance for growth potential and disease tolerance in the hybrid crosstype seed. For instance, we know that both growth potential and needle cast disease tolerance are inversely related to parental elevation. If either or both of these traits is inherited in a partial to full dominance fashion in the high elevation direction, then both high and low parent seed lots will contain a greater proportion of seed with lower growth potential and lower disease tolerance when planted at low to mid elevation sites. The reverse is true if dominance is in the low elevation direction, since the low parent seed lots will retain full growth potential and disease tolerance, and the high seed lots will have both greater growth potential and disease tolerance, wherever planted. If inheritance is of an additive mode, then levels of growth potential and disease tolerance will be intermediate between pure low (low × low) and pure high (high × high) fractions. For this orchard, assuming panmixia (random mating among all parents), approximately 13% of orchard seed will be of the hybrid crosstype.[1]

The objective of this study is to determine the modes of inheritance of height growth potential, needle cast susceptibility and other traits of adaptive significance such as growth timing and frost tolerance.

MATERIALS AND METHODS

Eight trees were selected from each the Thompson-Okanagan and Nelson seed planning zone parent tree collections. Four trees in each set of eight were from under 800 meters (L) elevation and four were from over 1,400 meters (H). Trees from each seed planning zone set were intercrossed in a partial factorial design resulting in 2 unrelated L × L crosses, 2 unrelated H × H crosses and 4 unrelated H × L crosses (Figure 2). In spring of 1992, seed of the resulting 16 families were sown and grown in styro 415 B containers (105 ml). Seedlings were lifted and cold stored in fall 1992. Three forest planting sites were prepared in the fall of 1992. Site elevations were 500 m (Skimikin), 950 m (Eagle Bay) and 1,500 m (Fly Hills), all located in the Nelson (NE) planning zone. All sites were planted in the spring of 1993. A randomized complete block design was used with a 4-tree row plot per family per each of 12 blocks. Tree heights were measured each fall, beginning in 1993. At the low elevation Skimikin site, *Lophodermella* needle cast infection was particularly heavy in 1996 and 1997. Needle cast was not observed at the other two sites. All trees at the Skimikin site were assessed for severity of infection in both years.

RESULTS

First year survival was excellent on all three sites and exceeded 90% at the end of the fifth growing season (1997) except for Fly Hills,

FIGURE 2. Partial factorial crossing scheme used for each seed planning zone parent tree collection.

	L_3	L_4	H_3	H_4
L_1	X		X	
L_2		X		X
H_1	X		X	
H_2		X		X

8 CROSSES:

2 L × L
4 L × H
2 H × H

the high elevation site, which suffered approximately 15% mortality due to bovine trampling. Mean plantation 6 year heights differed greatly among sites and varied inversely with elevation of site (Figures 3-6).

Crosstypes (L × L, H × L and H × H) within seed planning zones are strongly differentiated for 6 year heights at all sites (Figures 3-6). For both seed planning zone crossing sets, the hybrid crosses (X's) are

FIGURE 3. 6-year mean tree heights by cross on mean parent tree elevation at the Skimikin site.

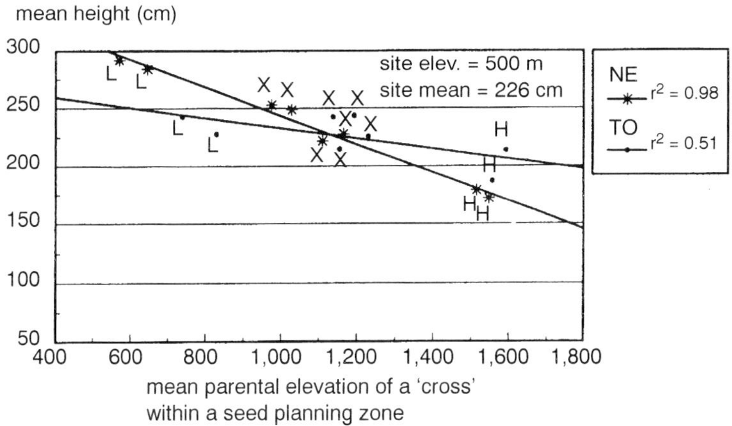

FIGURE 4. 6-year mean tree heights by cross on mean parent tree elevation at the Eagle Bay site.

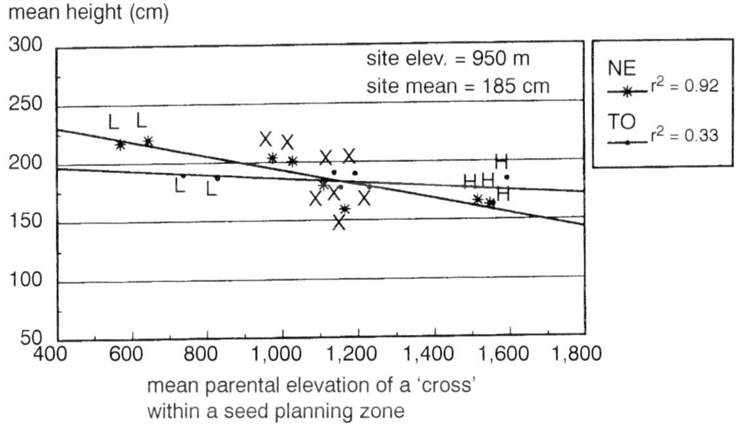

FIGURE 5. 6-year mean tree heights by cross on mean parent tree elevation at the Fly Hills site.

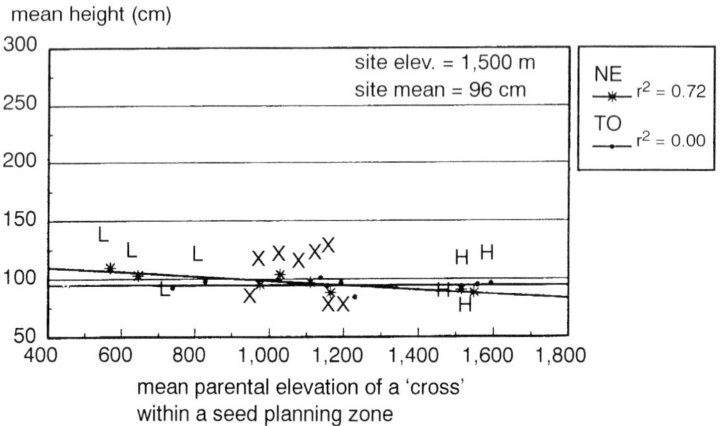

intermediate in height, between the low and high crosses (L's and H's), on all sites. Also of interest is the consistent difference between the seed planning zone crossing sets in the decline of productivity with increasing mean elevation of cross parents. Thompson-Okanagan zone crosstypes are less well differentiated for height growth than are their Nelson zone counterparts. The greatest high-versus-low crosstype difference (both absolute and relative differences for crossing sets combined) is on the highly productive low elevation Skimikin site and the least on the low productivity high elevation Fly Hills site (Figure 6).

The 1996 and 1997 *Lophodermella* needle cast surveys gave similar results for height growth. In 1997, the high crosstype had 82 percent of its trees with greater than 50 percent needle infection, while the low crosstype had 20 percent of its trees similarly infected (Figure 7). The hybrid crosstype was intermediate between the others, with 49 percent of trees heavily infected. Percentages of trees with no infection were 3, 17, and 47 for high, hybrid and low crosstypes, respectively.

DISCUSSION

These results suggest that both height growth potential and needle cast susceptibility are inherited in an additive fashion, i.e., intermediacy of the hybrid crosstype families, for both traits. Height growth is

FIGURE 6. 6-year mean tree heights for 3 crosstypes (seed planning zones combined) across 3 sites.

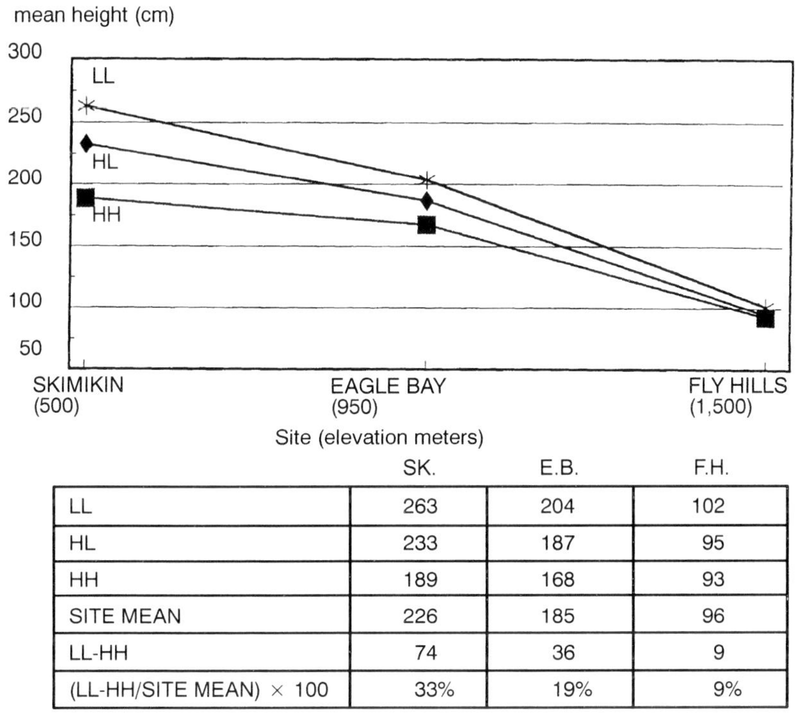

	SK.	E.B.	F.H.
LL	263	204	102
HL	233	187	95
HH	189	168	93
SITE MEAN	226	185	96
LL-HH	74	36	9
(LL-HH/SITE MEAN) × 100	33%	19%	9%

commonly assumed to be a quantitative trait, under the control of several small additive-effect loci. Populations at opposite ends of an environmental gradient (such as elevation) might be expected to differ in allelic frequencies at each of several controlling loci and, when intercrossed, produce intermediate hybrids. Intermediacy of hybrids for needle cast susceptibility is, on the other hand, somewhat unexpected. Most plant diseases for which there is some understanding of the underlying genetic mechanism are thought to be controlled by a single locus with dominant and recessive alleles. These results do not deny that a single locus may control needle cast susceptibility, but they do argue against a dominant/recessive gene action model. Also of interest is that only a weak crosstype by site interaction exists for height growth. The low elevation crosstype is superior on all three sites, although both the relative and absolute advantage of low cross-

FIGURE 7. Percent *Lophodermella* rust infected needles by crosstype (3) at the low elevation Skimikin site 1997.

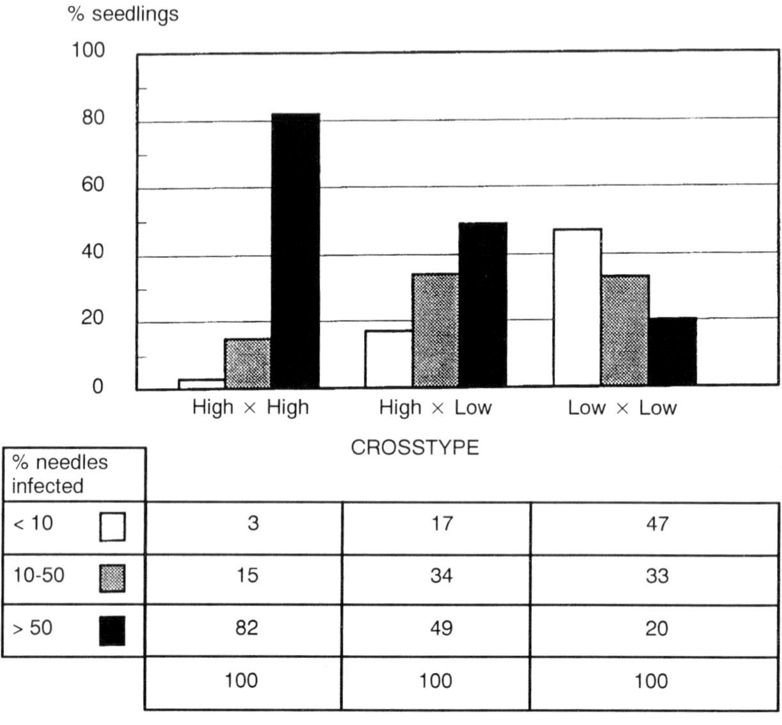

% needles infected	High × High	High × Low	Low × Low
< 10	3	17	47
10-50	15	34	33
> 50	82	49	20
	100	100	100

types over high crosstypes decrease with elevation of site (Figure 6). Trees at the high elevation Fly Hills site (1,500 meters) are not yet tall enough to be above the winter snow accumulation. When they are above the snow, winter frost tolerances will be tested. If the growth advantage of low elevation families is due to an extension of the growing season, then these families may suffer more from early or mid-winter cold than families with more conservative growth/growth timing habits. Greenhouse and nursery bed studies to characterize growth timing (flush and cessation) and frost hardiness acquisition of all three crosstypes are under way.

The consistently greater height growth differentiation between low and high crosstypes for the NE zone as compared to the TO zone may be due only to chance sampling of parents from the zones. On the other hand, steeper slopes of the NE zone crosstype growth functions (Figures 3-5) may be the result of differing selection pressures and genotypic norms in the two geographic areas. Growth potential of high elevation origin trees is probably limited by selection for growth timing so as to accommodate a relatively short frost-free growing period. NE zone high elevation trees have evolved under colder, wetter climatic regimes than their TO counterparts and might be expected to have correspondingly more conservative (shorter) growing periods at all sites, resulting in reduced growth. Growth potential of low elevation trees in the southern B.C. interior is thought to be closely associated with growing season moisture availability. The TO zone is characterized by hot, dry summers while the NE zone is cooler and has, on average, much more precipitation from April to September. Selection for drought tolerance has likely reduced the growth potential of low elevation trees from the TO compared with NE trees from similar elevations. Thus, differences in selective norms for low and high elevation trees from these zones may have contributed to the observed zone differences in crosstype differentiation.

The practical significance for orchard managers and silviculturists is that both height growth potential and needle cast susceptibility of seed lots from this and other lodgepole pine orchards can be considered as varying linearly (inversely) with the elevation mean of contributing seed and pollen parents. Growth potential and disease susceptibility of seed lots of known parental contribution will thus be more predictable in the future.

NOTE

1. Total possible crosstype proportions are: (17/65 low el. parents + 32/65 mid el. Parents + 16/65 high el. parents)^2 = $(17/65)^2$ low × low + $(32/65)^2$ mid × mid + $(16/65)^2$ high × high + 2(17/65 × 32/65) low × mid + 2(32/65 × 16/65 mid × high + 2(17/65 × 16/65) low × high. So low × high hybrid proportion = 2 (17/75 × 16/65) = 13%.

REFERENCES

Rehfeldt, G.E. 1994. Evolutionary genetics, the biological species, and the ecology of the Interior Cedar-Hemlock Forests. In Proceedings of Interior Cedar-Hemlock-White Pine Forests: Ecology and Management. March 2-3, 1993, Washington State University, Pullman, Washington.

Rehfeldt, G.E. 1988. Ecological genetics of *Pinus contorta*: a synthesis. Silvae Genetica 37:131-135.

Wu, H.X., and C.C. Ying. 1997. Genetic parameters and selection efficiencies in resistance to western gall rust, stalactiform blister rust, needle cast, and sequoia pitch moth in lodgepole pine. Forest Science. 43: 571-581.

Wu, H.X., Cheng C. Ying, and John A. Muir. 1996. Effect of geographic variation and jack pine introgression on disease and insect resistance in lodgepole pine. Canadian Journal of Forest Research. 26: 711-726.

Xie, Chang-Yi, and Cheng C. Ying. 1995. Genetic architecture and adaptive landscape of interior lodgepole pine (*Pinus contorta ssp. Latifolia*) in Canada. Canadian Journal of Forest Research 25: 2010-2021.

Ying, Cheng C., Keith Illingworth, and Michael Carlson. 1985. Geographic variation of Lodgepole pine and its implications for tree improvement in British Columbia. *In* Proc. Symp. Lodgepole Pine the Species and Its Management. May 14-16, 1984, Vancouver, B.C. In press.

Growth and Biomass Allocation of *Gliricidia sepium* Seed Sources Under Drought Conditions

M. García-Figueroa
J. J. Vargas-Hernández

INTRODUCTION

Gliricidia sepium (Jack.) Walp. is a multipurpose, nitrogen-fixing tree species distributed in the dry tropical regions of Mexico and central America. It is a valuable species for agroforestry systems (Stewart et al., 1996). The species grows well in low-fertility soils (Sanguinga et al., 1992) and it seems to be adapted to dry conditions (Glover, 1987). In several field trials established throughout the world, a broad growth variation has been shown within the species (Glover, 1987; Dunsdon and Simons, 1996). However, there is no information available about the ecophysiological basis of this variation. It is well known that under stress conditions, most tree species modify their carbon allocation to shoot and root (Bongarten and Teskey, 1987; Molchanov et al., 1994). In this study, we evaluated the effects of water

---

M. García-Figueroa is Research Assistant and J. J. Vargas-Hernández is Associate Professor, Forestry Department, Instituto de Recursos Naturales, Colegio de Postgraduados, Montecillo, Edo. de México. C.P. 56230, México.

Address correspondence to J. J. Vargas-Hernández at the above address.

This study received financial support from CONACYT through research project 1306A-9206.

[Haworth co-indexing entry note]: "Growth and Biomass Allocation of *Gliricidia sepium* Seed Sources Under Drought Conditions." García-Figueroa, M., and J. J. Vargas-Hernández. Co-published simultaneously in *Journal of Sustainable Forestry* (Food Products Press, an imprint of The Haworth Press, Inc.) Vol. 10, No. 1/2, 2000, pp. 45-50; and: *Frontiers of Forest Biology: Proceedings of the 1998 Joint Meeting of the North American Forest Biology Workshop and the Western Forest Genetics Association* (ed: Alan K. Mitchell et al.) Food Products Press, an imprint of The Haworth Press, Inc., 2000, pp. 45-50. Single or multiple copies of this article are available for a fee from The Haworth Document Delivery Service [1-800-342-9678, 9:00 a.m. - 5:00 p.m. (EST). E-mail address: getinfo@haworthpressinc.com].

stress on growth, nodulation and biomass allocation of seedlings from 12 seed sources of *G. sepium*. This information is useful to identify traits that could be important for adaptation of this species to dry sites.

METHODS

Open-pollinated seeds of *G. sepium* from 12 locations were used in this study. Eight seed sources were collected from the east coast of Mexico (18°15′-19°46′ N, 90°39′-96°25′ W, 15-120 m) and a further four from the west coast (16°40′-18°28′ N, 94°58′-102°34′ W, 20-1100 m). Seedlots were superficially disinfected and germinated in sand, under sterile conditions. Two weeks after emergence, seedlings similar in size from each seed source were transplanted into 1 L polyethylene bags filled with a mix of sand and local agricultural top-soil (2:3, vol.) sterilized at 120°C for 60 minutes. Seedlings were maintained in a greenhouse and inoculated with a *Rhizobium* strain obtained from *G. sepium* nodules collected in northern Veracruz (19°59′ N, 97°13′ W), at an elevation of 520 m above sea level. A factorial design arranged in split-plots with two replications was used. Soil moisture regimes (S_0 = Control; S_1 = drought) were assigned to main plots and seed sources to sub-plots, each sub-plot represented by 15 seedlings. Drought began when seedlings were three months old. Control seedlings were maintained with soil moisture near field capacity, whereas stressed seedlings were subjected to a 30-day drought period.

To determine seedling growth, dry weight partitioning, and nodulation, five seedlings per sub-plot were destructively sampled from seedling groups at the beginning, at the end, and 30 days after the drought period. Variables measured included stem diameter and height, number of nodules, leaf area, and dry weight of stem, foliage, roots and nodules. To test for differences in dry weight allocation patterns among seed sources in response to drought, allometric regressions were performed for each group of seedlings using the model:

$$\text{Ln (shoot mass)} = a + b \text{ Ln (root mass)}.$$

RESULTS AND DISCUSSION

Effect of Drought on Seedling Growth and Nodulation

Except for seedling height, all other traits were significantly affected ($p \leq 0.05$) by drought (Table 1). Number and dry weight of nodules were the traits most affected by water stress, showing over 90% reduction at the end of drought (Table 2). Water stress also severely affected leaf area and other associated variables, such as shoot and total weight. However, root growth was less affected by drought, implying shifts in photosynthate allocation towards the root caused were by water stress. Differences between treatments 30 days after the end of drought were even larger, indicating that seedlings were not able to recover yet from water stress (Table 2). Khuns and Gjerstad (1988) showed that water stress effects in *Pinus taeda* L.

TABLE 1. Analysis of variance (mean squares) for growth and nodule traits of *Gliricidia sepium* seedlings after a 30-day drought period.

	Source of variation[†]					
Trait	Drought (D)	Block	Error a	Seed source (S)	D × S	Error b
Stem diameter (mm)	13.890 ns	0.380	3.560	2.145 ns	4.610 ns	3.740
Stem height (cm)	121.620 *	69.35	0.426	81.51 *	18.45 *	7.951
Root length (cm)	154.580 **	18.84	0.001	64.19 **	11.50 ns	19.22
Leaf area (cm^2)	971863.2 *	45.14	562.1	15547.4 *	13772.5 *	5817.2
Root weight (mg)	1.602 *	0.056	0.005	0.098 **	0.006 ns	0.010
Stem weight (mg)	1.993 *	0.241	0.001	0.182 **	0.039 *	0.015
Leaf weight (mg)	0.693 *	0.491	0.004	0.808 *	0.779 *	0.329
Total weight (mg)	17.970 *	0.431	0.013	0.616 **	0.189 *	0.084
Number of nodules	271.120 **	0.671	0.003	33.04 *	35.25 *	14.11
Nodule weight (mg)	1.325 *	0.148	0.008	0.086 ns	0.065 ns	0.059

[†] For a given source: ns = non-significant at P = 0.05; * = significant at P = 0.05; and ** = significant at P = 0.01.

TABLE 2. Mean values for growth and nodule traits of *Gliricidia sepium* seedlings under two soil moisture conditions (S_0 = control; S_1 = drought).

	At the end of drought			30 days after drought		
Trait	S_0	S_1	S_1/S_0 (%)	S_0	S_1	S_1/S_0 (%)
Stem diameter (mm)	3.7	3.2	86	4.1	2.9	71
Stem height (cm)	24.7	23.3	94	24.9	23.5	94
Root length (cm)	24.9	23.4	94	24.2	22.8	94
Leaf area (cm^2)	189	58	31	116	9	8
Root weight (mg)	341	320	94	539	369	68
Stem weight (mg)	629	446	71	1021	563	55
Leaf weight (mg)	611	262	43	452	134	30
Total weight (mg)	1581	1028	65	2012	1066	53
Number of nodules	2.4	0.2	8	5.1	0.0	0
Nodule weight (mg)	2.5	0.2	8	4.4	0.0	0

seedlings remained several days after stress was relieved, because of time needed for recovery. A reduction in the shoot:root ratio is considered an adaptive response of tree seedlings to water stress (Pereira and Kozlowski, 1976; Molchanov et al., 1994).

Variation Among Seed Sources

Significant differences among seed sources were found for all seedling growth traits, except for stem diameter and nodule weight (Table 1). In addition, a significant seed source × environment interaction was detected for most traits, showing a broad variation among seed sources in growth and adaptation in response to water stress (Table 1). Rank changes in growth rate were common when seed sources were compared at both soil water regimes. However, no geographic pattern in total dry weight was observed. Seed sources also differed in dry matter partitioning between shoot and root in response to water stress (Figure 1). In well-watered plots, all seed sources had similar slopes in the allometric regression of shoot to root mass ($0.613 \leq b \leq 0.741$), even though there were differences in initial size of seedlings. Under drought conditions, slope of regression lines varied from 0.099 up to 0.473 (Figure 1). 'Barroso' (B), 'San Marcos' (SM), and 'Los Amates' (LA) showed the lowest slope, indicating lower biomass allocation to shoot as drought progressed. On the other hand, 'Escarcega' (E) and 'Playa Azul' (PA) had higher slope values. Differences in leaf retention among seed sources might account for some of the differences in slope. In fact, 'Barroso' and 'San Marcos' had the highest leaf fall. However, 'Los Amates', with a similar regression slope, had the lowest leaf fall of all seed sources. Bongarten and Teskey (1987) found similar changes in dry weight allocation patterns among seed sources of *Pinus taeda* in response to soil moisture. In that study, however, differences among seed sources were higher under favorable conditions. Thus, our results suggest that in *Gliricidia sepium*, differences in biomass allocation between shoot and root might be important for adaptation to water stress.

CONCLUSION

Water stress had a severe impact on seedling growth and nodule development in *Gliricidia sepium*. Even though differences in growth

FIGURE 1. Relationship between shoot weight and root weight for *Gliricidia sepium* seedlings from several seed sources (B = 'Barroso'; SM = 'San Marcos'; LA = 'Los Amates'; E = 'Escarcega'; PA = 'Playa Azul') grown under two soil moisture conditions (S_0 = Control; S_1 = Drought).

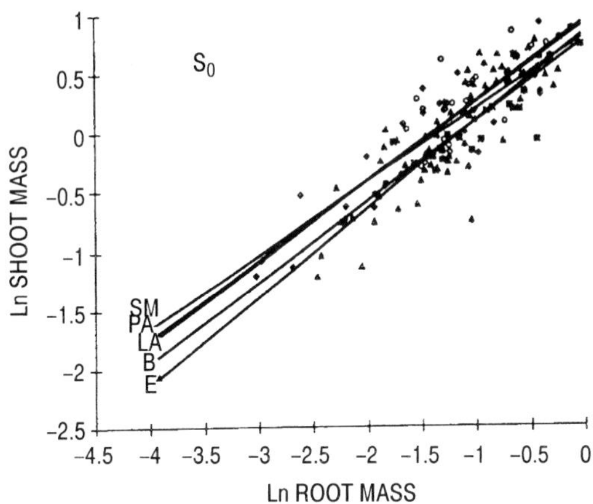

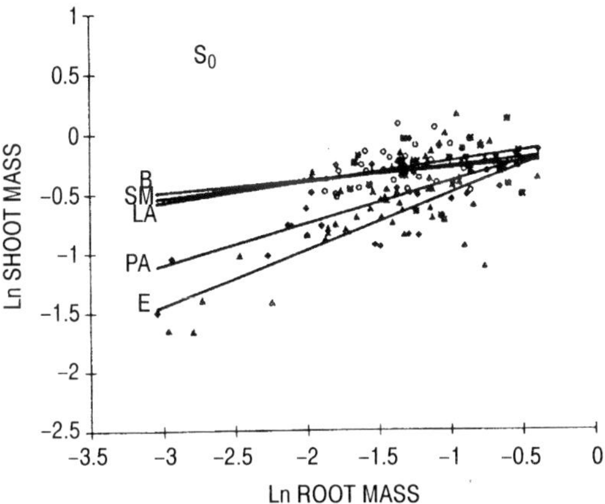

rate among seed sources decreased under drought, a significant seed source × environment interaction was detected, indicating a broad variation among them in adaptation and growth in response to water stress. In addition, seed sources differed in the ratio of biomass allocation to shoot and root under drought. No geographic pattern could be detected in growth rate or biomass allocation among seed sources.

REFERENCES

Bongarten, B.C. and R.O. Teskey. 1987. Dry weight partitioning and its relationship to productivity in loblolly pine seedlings from seven sources. For Sci 33: 255-267.

Dunsdon, A.J. and A.J. Simons. 1996. Provenance and progeny trials. In: pp. 93-118. J.L. Stewart, G.E. Allison, A.J. Simons (Eds.). *Gliricidia sepium*, Genetic Resources for Farmers. Oxford Forestry Institute, University of Oxford. Tropical Forestry Papers 33.

Glover, N. 1987. Variation among provenances of *Gliricidia sepium* (Jacq.) Walp. and implications for genetic improvement. In: pp. 168-173. D. Withington, N. Glover, J.L. Brewbaker (Eds.). *Gliricidia sepium* (Jacq.) Walp.: Management and Improvement. Nitrogen Fixing Tree Association, Waimanalo, Hawaii, USA. Special Publication 87-01.

Khuns, M.R. and D.H. Gjerstad. 1988. Photosynthate allocation in loblolly pine (*Pinus taeda*) seedlings as affected by moisture stress. Can J For Res 18: 285-291.

Molchanov, A.G., Mamaev V.V. and A.Y. Gopius. 1994. Fractional composition of the phytomass of oak seedlings in conditions of soil drought. Lesovedenie 1: 71-76.

Pereira, J.S. and T.T. Kozlowski. 1976. Leaf anatomy and water relations of *Eucalyptus camaldulensis* and *E. globulus* seedlings. Can J Bot 54: 2868-2880.

Sanguinga, N., Danso, S.K.A. and G.D. Bowen. 1992. Variation in growth, sources of nitrogen and N-use efficiency by provenances of *Gliricidia sepium*. Soil Biol Biochem 24: 1021-1026.

Stewart, J.L., Allison, G.E. and A.J. Simons (Eds.). 1996. *Gliricidia sepium*, genetic resources for farmers. Oxford Forestry Institute, University of Oxford. Tropical Forestry Papers 33. 125 p.

Microsatellite Markers for Alpine Larch and Western Larch

P. D. Khasa
B. Jaquish
B. P. Dancik

INTRODUCTION

In British Columbia, alpine larch (*Larix lyallii* Parl.) is a valuable tree for watershed protection, avalanche control, and wildlife habitat. Western larch (*Larix occidentalis* Nutt.) is a commercial tree that is highly valued for lumber and pulp production. Previous research has shown that western larch and alpine larch form putative natural hybrids in the zones of range overlap (Carlson et al., 1990). In a 4-month nursery test, western larch seedlings were ca. four times taller than the hybrids and 10 times taller than alpine larch but the stems of hybrids

P. D. Khasa is Research Assistant Professor, and B. P. Dancik is Professor, Department of Renewable Resources, University of Alberta, Edmonton, Alberta, Canada T6G 2H1.

B. Jaquish is Research Scientist, Research Branch, B.C. Ministry of Forests, Kalamalka Research Station, 3401 Reservoir Road, Vernon, British Columbia, Canada V1B 2C7.

Address correspondence to P. D. Khasa at the above address (E-mail: dkhasa@gpu.srv.ualberta.ca).

The authors are grateful to Dr. Craig Newton (BCR Inc., Vancouver) for preparing the enriched library.

Financial support provided by the Forest Renewal BC (FRBC Reference: HQ96090) is gratefully acknowledged.

[Haworth co-indexing entry note]: "Microsatellite Markers for Alpine Larch and Western Larch." Khasa, P. D., B. Jaquish, and B. P. Dancik. Co-published simultaneously in *Journal of Sustainable Forestry* (Food Products Press, an imprint of The Haworth Press, Inc.) Vol. 10, No. 1/2, 2000, pp. 51-56; and: *Frontiers of Forest Biology: Proceedings of the 1998 Joint Meeting of the North American Forest Biology Workshop and the Western Forest Genetics Association* (ed: Alan K. Mitchell et al.) Food Products Press, an imprint of The Haworth Press, Inc., 2000, pp. 51-56. Single or multiple copies of this article are available for a fee from The Haworth Document Delivery Service [1-800-342-9678, 9:00 a.m. - 5:00 p.m. (EST). E-mail address: getinfo@haworthpressinc.com].

were significantly thicker than those of either parent (Carlson, 1994). Since hybrid seedlings are robust and stocky, they may be useful in revegetating cold, moist sites between the elevational ranges of alpine and western larches (Carlson, 1994).

As yet, there is little information available on genetic variation in these species for assisting decision-making in programs of gene resource management and tree improvement. Indeed, programs of larch genetics research and tree improvement are of recent origin in B.C. (Jaquish et al., 1995). In the U.S., studies of genetic variation of western larch began in the mid-1970s and have focused on populations in the central and southern portion of the species' range. A study by Fins and Seeb (1986), examined the average levels of allozyme variation in 19 U.S. western larch populations from the Inland Empire and found low values of genetic variation, suggesting that genetic drift has played a major role in the genetic history of the species. A recent study on allozyme variation of western larch in B.C., however, has indicated that western larch harbors levels of genetic variation comparable to other wide-ranging, long-lived, outcrossing-wind-pollinated species (Jaquish and El-Kassaby, 1998).

Microsatellite markers or simple sequence repeats (SSRs), consisting of tandem repeats of relatively short sequences (1-10 bases long), are useful genetic markers in tree population genetic studies (Powell et al., 1995). These hypervariable markers are preferred (Hughes and Queller, 1993; Queller et al., 1993), especially in populations with little variation such as the U.S. western larch populations (Fins and Seeb, 1986); and alpine larch populations which are distributed in isolated patches (Arno, 1990). Two advantages of microsatellite loci are that they are generally highly variable codominant loci and that they are PCR-based markers. Attempts to detect losses of variation in isolated patches or to make comparisons of genetic variability between populations would be enhanced through the use of more variable loci. Therefore, microsatellites are more informative than allozymes in population genetic studies, taking advantage of their length polymorphism (Powell et al., 1995). A disadvantage, however, at least in plants, has been the considerable time taken in development of microsatellite loci.

METHODS

Two strategies (database searches and cloning) were used to develop microsatellite markers for alpine larch and western larch.

Database Searches

Database searches (in GenEMBLPlus) for larch were performed using the Genetics Computer Group (GCG, 1994). We searched for all types of mononucleotide repeats with a minimum number of seven repeats; and for all types of di-, tri-, and tetranucleotide repeats with a minimum number of four repeats without mismatch by using the FINDPATTERNS program. Specific set of primers (18 to 24 bp) complementary to their flanking regions were designed with the aid of the program PRIME of the GCG sequence analysis software.

Cloning of Microsatellite Loci

Briefly, alpine larch genomic DNA was extracted and purified (Khasa et al., 1999). Both non-enriched and enriched library strategies were used. In the non-enriched library strategy, genomic DNA of alpine larch was digested with *Sau*3AI. Size-selected (approximately 300-700 bp) fragments were excised from the agarose gel, purified with 'prep A gene' kit (BIO-RAD) and then cloned into *Bam*HI digested and dephosphorylated pUC18 (Pharmacia Biotech). Following transformation of competent *Escherichia coli* DH5(cells, the size-selected library was screened by colony hybridization using ^{32}P-labeled poly (GA+CA) in 6 × SSPE, 5 × Denhardts, 1% SDS, with 3 post-hybridization washes in 0.5 × SSC, 0.1% SDS (Sambrook et al., 1989). Putative positive colonies were replated and a second round screening carried out to confirm their positive status by the polymerase chain reaction (PCR) amplification of the inserts from forward and reverse primers flanking the cloning site. Enriched library with CA repeats was constructed by Dr. Craig Newton (BCR Inc., Vancouver, B.C.). Sequencing of positive clones from both non-enriched and enriched libraries, was performed using an ABI 373 automated DNA sequencer. Microsatellite primers were also designed with the aid of the program PRIME of the GCG sequence analysis software.

Screening of Microsatellite Primers

Microsatellite primers developed from both strategies were used to amplify microsatellite loci using the PCR technique. PCR cocktails and cycling conditions are presented elsewhere (Khasa et al., 1999).

The PCR products along with 100 and 20 base pair ladder-DNA sizing markers (Gensura, Bio/Can Scientific), were electrophoresed for 3 h in 2% agarose gel in 1 × TBE (tris-borate-EDTA) buffer into which the ethidium bromide dye was incorporated. The PCR products that amplified cleanly at or near the expected size were tested for their polymorphism on 6% denaturing polyacrylamide gels in 8 M urea and 1 × TBE buffer run at 55 W constant power for 3 h. Gels were fixed, dried and stained with silver nitrate using a 'DNA Silver Staining System' kit (Promega Silver Sequence™ Staining protocol) as modified by Echt et al. (1996).

RESULTS AND DISCUSSION

For larch, very little DNA sequence information is available. The search in GenEMBLPlus database for larch revealed 2 mononucleotides with a minimum number of 6 repeats and 1 trinucleotide with a minimum number of 5 repeats from the sequences of the *rbc*L gene (Table 1). The remaining repeats conveyed in Table 1 were found by the cloning/sequencing strategy. Enriched library strategy yielded clones that contained more microsatellites than the non-enriched one. The simple sequence repeats found were divided into three categories: perfect repeat sequences without interruptions, imperfect repeat sequences with interruption(s), and compound repeat sequences with adjacent tandem simple repeats of a different sequence. In general and more specifically in animals, the longest run of uninterrupted simple sequence repeats (i.e., minimum number of 10 repeats) is found to be the best predictor of informativeness (Webber, 1990). However, in plants, a shorter run of uninterrupted simple sequence repeats can also be informative even though less polymorphic (Lagercrantz et al., 1993; Terauchi and Konuma, 1994). Therefore, we screened and retained for testing short simple sequence repeats, having at least 6 and 5 repeats for dinucleotide and trinucleotide repeats, respectively.

In total, fourteen microsatellite regions have been isolated and characterized and 7 out of 14 are useful markers (Table 1). Most of the identified repeat units found were dinucleotide repeats, within many of them being compound motifs consisting of two different dinucleotide repeats. The dinucleotide repeats were long repeats (4 to 20 repeat units). Trinucleotides were rarer than dinucleotides. Although some microsatellites displayed extremely high levels of allelic variation as

TABLE 1. Larch simple sequence repeats.

Repeats*	GenBank accession No/library clone	Locus name	Optimal annealing temperature (°C)	Expected PCR product length (bp)	Remarks
$(TCT)_5$	X54464, *L. laricina*	*UAKLLA1*	53	181	Polymorphic
$(C)_6$	X63663, *L. occidentalis*	*UAKLOC1*	57	226	Monomorphic
$(A)_7$	X16039, *L. laricina*	*UAKLLA2*	52	77	Monomorphic
$(CA)_{17}$	LLY1 (NEL)	*UAKLLY1*	47	100	ND
$(CA)_5GA(CA_4)$	LLY2(EL)	*UAKLLY2*	55	260	Polymorphic
$(TA)_9(TG)_{16}$	LLY3 (EL)	*UAKLLY3*	55	500	ND
$(TC)_6(AC)_8NNN(AC)_{12}$	LLY4 (EL)	*UAKLLY4*	65	250	Multi-locus
$(GT)_{17}$	LLY6 (EL)	*UAKLLY6*	48	247	Polymorphic
$(TG)_8$	LLY7 (EL)	*UAKLLY7*	50	186	Polymorphic
$(CA)_5AA(CA)_7$	LLY8 (EL)	*UAKLLY8*	60	401	ND
$(T)_8 \ldots (TG)_9$	LLY9 (EL)	*UAKLLY9*	54	219	Polymorphic
$(CA)_5AA(CA)_7$	LLY10 (EL)	*UAKLLY10a*	54	281	Polymorphic
$(CA)_{10}$	LLY10 (EL)	*UAKLLY10b*	60	119	ND
$(AT)_5(GT)_{20}(GA)_6(A)_7$	LLY13 (EL)	*UAKLLY13*	60	166	Polymorphic
$(CA)_{10}TA(CA)_4TA(CA)_4$	LLY14 (EL)	*UAKLLY14*	60	254	Polymorphic

* Information on primer sequences may be obtained from the authors. NEL, non-enriched library; EL, enriched library. The nomenclature of microsatellite loci follows this convention: the name of institute or location of origin and author, and the species, followed by a numerical designator, ND, not determined yet.

compared to allozymes, not all did. The polymorphic microsatellite loci found exhibited from 2 to 7 alleles which make them suitable for population genetic studies. In our case, microsatellite markers developed for one species (alpine larch) successfully amplify polymorphic loci in its related species (western larch). These results indicate that SSR loci are conserved among closely related species.

CONCLUSIONS

Microsatellite markers are co-dominant markers that exhibit high levels of allelic variation attributes making them suitable as genetic markers for the genetics of larch species. The practical implications of these findings are to: (i) rapidly and easily identify seed lots in operational nursery practices, (ii) depict the patterns of mating system and genetic structure, (iii) conduct pedigree studies, and (iv) facilitate a program in genetic improvement and gene management of alpine larch, western larch and related species.

REFERENCES

Arno, S.F. 1990. Alpine larch. Pp. 152-159. *In*: Sylvics of North America. Vol. 1., Conifers. R.M. Burns and B.H. Honkala (Technical coordinators). U.S. Dep. Agric. Agric. Handb. 654.

Carlson, C.E. 1994. Germination and early growth of western larch (*Larix occidentalis*), alpine larch (*Larix lyallii*), and their reciprocal hybrids. Can. J. For. Res. 24: 911-916.

Carlson, C.E., S.F. Arno and J. Menakis. 1990. Hybrid larch of the Carlton Ridge research natural area in western Montana. Natural Areas Journal 10: 134-139.

Echt, C.S., P. May-Marquardt, M. Hseih and R. Zahorchak. 1996. Characterization of microsatellite markers in eastern white pine. Genome 39:1102-1108.

Fins, L. and L.W. Seeb. 1986. Genetic variation in allozymes of western larch. Can. J. For. Res. 16: 1013-1018.

Genetics Computer Group, Inc. 1994. The Wisconsin Sequence Analysis Package, Version 8, University Research Park, Madison, Wisconsin, USA.

Hughes, C.R. and D.C. Queller. 1993. Detection of highly polymorphic microsatellite loci in a species with little allozyme polymorphism. Mol. Ecol. 2: 131-137.

Jaquish, B. and Y. A. El-Kassaby. 1998. Genetic variation of western larch in British Columbia and its conservation. J. Heredity 89:248-253.

Jaquish, B., G. Howe, L. Fins, and M. Rust. 1995. Western larch tree improvement programs in the Inland Empire and British Columbia. Pp. 452-460. *In*: Ecology and management of *Larix* forests: a look ahead. Proc. of an International Symposium. Schmidt, W.C. and K.J. McDonald (eds.), October 1992, Whitefish, MT, USA. Gen. Tech. Trp. INT-GTR-319. USDA Forest Service, Intermountain Research Station, Ogden, UT.

Khasa, P.D., B. Jaquish, S. Pluhar and B. P. Dancik. 1999. Isolation, characterization and inheritance of microsatellite loci in alpine larch and western larch. Genome (in press).

Lagercrantz, U., H. Ellegren and L. Andersson. 1993. The abundance of various polymorphic microsatellite motifs differs between plants and vertebrates. Nucleic Acids Res. 21: 1111-1115.

Queller, D.C., J.E. Strassmann, and C.R. Hughes. 1993. Microsatellites and kinship. Tree 8: 285-290.

Powell, W., Morgante, M., McDevitt, R., Vendramin, G.G. and Rafalski, J.A. 1995. Polymorphic simple sequence repeat regions in chloroplast genomes: applications to the populations genetics of pines. Proc. Natl. Acad. Sci. USA 92: 7759-7763.

Sambrook, J., E.F. Fritsch and T. Maniatis. 1989. Molecular cloning: a laboratory manual. 2nd ed. Cold Spring Harbor Laboratory Press, New York, USA.

Terauchi, R. and A. Konuma. 1994. Microsatellite polymorphism in *Dioscorea tokoro*, a wild yam species. Genome 37: 794-801.

Webber, J.L. 1990. Informativeness of human (dC-dA)n.(dG-DT)n polymorphisms. Genomics 7: 524-530.

Range-Wide Genetic Variation in Port-Orford-Cedar (*Chamaecyparis lawsoniana* [A. Murr.] Parl.): I. Early Height Growth at Coastal and Inland Nurseries

Jay H. Kitzmiller
Richard A. Sniezko

INTRODUCTION

Public resource managers and scientists in Oregon and California are conducting a species conservation program to protect Port-Orford-

---

Jay H. Kitzmiller is Regional Geneticist, Pacific Southwest Region, USDA Forest Service, Genetic Resource Center, 2741 Cramer Lane, Chico, CA 95928.

Richard A. Sniezko is Geneticist, Pacific Northwest Region, USDA Forest Service, Dorena Tree Improvement Center, Cottage Grove, OR 97424.

The authors are grateful to Lee Riley and crew at Dorena Tree Improvement Center; Lavelle Frisbee, Ruth Trimble, Bill Jones, and Jim Nelson at Humboldt Nursery for measuring thousands of trees and supporting this study; Jim Jenkinson and Roger Stutts, Institute of Forest Genetics, Placerville, CA for initial project coordination (1994-95); and Bill Powers and Chuck Frank, Klamath NF for their technical support.

Funding was provided by Forest Pest Management, USDA Forest Service.

[Haworth co-indexing entry note]: "Range-Wide Genetic Variation in Port-Orford-Cedar (*Chamaecyparis lawsoniana* [A. Murr.] Parl.): I. Early Height Growth at Coastal and Inland Nurseries." Kitzmiller, Jay H., and Richard A. Sniezko. Co-published simultaneously in *Journal of Sustainable Forestry* (Food Products Press, an imprint of The Haworth Press, Inc.) Vol. 10, No. 1/2, 2000, pp. 57-67; and: *Frontiers of Forest Biology: Proceedings of the 1998 Joint Meeting of the North American Forest Biology Workshop and the Western Forest Genetics Association* (ed: Alan K. Mitchell et al.) Food Products Press, an imprint of The Haworth Press, Inc., 2000, pp. 57-67. Single or multiple copies of this article are available for a fee from The Haworth Document Delivery Service [1-800-342-9678, 9:00 a.m. - 5:00 p.m. (EST). E-mail address: getinfo@haworthpressinc.com].

cedar (*Chamaecyparis lawsoniana*) against a major threat from a highly virulent, exotic root disease fungus, *Phytophthora lateralis*. Port-Orford-cedar (POC) is narrowly restricted to northwestern California and southwestern Oregon where it grows in a variety of habitats, including those with ultramafic soils, from sea level to 1950 m. The genetic phase of this concerted effort includes genetic testing, selection, and breeding to:

a. determine the genetic structure of populations,
b. develop disease resistant strains, and
c. identify adaptive zones for use of resistant stock in restoration plantings.

Genetic conservation of POC is justified by:

a. the species' ecological amplitude and importance,
b. its high commercial value for both timber and ornamental markets,
c. the increasing mortality from root rot in natural stands, and
d. the apparently low disease resistance.

In 1995, POC nursery stock was grown from range-wide seed collections for three types of genetic tests. Root disease resistance testing on a large scale began using 344 wind-pollinated (W-P) families at Oregon State University (Sniezko et al., 1996). A long-term common garden study was outplanted on four forest sites in 1996. A short-term common garden study was established in 1996 at a coastal (Humboldt Nursery) and an inland (Dorena Tree Improvement Center) site using 298 W-P families. This paper presents the two-year height growth results from the short-term "raised bed" study.

Much is known about the allozyme diversity, ecology, and physiology of the species (Millar and Marshall, 1991; Millar et al., 1992, unpublished; Zobel and Lui, 1980; Zobel, 1983; Zobel et al., 1985; Zobel, 1998). These previous studies found relatively low genetic differentiation among populations. However, the genetic structure of growth, survival, and disease resistance traits from range-wide common garden tests is not known.

METHODS

Seed Collection and Sample Populations

Seed was collected from 344 healthy parent trees in 52 native stands during 1991-94 by the United States Forest Service (USFS) and Bureau of Land Management (BLM). Two bulk seed lots from Humboldt County, California were also included. Stands were sampled throughout much of the ca 350 km NW-SE range from the extreme northwestern portion (Oregon Dunes, a few meters from the ocean) to the extreme southeastern stands (Pond Lily Creek, Upper Trinity River at 1630 m elevation and 144 km inland). Sample trees were grouped into 10 regional watersheds: six in Oregon and four in California, and into 52 stands: 36 in Oregon and 16 in California (Table 1). Within stands, most sample trees were located within 1 km distance and 100 m in elevation. However, the distribution of watersheds, stands within watersheds, and trees within stands was uneven.

Raised Bed Study Design

In 1995, 300 seed lots were grown in 4 cm^3 containers at the Simpson Korbel Nursery, Korbel, California. In spring 1996, the 1-0 seedlings were transplanted to two locations: Dorena Tree Improve-

TABLE 1

Regional Watershed	No. Stands	No. Trees	Elev. Range (m)	Latitude Range (deg)	Longitude Range (deg)	2-Yr Ht (cm)
Trinity	2	9	1585-1615	41.0885-.1255	122.4720-.5301	80.0
Sacramento	3	30	1143-1585	41.2200-.2500	122.3959-.4600	85.9
Klamath	3	24	914-1372	41.0000-.8234	123.4651-.9000	99.0
Smith	8	40	402-1585	41.7237-.9657	126.6493-124.0690	99.4
Illinois	2	13	1024-1067	42.0332-.1250	123.3553-.5535	102.2
Applegate	4	29		42.1188-.2073	123.2789-.4057	104.1
Rogue	6	28	664-1097	42.4277-.6917	123.7248-124.2843	105.4
Coquille	18	82	122-838	42.7083-.2600	123.7800-124.1333	115.9
Dunes	4	26	15-59	43.3400-.4500	124.2500-.3400	124.4
Sixes and Elk	2	19	61-246	42.7799-.8333	124.4315-.4833	127.9

ment Center (DTIC), Cottage Grove, Oregon, and Humboldt Nursery (HN), McKinleyville, California. The HN site is only 3.2 km from the ocean at 76 m elevation near the southern end of the range. DTIC is 100 km from the coast at the northernmost latitude of the species and at 245 m elevation. At DTIC the experimental design was a randomized, complete block with six blocks and 300 seed lots (includes two bulk lots) arranged in 3-tree family row plots. At DTIC all six blocks were artificial raised beds with organic rooting medium 40 cm deep over a gravel base. At DTIC, three of the blocks were shaded with 47% shade-cloth during the growing season and three were unshaded. At HN, the design was similar, except three of the blocks were raised beds (HN-RB) and about 20 m away were three conventional nursery beds (HN-NB). All HN blocks had native mineral soil and were transplanted one month later than DTIC. In addition, HN-RB were sheltered from wind and partially shaded by a mature spruce canopy, while the HN-NB were fully exposed to sun and wind. Thus, HN treatments represented two contrasting microenvironments confounding differences in exposure to wind, sun, and temperature extremes. Spacing of trees was 210, 204, and 160 cm^2 per seedling, respectively, for DTIC, HN-NB, and HN-RB. All trees were irrigated and fertilized at maintenance levels.

Traits Measured and Statistical Analysis

Treatment design inconsistencies (spacing, type of bed, and shading) at HN resulted from lack of funding and materials for constructing additional raised beds. Confounding effects from other factors with treatment and treatment interaction effects at HN necessitated analyzing the treatments first as separate experiments (by location-treatment), then by location combined across treatment, and finally, combined across both treatment and location. This approach aided interpretation of results.

Seed weight and percent filled seed were available for 223 families. Total height (longest leader), mortality (summarized as arcsin [proportion dead]$^{1/2}$), and shoot condition (dieback, foliage discoloration) were taken at the end of the first and second year after transplanting. Simple correlations among traits were made using 223 families with both seed and seedling data. Regressions for height on location variables were based on all 298 families sown.

For height traits, dead and unhealthy trees were removed prior to analysis. Means for missing plots (15 at DTIC, 12 from shaded blocks, 2 at HN) were estimated by adjusting location means for block and family means. All analyzes involving height were based on plot means and were balanced at the family level. SAS (ver. 6.12) was employed for all analyses. Where necessary, F-tests were synthesized from expected mean squares, and degrees of freedom were calculated (Satterthwaite, 1946).

Expected mean squares for a "combined experiment" analysis across locations and treatments are shown in Table 2. Locations and treatments were considered as fixed effects, blocks and families as random. For the purposes of examining genotype × environment interactions (G- × -E) and estimating variance components, the family effect was partitioned into the nested genetic sources: watersheds, stands within watersheds, families within stands, along with their associated interactions with locations.

RESULTS

Locations and Bed Treatments

Trees at Dorena grew rapidly, averaging 117 cm for total height (2-yrs after transplanting) which was 21% taller than trees at Hum-

TABLE 2. ANOVA and expected mean squares for combined experiments across sites and treatments

	Source	Deg Freedom	MS	Expected Mean Squares
F	Location	p – 1 = 1	1	$\sigma^2 + 298\ \sigma^2_B + 3\ \sigma^2_{PF} + 1788\ \Sigma P^2/1$
F	Trt	t – 1 = 1	2	$\sigma^2 + 298\ \sigma^2_B + 6\ \sigma^2_{TF} + 1788\ \Sigma T^2/1$
	Loc × Trt	(p – 1)(t – 1) = 1	3	$\sigma^2 + 298\ \sigma^2_B + 3\ \sigma^2_{PTF} + 894\ \Sigma PT^2/1$
	Blocks (Loc × Trt)	pt(b – 1) = 8	4	$\sigma^2 + 298\ \sigma^2_B$
	Families	f – 1 = 297	5	$\sigma^2 + 12\ \sigma^2_F$
	Loc × Fam	(p – 1)(f – 1) = 297	6	$\sigma^2 + 6\ \sigma^2_{PF}$
	Trt × Fam	(t – 1)(f – 1) = 297	7	$\sigma^2 + 6\ \sigma^2_{TF}$
	Loc × Trt × Fam	(p – 1)(t – 1)(f – 1) = 297	8	$\sigma^2 + 3\ \sigma^2_{PTF}$
	Error	pt(b – 1)(f – 1) = 2376	9	σ^2
	Corr. Total	ptbf – 1 = 3575		

p = 2, t = 2, b = 3, f = 298, MS = sample mean square, F-test for Locations = MS1/(MS4 + MS6 – MS9), F-test for Treatments = MS2/(MS4 + MS7 – MS9), F-test for Loc × Trt = MS3/(MS4 + MS8 – MS9) boldt. At DTIC, shaded trees grew only slightly taller than trees in open sun the first year, but the second year, trees in open sun grew 26% taller ($p < 0.01$) than shaded trees. At HN, the response pattern was different. Sheltered trees were 13% taller the first year ($p = 0.06$), but subsequently trees grew at the same rate in both treatments (Table 3). This location-×-treatment interaction was significant ($p < 0.01$) for second year height growth and for total two-year height ($p = 0.04$). A delay of 1-month in planting date in HN-NB may account for its slower first-year growth. Overall mortality was very low (1.2%) at HN, but was 6.7% at DTIC, where most mortality occurred in shade. At HN, the reverse was true. Gophers caused most mortality in open, barrier-free nursery beds (HN-NB).

Thus, POC benefited slightly from shading and shelter initially before heavy inter-tree shading and root competition began. Then, as inter-tree competition intensified, trees in open sun grew taller and had less mortality at DTIC. However, at HN, mean final height of treatments were similar, perhaps because conditions for growth became less favorable in the sheltered, raised bed where tree spacing was closest and rooting space was restricted. Typical coastal cloudiness reduced the sunlight in both treatments.

TABLE 3. ANOVA for combined experiments across treatments by site and height variable.

		1st Yr Ht		2nd Yr Ht		2nd Yr Ht	Growth
Source	DF	MS	Prob	MS	Prob	MS	Prob
Humboldt:							
Treatments	1	17703.10	.0646	21956.51	.1140	228.74	.6427
Blocks (Trt)	4	2763.39	.0001	5399.85	.0001	911.65	.0001
Families	299	275.73	.0001	1038.35	.0001	343.01	.0001
Trt × Fam	299	36.49	.0158	166.19	.4914	104.74	.2863
Error	1196	30.14		166.26		99.66	
Dorena:							
Treatments	1	1679.14	.8225	52880.56	.1672	73470.43	.0036
Blocks (Trt)	4	29278.45	.0001	18612.34	.0001	1973.22	.0001
Families	299	303.16	.0001	2213.10	.0001	931.66	.0001
Trt × Fam	299	43.65	.0910	255.57	.3207	124.88	.3896
Error	1196	38.75		245.37		121.93	

Genetic Sources and G-×-E Interactions

Differences among families and the location-×-family interaction were significant ($p < 0.01$) for height both years (Table 4). The family mean phenotypic correlation for total height between locations was 0.81, which reflects high overall stability across locations. The Type B genetic correlation adjusted for heterogeneity of family variances (Surles et al., 1995) was 0.94 for total height, which indicates the interaction was caused by scale effects and not by true rank changes among family genotypes. Significant differences were also evident in 1-yr height for treatment-×-family and location-×-treatment-×-family interactions (Table 4), but these subsequently became non-significant. Also, their underlying cause was examined by conducting a joint regression analysis (see Westfall, 1992) for family height regressed across sequential location-treatment means. Eberhardt and Russell's slope coefficient and residual variance parameters were computed and then correlated with Wricke's ecovalence. This showed a significant positive correlation ($r = 0.81$) for ecovalence with heterogeneity of residual variances, but not with heterogeneity of slopes ($r = 0.07$). Thus, scale effects again caused the G-×-E interactions.

Variance Components for Height Traits

The genetic structure of POC for early height is described in the proportion of total variance residing at various levels and from differ-

TABLE 4. ANOVA for combined experiments across sites and treatments by height variable.

		1st Yr Ht		2nd Yr Ht		2nd Yr Ht	Growth
Source	DF	S	Prob	MS	Prob	MS	Prob
Locations	1	118349.30	.0263	363448.82	.0006	66968.30	.0001
Treatments	1	15143.27	.3594	3344.02	.6120	32750.14	.0014
Loc × Trt	1	4238.27	.6209	71493.04	.0405	40949.03	.0007
Blocks (Loc × Trt)	8	16020.92	.0001	12006.10	.0001	1442.43	.0001
Families	299	532.51		2848.30	.0001	1023.34	.0001
Loc × Fam	299	46.38	.0002	403.15	.0001	251.33	.0001
Trt × Fam	299	40.46	.0275	204.21	.5268	112.71	.4129
Loc × Trt × Fam	299	39.67	.0464	217.56	.2525	116.91	.2590
Error	2392	34.44		205.81		110.79	

ent sources (Table 5). Site location effects were assumed fixed and thus excluded from the SAS Restricted Maximum Likelihood model. Genetic main effects of watersheds, stands within watersheds, and families within stands accounted for 47.5% of the total variability in total height, and most (37.4%) of this resided at the watershed level. Families accounted for twice as much as stands. The difference between the tallest and shortest stand means within watersheds varied by 2-31% (ave. = 12%), while that between family means within stands varied by 2-47% (ave. = 22%).

G-×-E interaction components accounted for only 6.1%. These were largely due to location × family "scale" effects. Environmental block effects, which included shade treatment effects here, accounted for 9.9%. The remaining 36.5% was due to random plot effects.

Compared to the first year height components, a shift occurred in second year growth toward a lower component for family, higher for interaction, lower for block, and higher for plot error (Table 5). Intense competition among trees in different plot neighborhoods apparently resulted in inconsistent performance across blocks and locations, especially at the family (plot) level.

Correlations Among Traits and with Geographic Origin

Using the 134 families and 32 sub-stands (spatially separate groups within stands) having both seed and seedling data from only the 1993 seed collections, stands with lighter seed had lower filled seed, shorter trees, and greater mortality at DTIC (Table 6). Using the 223 families and 57 sub-stands having both seed and seedling traits from all collection years, watersheds and sub-stands from higher elevations, more

TABLE 5. Distribution of variance components for height traits

Varcomp Trait	Watershed[1]	Stand/W	Family/S	Loc × W	Loc × S	Loc × F	Block	Error
1st Yr Ht	26.5**	2.8**	9.1**	0.1	0.0	1.2**	35.0**	25.4
HG 2nd Yr	28.0**	2.6**	2.8**	3.6**	0.1	6.7**	11.2**	45.0
2nd Yr Ht	37.4**	3.4**	6.7**	1.6**	0.0	4.5**	9.9**	36.5
Total 2nd Ht	<------	47.5	------>	<------	6.1	------>	<----- 46.4----->	

[1] ** = significant at $p < 0.01$

southern latitudes, and more eastern longitudes had higher mortality at DTIC and had shorter trees at both locations (Table 7). Also, sub-stands with taller trees after the first year had greater height growth during the second year. Regressions for height on location variables were highly significant for latitude, longitude, and elevation (Table 7), but were strongest for watershed elevation (Figure 1). Height decreased 28.2 cm per 1000 m increase in source elevation. Trees from the low Sixes and Elk watershed averaged 60% taller than those from the high Trinity watershed.

CONCLUSIONS

The environmental components–transplanting location and "shade" treatments–had significant and major effects on height growth. Sur-

TABLE 6. Simple correlations between seed traits and mortality and height traits for the 1993 seed collections based on sub-stand means. Critical values are 0.31 (p = 0.05) and 0.41 (p = 0.01).

Traits[1]	% Filled	Dead_Hn	Dead_DTIC	1st Yr Ht	2nd Yr Ht	2nd Yr Ht Gr
Seed Weight	0.60	−0.18	−0.56	0.49	0.43	0.46
% Filled		−0.18	−0.49	0.20	0.18	0.19

[1] % filled seed, arcsin (proportion dead)$^{1/2}$ at HN and DTIC, total 1-, 2-yr height and height growth 2nd yr

TABLE 7. Simple correlations between height, mortality, and source location variables for all seed collection years based on sub-stand means. Critical values are 0.22 (p = 0.05) and 0.31 (p = 0.01).

Traits[1]	2nd Yr Ht	2nd Yr Ht Gr	Dead_HN	Dead_DTIC	El	Lat	Lon
1st Yr Ht	0.98	0.95	−0.25	−0.57	−0.87	0.77	0.77
2nd Yr Ht		0.99	−0.31	−0.60	−0.89	0.81	0.82
2nd Yr Ht Gr			−0.34	−0.61	−0.88	0.81	0.83
Dead_HN				0.26	0.19	−0.31	−0.29
Dead_DTIC					0.37	−0.37	−0.43
El						−0.78	−0.75
Lat							0.80

[1] arcsin (proportion dead)$^{1/2}$) at HN and DTIC, total 1-, 2-yr height and height growth 2nd yr

FIGURE 1. Watershed mean total height by elevation of origin.

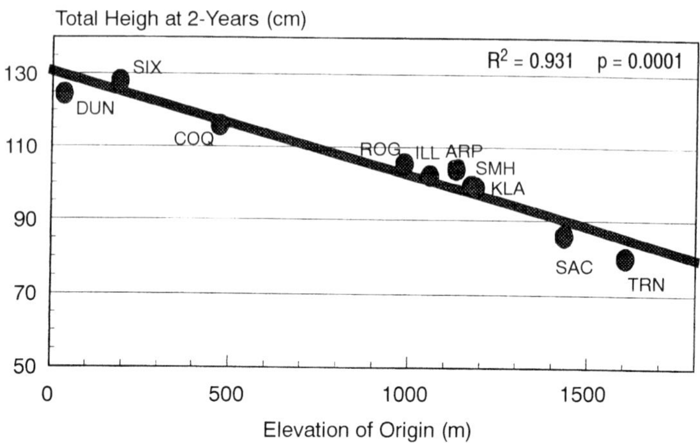

prisingly, the inland location had superior height to the coastal location both years, and "shading" was inferior to open sun the second year. The height growth response of POC families from different geographic regions and stands revealed a strong genetic structure with a well-defined geographic pattern. Height potential was highly related to genetic source at the watershed, stand, and family levels. Strong clinal patterns were found for height potential with source elevation, latitude, and longitude. G-×-E interactions, though statistically significant at watershed and family levels, were minor sources of variability in height, and were due to scale effects rather than rank changes. Southern and high elevation inland sources had low growth potential at both locations, while northern and low elevation coastal sources had high growth potential. Trees from low elevation, northern, and coastal sites had less mortality, higher seed weight and higher filled seed percent.

These tentative results show similar, though much stronger, population structure and geographic patterns as allozymes (Millar et al., 1992, unpublished report). Current results suggest that gene conservation practices should encompass: (a) seed zoning by watershed, subdivided by elevation bands, and (b) protecting the broad gene base for growth, including the adaptive extremes near the northern and southern limits. Long-term field studies and disease screening trials already underway will complement these nursery results and guide restoration planting of resistant stock.

REFERENCES

Millar, C.I., Delany, D.L., and R.D. Westfall. Genetic diversity in Port-Orford-cedar: range-wide allozyme study. Unpublished report, August 22, 1992.

Millar, C.I. and K.A. Marshall. 1991. Allozyme variation of Port-Orford-cedar (*Chamaecyparis lawsoniana*): implications for genetic conservation. Forest Science 37(4): 1060-1077.

Satterthwaite, F.E. 1946. An approximate distribution of estimates of variance components. Biometrics 2:110-114.

Sniezko, R.A., Hansen, E., and J.H. Kitzmiller. 1996. Genetic variation in *Phytophthora lateralis* resistance in Port-Orford-cedar: results of artificial inoculation of 344 families from a range-wide collection. Western Forest Genetics Association. Abstract.

Surles, S.E., T.L. White, and G.R. Hodge. 1995. Genetic parameter estimates for seedling dry weight traits and their relationships with parental breeding values in slash pine. Forest Science 41(3):546-563.

Westfall, R.D. 1992. Developing seed transfer zones. In L. Fins et al. (eds) Handbook of Quantitative Forest Genetics, pp. 313-398. Kluwer Academic Publishers. Netherlands.

Zobel, D.B. and V.T. Liu. 1980. Effects of environment, seedling age, and seed source on leaf resistance of three species of *Chamaecyparis* and *Tsuga chinensis*. Oecologia 46:412-419.

Zobel, D.B. 1983. Twig elongation patterns of *Chamaecyparis lawsoniana*. Bot. Gaz. 144:92-103.

Zobel, D.B., L.F. Roth, and G.M. Hawk. 1985. Ecology, pathology, and management of Port-Orford-cedar *(Chamaecyparis lawsoniana*). USDA For. Serv. Gen. Tech. Rep. PNW-184. 161 p.

Zobel, D.B. 1998. Chamaecyparis forests: a comparative analysis. In A.D. Laderman, editor. Coastally Restricted Forests, pp. 39-53, Oxford University Press, New York.

Genetic Linkage Mapping of Genomic Regions Conferring Tolerance to High Aluminum in Slash Pine

T. L. Kubisiak
C. D. Nelson
J. Nowak
A. L. Friend

INTRODUCTION

Reports of reduced growth and vigor of forest trees in Europe and North America have been accumulating in recent years. In eastern

---

T. L. Kubisiak is Research Geneticist, USDA Forest Service, Southern Research Station, Southern Institute of Forest Genetics, 23332 Highway 67, Saucier, MS 39574 USA.

C. D. Nelson is Project Leader, International Paper Company, Southlands Experiment Forest, 719 Southlands Road, Bainbridge, GA 31717 USA.

J. Nowak is Graduate Student and A. L. Friend is Associate Professor, Department of Forestry, Mississippi State University, Box 9681, Mississippi State, MS 39762-9681 USA.

Address correspondence to T. L. Kubisiak at the above address.

The authors would like to thank Clay D. Ware for his skilled technical assistance with the aluminum screening, and Kristel B. Davis and Glen N. Johnson for their skilled technical assistance with the polymerase chain reaction.

This paper is Journal Article No. FO101 of the Forest and Wildlife Research Centre, Mississippi State University.

[Haworth co-indexing entry note]: "Genetic Linkage Mapping of Genomic Regions Conferring Tolerance to High Aluminum in Slash Pine." Kubisiak, T. L. et al. Co-published simultaneously in *Journal of Sustainable Forestry* (Food Products Press, an imprint of The Haworth Press, Inc.) Vol. 10, No. 1/2, 2000, pp. 69-78; and: *Frontiers of Forest Biology: Proceedings of the 1998 Joint Meeting of the North American Forest Biology Workshop and the Western Forest Genetics Association* (ed: Alan K. Mitchell et al.) Food Products Press, an imprint of The Haworth Press, Inc., 2000, pp. 69-78. Single or multiple copies of this article are available for a fee from The Haworth Document Delivery Service [1-800-342-9678, 9:00 a.m. - 5:00 p.m. (EST). E-mail address: getinfo@haworthpressinc.com].

North America, increased mortality and reduced radial growth rates have been noted for red spruce, frasier fir, and sugar maple. USDA Forest Service inventory data from permanent survey plots has revealed an unexpected reduction of radial growth (~50%) in natural pine forests over the past 30 years (Sheffield and Cost, 1987). Aluminum (Al) ions have been implicated as one of the main factors contributing to this decline on mineral soils at a pH below 5.5 (Johnson and Siccama, 1983). Elevated Al in the soil solution has been suggested as a possible result of acid rain that acts either directly at toxic levels to limit root development, or through selective inhibition of nutrient uptake at sub-toxic levels that results in nutrient imbalances and reduced growth (Taylor, 1991). Ameliorating extremely troublesome areas with lime or nutrient treatments is possible but difficult and expensive. A more promising solution is to breed Al-tolerant trees.

In contrast to annual crop species, little is known about the genetics and physiology of Al tolerance in perennial forest tree species. A number of forest tree species have been classified according to their ability to tolerate Al (Hutchinson et al., 1986; Kelly et al., 1990; Raynal et al., 1990). However, the extent of intraspecific variation in Al tolerance is only now being more fully explored. Geburek and Scholz (1992) noted significant differences in root and shoot growth among six provenances of Norway spruce (*Picea abies* Karst.) subjected to 1.68 mM Al, and suggest that the genetics of the chosen plant material strongly influenced the results of their experiment. Nowak and Friend (1995) report significant differences in various growth parameters among six full-sib loblolly pine (*Pinus taeda* L.) and six full-sib slash pine (*Pinus elliottii* var. *elliottii* Engel.) families subjected to 4.5 mM Al in solution culture. The fact that conifer families often exhibit considerable intraspecific variation in growth responses when grown under Al stress conditions make it a promising candidate trait for genetic dissection using molecular markers.

The objective of this study was to employ DNA-based molecular markers to examine the inheritance of Al tolerance in a single full-sib family of slash pine known to exhibit substantial variation for various growth parameters when grown under high Al conditions (Nowak and Friend, unpublished data).

METHODS

Plant Material

The slash pine family for which results are reported was derived from a cross between the genotypes 18-62(f) × 8-7(m). Both parents are part of a five-tree diallel at the Southern Institute of Forest Genetics (SIFG) in Saucier, Mississippi. These trees were originally selected for growth rate, form, and disease tolerance. Of six full-sib slash pine families originally evaluated for Al tolerance, this family exhibited the greatest variation in percent weight gain when subjected to 4.5 mM Al (Nowak and Friend, unpublished data). Open-pollinated (OP) seeds were also collected from 18-62 and 8-7 to assess the relative Al tolerance of the parents.

Aluminum Tolerance of Parents, and Effect of Aluminum on Family 18-62(f) × 8-7(m)

In order to determine the effect of Al on family 18-62(f) × 8-7(m), 96 progenies were screened, 48 under low (0.01 mM) and 48 under high (4.5 mM) Al. To assess the relative Al tolerance of the parents, a total of 108 and 106 OP progenies from parents 18-62 and 8-7, respectively, were screened under high Al. All seeds were stratified for 18 days at 4°C, planted at a depth of 5 mm in silica sand in leach tubes, and grown for seven weeks. The seedlings were subsequently transferred to hydroponic solutions of Nowak and Friend (1995), minus Al, for 10 days of acclimation. Al tolerance was assessed by replacing these solutions with one containing the addition of 4.5 mM $AlCl_3$. All seedlings were weighed prior to pretreatment, prior to Al treatment, and 21 days after Al treatment. The percent weight gain over the treatment period was determined and used as a metric of Al tolerance.

Mapping Population

A total of 235 additional progenies from the family 18-62(f) × 8-7(m) were screened under high Al. The protocol used to assess Al tolerance was similar to that stated above, except that the seedlings were acclimated for 18 days and then exposed to Al for 28 days. In an attempt to augment the detection of genomic regions conditioning Al tolerance, we initially employed the method of selective genotyping described by Lander and Botstein (1989). A total of 27 progenies were selected from each tail of the percent weight gain distribution.

DNA Extraction and Random Amplified Polymorphic DNA (RAPD) Amplification

DNA was isolated from slash pine needles using a modification of the CTAB-based procedure outlined in Wagner et al. (1987). Oligonucleotide 10-mer primers were obtained from either Operon Technologies (Alameda, Calif., USA) or J.E. Carlson (Univ. of British Columbia, Vancouver, B.C., Canada). DNA amplification followed the protocol outlined in Nelson et al. (1994), modified by doubling the template DNA to 6.25 ng per reaction. Three hundred thirty-six primers were screened against both parents and six full-sib progenies (three from each tail of the percent weight gain distribution). A total of 120 primers were selected for mapping and data were collected for the remaining 48 individuals.

Segregation Analysis

JoinMap (version 2.0, Stam and Van Ooijen, 1995) was utilized to produce a comprehensive genetic linkage map for slash pine. Each polymorphism was tested for goodness of fit to its expected Mendelian segregation ratio using the chi-square (χ^2) test in the JoinMap single locus analysis (JMSLA) module. Two-point linkages were investigated using the JoinMap linkage group assignment (JMGRP) module and a log of the odds ratio (LOD) $\geq$ 4.0. For each of the suggested linkage groups, marker orders were determined using the JoinMap recombination estimation (JMREC) and map construction (JMMAP) modules.

Single-locus analysis of variance (ANOVA) models, in which the individual marker genotypes were used as class variables, were employed to investigate the co-inheritance of each marker locus with percent weight gain (Keim et al., 1990). The model for a straight line, $y = ax + b$ (where y = percent weight gain and x = marker genotype), was employed. An association between a marker and percent weight gain was considered significant if the probability of observing an F-value as large or larger than the observed value was $\leq$ 0.05. The proportion of the phenotypic variance explained by segregation of the marker was determined by the r-square (R^2) value. Marker loci found to be significantly associated with percent weight gain using the selective genotyping approach were further characterized on an additional 132 randomly selected progenies. In an attempt to increase the probability of detecting secondary QTL, multiple marker models were constructed. A separate model was constructed for each linkage group allowing for the removal of variation associated with any other QTL that might be present on the same linkage group. The best multi-variable ANOVA model was determined by using both the stepwise and maximum R^2 improvement methods available in the statistical analysis software SAS® (version 6.12). A set of genome-wide models were then constructed that included all the significant markers located on different linkage groups.

RESULTS

Effect of Aluminum on Family 18-62(f) × 8-7(m) and Relative Aluminum Tolerance of Parents

Results for family 18-62(f) × 8-7(m), an OP family from parent 18-62, and OP family from parent 8-7, grown under low and/or high Al, are presented in Table 1. The effect of Al on the percent weight gain of progenies from family 18-62(f) × 8-7(m) was found to be highly significant (Prob. > F < 0.0001). No significant difference in

TABLE 1. Mean, standard deviation, and range of percent weight gain for slash pine family 18-62(f) × 8-7(m), OP family 18-62, and OP family 8-7 subjected to either low and/or high levels of Al for 21 days.

			Percent Weight Gain		
Family	Treatment[a]	NPA[b]	MPWTGN[c]	SD[d]	Range
18-62(f) × 8-7(m)	Low	48	95.6	12.5	67.5 to 120.7
18-62(f) × 8-7(m)	High	48	67.5	13.3	39.4 to 95.6
18-62 OP	High	108	80.7	17.6	40.8 to 128.6
8-7 OP	High	106	85.4	22.0	3.3 to 167.2

[a]High = 4.5 mM Al, Low = 0.01 mM Al
[b]NPA = number of progeny analyzed
[c]MPWTGN = mean percent weight gain
[d]SD = standard deviation of mean percent weight gain

mean percent weight gain was detected between OP families from the two parents (Prob. > F = 0.297). Bartlett's test of homogeneity of variances did, however, suggest that the variances of percent weight gain for the parents were significantly heterogeneous ($\chi^2_{obs} = 5.296 > \chi^2_{(0.05, 1)} = 3.84$).

Molecular Markers and Linkage Map Construction

A total of 159 RAPD markers were scored on 54 progenies from the mapping population. Of the 159 marker loci, 126 were segregating 1:1 (62 were heterozygous in parent 18-62, and 64 were heterozygous in parent 8-7) and 33 were segregating 3:1 (heterozygous in both parents). Only nine markers (5.6%) were found to deviate significantly from their expected Mendelian inheritance ratio based on χ^2 analyses ($p \leq 0.05$). Of the 159 marker loci analyzed, 129 (81.1%) were linked to 17 groups (three or more loci) and 12 linked pairs at LOD > 4.0.

Molecular Evaluation of Aluminum Tolerance

The distribution of percent weight gain in the mapping population was determined to be normally distributed (n = 235) based on the Wilk-Shapiro test (Prob. < W = 0.7086). Of the 159 RAPD markers, a total of 14 were found to be significantly associated with percent weight gain based on single-marker ANOVAs and the selective genotyping approach. Only 13 of these markers were still significant after an additional 132 randomly selected progenies were included in the analyses (Table 2). Linkage group specific models support the existence of a single region on each of the groups C and E that appear to be conditioning an Al tolerance response. Genome-wide models support the existence of three putative regions; one on group C near marker 351_{2150}, one on group E near marker $Y10_{1125}$, and a third region near the unlinked marker $B08_{0850}$ (Figure 1). A model including all three marker loci as independent variables explained as much as 41.3% of the variation associated with percent weight gain using the selective genotyping approach (n = 50), but only 15.6% after an additional 132 randomly selected individuals were added to the analyses (n = 179). The mean and standard deviation of percent weight gain for individuals harboring different numbers of putative alleles for Al tolerance based on single marker and multiple marker genotypic classes are displayed in Table 3.

TABLE 2. Molecular markers significantly associated with tolerance to high levels of Al in the cross 18-62(f) × 8-7(m) based on single-marker ANOVA models.

Marker	LG[1]	NPA[2]	Prob. > F[3]	R-square[4]	NPA	Prob. > F	R-square
$G09_{0925}$	C	50	0.0131	0.119	179	0.0124	0.035
351_{2150}	C	53	0.0164	0.110	181	0.0067	0.040
531_{1050}	C	53	0.0266	0.091	182	0.0089	0.037
667_{1600}	C	53	0.0424	0.077	179	0.0307	0.026
$Y10_{1125}$	E	53	0.0013	0.182	185	0.0006	0.062
187_{1075}	E	53	0.0019	0.171	183	0.0004	0.067
269_{0700}	E	53	0.0019	0.171	184	0.0007	0.053
597_{1025}	E	51	0.0019	0.177	179	0.0029	0.049
$J01_{1050}$	E	50	0.0039	0.158	181	0.0009	0.059
471_{0800}	E	51	0.0476	0.076	178	0.0148	0.033
110_{2900}	LP	53	0.0384	0.080	181	0.0687	0.013
$B08_{0850}$	UL	52	0.0035	0.155	182	0.0004	0.067
269_{1850}	UL	53	0.0332	0.084	179	0.0199	0.030
452_{1700}	UL	52	0.0466	0.075	180	0.0272	0.027

[1]LG = linkage group designation according to Doudrick 1996
[2]NPA = number of progeny analyzed
[3]Prob. > F = probability of a greater F value
[4]R-square = proportion of the phenotypic data explained by the marker locus

DISCUSSION AND CONCLUSIONS

Analyses of the RAPD marker and Al tolerance data suggest that three unlinked genomic regions appear to condition a response to high Al in the slash pine family 18-62(f) × 8-7(m). A partial RAPD linkage map for the parent 8-7 has been published affording us the opportunity to use the same linkage group designations reported previously for slash pine 8-7 (Doudrick, 1996). Results of this study are in accordance with studies of Al tolerance in other species, such as wheat, where much of the observed variability in Al tolerance can be explained by the existence of as few as one to three genes (Aniol and Gustafson, 1984; Aniol, 1990).

The fact that only three genomic regions were identified in slash pine, however, does not preclude the possibility that additional genomic regions influencing Al tolerance may exist. Genomic regions influencing Al tolerance may have eluded detection because they were not segregating in family 18-62(f) × 8-7(m), were located in genomic regions devoid of molecular markers, or had an effect below the threshold for detection. To address the latter two concerns, we are

Figure 1. Marker orders for slash pine linkage groups C and E. Markers IDs are provided on the right side of each group and genetic distance in Haldane cM are provided on the left. RAPD markers are identified by the manufacturer primer code corresponding to the 10-mer primer responsible for amplification followed by a subscript four digit number indicating the approximate product size in base pairs. Those markers followed by asterisks were associated with Al tolerance; * = P > F < 0.05, ** = P > F < 0.01, and *** = P > F < 0.001.

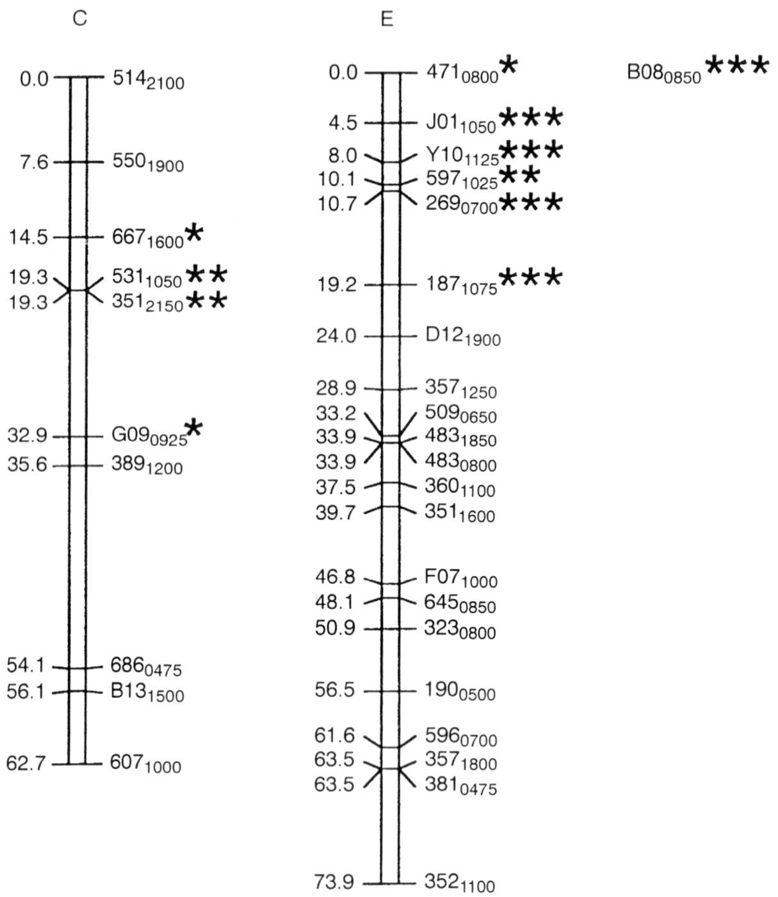

continuing to add markers to the 18-62(f) × 8-7(m) map, and have established all the full-sib progenies used in these analyses in a hedging orchard so that clonal propagules can be screened for Al tolerance. This additional data should afford us better genome coverage and allow us to obtain a more precise estimate (average) of percent weight gain.

TABLE 3. Mean and standard deviation of percent weight gain for individuals from the cross 18-62(f) × 8-7(m) harboring different numbers of putative alleles for Al tolerance based on molecular marker genotypes.

Marker genotype[a]	PNTA[b]	MNTA[c]	N[d]	MPWTGN[e]	SD[f]
Single markers					
aa	0	0	41	96.0	32.2
A_	1 or 2	1.5	141	113.5	37.1
bb	2	2	101	117.6	37.4
Bb	1	1	85	99.2	33.6
cc	2	2	49	124.9	42.0
C_	0 or 1	0.5	134	103.3	33.2
Three markers combined					
A_, bb, cc	5 or 6	5.5	26	133.9	41.4
A_, Bb, cc	4 or 5	4.5	9	128.8	48.9
aa, bb, cc	4	4	5	118.4	39.9
A_, bb, C_	3 or 5	4	52	116.6	33.6
aa, Bb, cc	3	3	8	103.0	28.8
A_, Bb, C_	2 or 4	3	52	97.8	29.9
aa, bb, C_	2 or 3	2.5	14	90.8	32.9
aa, Bb, C_	1 or 2	1.5	14	89.1	29.4

[a]A = tolerance-associated allele at marker 351_{2150}. b = tolerance-associated allele at marker $Y10_{1125}$. c= tolerance-associated allele at marker $B08_{0850}$
[b]PNTA = potential number of tolerance alleles
[c]MNTA = mean number of tolerance alleles
[d]N = number of individuals with particular marker genotype
[e]MPWTGN = mean percent weight gain of individuals with particular marker genotype
[f]SD = standard deviation of mean percent weight gain

Results of this study suggest that high levels of Al have an inhibitory effect on the growth of slash pine, but that tolerance mechanisms exist and appear to be genetically controlled. Potential gains in Al tolerance using marker-aided selection might be possible, provided that the effects of these genomic regions are indeed heritable. Our future goal is to confirm the inheritance of each of the three genomic regions in advanced generations.

REFERENCES

Aniol, A. 1990. Genetics of tolerance to aluminum in wheat (*Triticum aestivum* L. Thell). Plant Soil 123:223-227.

Aniol, A., and J.P. Gustafson. 1984. Chromosome location of genes controlling aluminum tolerance in wheat, rye, and triticale. Plant Physiol. 104:701-705.

Doudrick, R.L. 1996. Genetic recombinational and physical linkage analyses on slash pine. *In:* J.S. Heslop-Harrison (ed). Unifying plant genomes. Symposia of the Society for Experimental Biology, Number L. Society for Experimental Biology, Cambridge, UK. pp 78.

Geburek, T., and F. Scholz. 1992. Response of *Picea abies* (L.) Karst. Provenances to sulfur dioxide and aluminum: a pilot study. Water Air Soil Pollut. 62:227-232.

Hutchinson, T.C., Bozic, L., and G. Munoz-Vega. 1986. Responses of five species of conifer seedlings to aluminum stress. Water Air Soil Pollut. 31:283-294.

Johnson, A.H., and T.G. Siccama. 1983. Acid deposition and forest decline. Environ. Sci. Technol. 17:294-305.

Keim, P., Diers, B.W., Olson, T.C., and R.C. Shoemaker. 1990. RFLP mapping in soybean: association between marker loci and variation in quantitative traits. Genetics 126:735-742.

Kelly, J.M., Schaedle, M., Thorton, F.C., and J.D. Joslin. 1990. Sensitivity of tree seedlings to aluminum. II. Red oak, sugar maple, and European beech. J. Environ. Qual. 19:172-179.

Lander, E.S., and Botstein, D. 1989. Mapping Mendelian factors underlying quantitative traits using RFLP linkage maps. Genetics 121:185-199.

Mason, M.E., and J.M. Davis. 1997. Defense response in slash pine: chitosan treatment alters the abundance of specific mRNAs. Mol. Plant Microbe Interaction 10:135-137.

Nelson, C.D., Kubisiak, T.L., Stine, M., and W.L. Nance. 1994. A genetic linkage map of longleaf pine (*Pinus palustris* Mill.) based on random amplified polymorphic DNAs. J. Heredity 85:433-439.

Nowak, J., and A.L. Friend. 1995. Aluminum sensitivity of loblolly pine and slash pine seedlings grown in solution culture. Tree Physiology 15:605-609.

Raynal, D.J., Joslin, J.D., Thorton, F.C., Schaedle, M., and G.S. Henderson. 1990. Sensitivity of tree seedlings to aluminum: III. red spruce and loblolly pine. J. Environ. Qual. 19:180-187.

Sheffield, R.M., and N.D. Cost. 1987. Behind the decline. J. For. 85:29-33.

Stam, P., and J.W. Van Ooijen. 1995. JoinMap™ version 2.0: software for the calculation of genetic linkage maps. CPRLO-DLO, Wageningen, The Netherlands, 59 pp.

Taylor, G.J. 1991. Current views of the aluminum stress response; the physiological basis of tolerance. Current Topics in Plant Biochem. Physiol. 10:57-93.

Wagner, D.B., Furnier, G.R., Saghai-Maroof, M.A., Williams, S.M., Dancik, B.P., and R.W. Allard. 1987. Chloroplast DNA polymorphisms in lodgepole and jack pines and their hybrids. Proc. Natl. Acad. Sci. USA 84:2097-2100.

Impact of Forest Genetics on Sustainable Forestry– Results from Two Cycles of Loblolly Pine Breeding in the U.S.

Bailian Li
Steve McKeand
Robert Weir

INTRODUCTION

Investment in genetic improvement offers increased forest productivity and an enhanced timber supply. Forests comprise more than 50% of the land cover in the southern U.S. and supply 53% of the timber harvested in the U.S. The southern pines are the most commonly planted species, about 11 million hectares in plantations. The South plants approximately 1.2 billion seedlings annually, 80% of which are loblolly pine seedlings and 20% are slash pine seedlings, and virtually all planting stocks are from genetically improved seedlings from seed orchards. The North Carolina State University-Industry Cooperative

---

Bailian Li is Associate Professor, Steve McKeand is Professor, and Robert Weir is Director, NC State University Tree Improvement Program, Department of Forestry, Box 8002, North Carolina State University, Raleigh, NC 27695-8002 USA.

Support for this work was provided by the North Carolina State University-Industry Cooperative Tree Improvement Program, the Department of Forestry at NCSU, and the North Carolina Agricultural Research Service.

[Haworth co-indexing entry note]: "Impact of Forest Genetics on Sustainable Forestry–Results from Two Cycles of Loblolly Pine Breeding in the U.S." Li, Bailian, Steve McKeand, and Robert Weir. Co-published simultaneously in *Journal of Sustainable Forestry* (Food Products Press, an imprint of The Haworth Press, Inc.) Vol. 10, No. 1/2, 2000, pp. 79-85; and: *Frontiers of Forest Biology: Proceedings of the 1998 Joint Meeting of the North American Forest Biology Workshop and the Western Forest Genetics Association* (ed: Alan K. Mitchell et al.) Food Products Press, an imprint of The Haworth Press, Inc., 2000, pp. 79-85. Single or multiple copies of this article are available for a fee from The Haworth Document Delivery Service [1-800-342-9678, 9:00 a.m. - 5:00 p.m. (EST). E-mail address: getinfo@haworthpressinc.com].

Tree Improvement Program (NCSU-ICTIP) has completed 42 years of genetic improvement for loblolly pine in the southeastern U.S. The 17 industries and five states who are currently members of the Cooperative, plant more than 600,000,000 trees on 350,000 hectares, annually, accounting for 37% of the annual tree planting in the country.

The impact of the tree improvement on forest productivity has been substantial through the two cycles of breeding, testing and selection completed by the NCSU-ICTIP. Trees grown from seeds of first-generation seed orchards have produced 7-12% more volume per acre at harvest than trees grown from wild seed (Talbert, 1982). With additional improvement in value from quality traits (stem straightness, disease resistance, wood density), the estimated genetic gain in value from first-generation breeding is about 20% (Talbert et al., 1985). Second-generation seed orchards are now producing more than 50% of the total seed harvest in the region. Progeny test data from second-generation seed orchards are now available to provide genetic gain estimates. In this paper, we summarize the genetic gains from two cycles of loblolly pine breeding by the NC State Tree Improvement Program and discuss the impact on stand productivity and sustainable forestry.

METHODS

The first generation genetic tests were established with full-sib families generated with an incomplete factorial mating design (Talbert, 1982; Li et al., 1996). Row plots were used to evaluate family variation and compare the selected stock with unimproved check lots. Tests were measured at age 4, 8 and 12 years to estimate average percentage gains in height growth over unimproved check lots (Talbert et al., 1985). Second-generation selections were made in first-generation progeny tests and grafted to establish second-generation seed orchards.

Open-pollinated progeny tests from those second-generation seed orchards were established throughout the Southeast. The number of families in each test series ranged from 19 to 44, and included several unimproved check lots. Each test series generally included four tests established over a two-year period at two locations. The experimental design was a randomized complete block with six blocks and 6-tree row plots. All tests we:e measured for tree height, and some were measured for DBH, stem straightness and rust infection (Li et al.,

1997). The tests were grouped into four general geographic regions, Virginia and northern North Carolina, Atlantic Coastal Plain, Lower Gulf, and Piedmont, for genetic gain estimates (Table 1). The numbers of families ranged from 83 to 285 per region.

Details of the data analysis and genetic gain estimates for those second-generation open-pollinated progeny tests were given by Li et al. (1997). Briefly, a method of the best linear unbiased predictions (BLUP) (Huber, 1993) was used to estimate parental breeding values for 8-year height, measuring the amount of genetic superiority that can be inherited by progeny of parents. The expected genetic gain was then calculated from the predicted breeding values for height. Using unimproved checklots in progeny tests as the baseline, genetic gains were estimated as the percentage over local unimproved check lots. Breeding values for rust infection were calculated as the expected rust infection percentages for parents at a 50% infection level (R-50) in the stands.

Using the methods described by Talbert (1982) and Talbert et al. (1985), volume gains at rotation (age 25) were estimated from 8-year height gains. Briefly, percentage height gains at age eight years were assumed to equate to percentage gains at age 12, and the 8-year differences in height were assumed to equal site index value changes at stand age 12. The site index values were then used with the growth and yield model first developed by Hafley et al. (1982) to estimate the volume in unthinned plantations at age 25 years. The simplifying assumption was that the shape of height over age curves is essentially equivalent for all families and that selection has little impact on other

TABLE 1. Predicted average growth gains (%) for 8-year height gains and 25-year volume gains of second-generation loblolly pine seed orchards for different geographic regions, Virginia and northern North Carolina (VA/NC), Atlantic Coastal Plain, Lower Gulf, and Piedmont. Gains are estimated as the percent difference over unimproved check lots.

	Average for All Families in Region		Average for Top 30% of Families	
Region	% Height Gain	% Volume Gain	% Height Gain	% Volume Gain
VA/NC	8.1	17.0	12.7	27.0
Atlantic Coast	6.1	12.8	12.4	26.3
Lower Gulf	7.7	16.0	14.6	31.2
Piedmont	9.9	20.7	16.0	34.9
Average	8.0	16.5	13.9	29.9

parameters of stand growth and yield such as mortality functions and height-diameter relationships. These assumptions were found to be reasonable in most situations (Buford and Burkhart, 1987).

RESULTS

Genetic gains for 25-year volume from an unthinned plantation are approximately 7% on average for unrogued first-generation seed orchards and 12% on average for rogued seed orchards (Talbert, 1982). Genetic gains over unimproved checklots from second-generation seed orchards for growth are substantially higher than those from the first-generation seed orchards (Table 1). The genetic gains for all families in a region are representative of the gains from unrogued second-generation seed orchards, while gains for the top 30% of families are representative of the gains from intensively rogued second-generation seed orchards. Eight-year height gain averaged 8% above the local checklots for Virginia/North Carolina, 8% for the Lower Gulf, 6% for the Atlantic Coastal Plain, and 10% for the Piedmont region. The estimated rotation volume (25-year) gains over unimproved check lots ranged from 13% to 21% for unrogued orchards and 26% to 35% for the top 30% of families in rogued orchard. Since these estimates represent genetic gains from two cycles of breeding, it is clear that by subtracting the first-generation gains, second-generation breeding and selection, on average over regions, has produced additional 7% and 18% volume gains for unrogued and rogued seed orchards, respectively, over the first-generation seed orchards (Table 1).

Improvement in resistance to fusiform rust is apparent based on the significantly lower R-50 for second-generation families than for unimproved check lots (Li et al., 1997). For example, in the Atlantic Coastal plains, about 80% of the families had lower R-50 breeding values than the checklots. The top ranked 30% of families for rust in this region had an R-50 of 29.6%, significantly lower than the checklot average (above 63%). Similar differences in R-50 were observed for the Piedmont population, which averaged 28% for the best 30% families and 56% for checklots.

Much greater genetic gain can be expected from utilizing the best families since large differences were observed among second-generation families. The best Atlantic Coastal family had over 38% volume gain over the unimproved checks, while the best Piedmont family had

66% volume gain over the unimproved checks (Li et al., 1997). Although genetic gain for stem straightness is difficult to quantify based on arbitrary scoring systems, it is evident that most of the second-generation families had much better stem/crown quality than the unimproved checks. The unimproved checklots are usually ranked at or near the bottom for stem straightness.

DISCUSSION

Loblolly pine, already the most significant commercial tree species in the South, will become an increasingly important source for softwood fiber for pulp and timber. To meet future demands without increasing pressures on old-growth and ecologically sensitive forests, timber productivity per acre must increase, rather than by managing more acres of forest (Gladstone and Ledig 1990). Intensively managed plantations of loblolly pine, employing the best genetically improved planting stock and best silvicultural practices, are believed to be the most effective strategies to meet these demands. With the 7-12% more volume per hectare at harvest from first-generation and 17-30% more volume per hectare at harvest from second-generation than trees grown from wild seed, the impact of the tree improvement on forest productivity has been substantial through the two cycles of breeding by the NCSU-ICTIP. Genetically improved stock has not only demonstrated outstanding growth. It also has lower infection from fusiform rust, typically 20%-25% below the unimproved checklots. With additional improvements in value from quality traits (stem straightness and wood quality), the realized genetic gains in value should be much greater. With the current large scale planting program, these fast-growing plantations of loblolly pine will have much more significant impact on the future wood supply and the sustained management of forest resources in the southern U.S. Although only 15 percent of the commercial forests are currently in plantations (11 million hectares), almost 50 percent of the South's timber supply will soon come from them (Kellison, 1997). Improved wood production on limited commercial lands will reduce the logging pressures on natural forests and provide better opportunities for the use of natural forests and forestlands for conservation and other recreational purposes. Clearly, results from two-cycles of loblolly pine breeding strongly suggest that genetic improvement, as an integrated part of intensively managed plantations, can contribute significantly to the sustained use of forest resources.

CONCLUSIONS

Considering the large scale tree planting program in the South, the impact of the tree improvement on forest productivity has been substantial through the two cycles of breeding by the NC State University-Industry Cooperative Tree Improvement Program:

- 7% volume gain over unimproved stock from unrogued first-generation seed orchards,
- 12% volume gain over unimproved stock from rogued first-generation seed orchards,
- 17% volume gain over unimproved stock from unrogued second-generation seed orchards, and
- 30% volume gain over unimproved stock from rogued second-generation seed orchards.

The future impact will be even more dramatic as the tree improvement program moves to advanced generations (3rd and 4th). Together with intensive silvicultural practices, tree improvement will contribute significantly to sustainable forestry in the future.

REFERENCES

Buford, M.A. and H.E. Burkhart. 1987. Genetic improvement effects on growth and yield of loblolly pine plan.ations. For. Sci. 33:707-724.

Gladstone, W.T., and F.T. Ledig. 1990. Reducing pressure on natural forests through high-yield forestry. Forest Ecology Management 35:69-78.

Hafley, W.L., W.D. Smith, and M.A. Buford. 1982. A new prediction model for unthinned loblolly pine in plantations. Tech. Rep. No. 1, South. For. Res. Center, North Carolina State Univ., Raleigh, NC. 65 p.

Huber, D.A. 1993. Optimal mating designs and optimal techniques for analysis of quantitative traits in forest genetics. Ph.D. Dissertation, Dept. of Forestry, University of Florida, Gainesville. 151 p.

Kellison, R.C. 1997. Production forestry into the 21st century, a world view. *In*: Proc. 24th South. For. Tree Impr. Conf., pp. 3-10. June 9-12, 1997. Orlando, FL.

Li, B., S.E. McKeand, A.V. Hatcher, and R.J. Weir. 1997. Genetic gains of second-generation selections from the NCSU-Industry Cooperative Tree Improvement

Program. Proc. 24th South. For. Tree Impr. Conf., pp. 234-238. Orlando, Florida, June 9-12, 1997.

Li, B., S.E. McKeand and R.J. Weir. 1996. Genetic parameter estimates and selection efficiency or the loblolly pine breeding the southeastern U.S. *In* Dieters, M.J., Matheson, A.C., Nikles, D.G., Harwood, C.E., and Walker, S.M. (eds.). Tree Improvement for Sustainable Tropical Forestry. Proc. QFRI-IUFRO Conf., pp. 164-168, Caloundra, Queensland, Australia. 27 Oct.-1 Nov., 1996.

Talbert, J.T. 1982. One generation of loblolly pine tree improvement: results and challenges. *In* Pollard, D.F.W., D.G. Edwards, and C.W. Yeatman (eds.). Proc. 18th Meeting of the Can. Tree Impr. Assoc., Part 2, pp. 106-120, Duncan, B.C.

Talbert, J.T., R.J. Weir and R.D. Arnold. 1985. Costs and benefits of a mature first-generation loblolly pine tree improvement program. Journal of Forestry 83:162-166.

Responsiveness of Diverse Provenances of Loblolly Pine to Fertilization–Age 4 Results

S. E. McKeand
J. E. Grissom
J. A. Handest
D. M. O'Malley
H. L. Allen

S. E. McKeand is Professor, J. E. Grissom and J. A. Handest are graduate students, D. M. O'Malley is Associate Professor, and H. L. Allen is Professor, Department of Forestry, College of Forest Resources, North Carolina State University, Raleigh, NC, USA.

Address correspondence to: S. E. McKeand, Professor, Department of Forestry, College of Forest Resources, North Carolina State University, Campus Box 8002, Raleigh, NC 27695-8002 USA (Email: McKeand@cfr.ncsu.edu).

The authors appreciate the Texas Forest Service providing the seeds from the Lost Pines provenance. The authors are grateful to many graduate students, faculty, and staff at NCSU for their technical assistance, but in particular to Paula Otto Zanker, Chris Hunt, and Matt Workman for their many hours of work in Scotland County.

Financial support for this research has been provided by the Department of Forestry, NCSU; McIntire-Stennis Project NCZ04149; Agricultural Research Service, NCSU; North Carolina Biotechnology Center grant 9413-ARG-0035; USDA Forest Service; Tree Improvement Program, NCSU; Forest Biotechnology Program, NCSU; Forest Nutrition Cooperative, NCSU; Bowater Inc.; Champion International; Georgia-Pacific Corp.; Rayonier; Westvaco Corp.; and Weyerhaeuser Company.

[Haworth co-indexing entry note]: "Responsiveness of Diverse Provenances of Loblolly Pine to Fertilization–Age 4 Results." McKeand, S. E. et al. Co-published simultaneously in *Journal of Sustainable Forestry* (Food Products Press, an imprint of The Haworth Press, Inc.) Vol. 10, No. 1/2, 2000, pp. 87-94; and: *Frontiers of Forest Biology: Proceedings of the 1998 Joint Meeting of the North American Forest Biology Workshop and the Western Forest Genetics Association* (ed: Alan K. Mitchell et al.) Food Products Press, an imprint of The Haworth Press, Inc., 2000, pp. 87-94. Single or multiple copies of this article are available for a fee from The Haworth Document Delivery Service [1-800-342-9678, 9:00 a.m. - 5:00 p.m. (EST). E-mail address: getinfo@haworthpressinc.com].

INTRODUCTION

Tree improvement has made significant contributions to forestry and plantation management the last 40 years. In the southeastern United States, managers of wood-based manufacturing facilities have realized that the future of their industry depends upon a reliable, ecologically sustainable and economically affordable supply of wood. Plantations of genetically improved forest trees are critical to maintaining this supply.

Loblolly pine (*Pinus taeda* L.) is by far the most important forest tree species in the South, with over 800,000,000 seedlings planted annually by forest industry and non-industrial private forest landowners. Genetic gains from tree improvement programs have been large (e.g., Li et al., 1998), since geographic and within-provenance variation for growth and adaptive traits in loblolly pine is very large. General trends in productivity variation are that families from southern and eastern coastal sources grow faster than families from northern, western, and interior sources (e.g., McKeand et al., 1989; Wells, 1983; Wells and Lambeth; 1983; Schmidtling, 1994). Contrasting the response to nutrient stress of two very different provenances of loblolly pine such as from the "Lost Pines" region of Texas and the Atlantic Coastal Plain may give us insight into the adaptive significance of different ecophysiological traits.

Previous work indicates that the Lost Pines Texas (LPT) sources are generally more stable across environments, while productivity of eastern sources depends more on the environment (van Buijtenen, 1978). Eastern sources were very responsive to environmental enhancement, since productivity was high on the better sites, but very low on the droughty sites. In this paper, we describe a study designed to assess spatial and temporal variation in response of loblolly pine genotypes to environmental stress. Trees have completed four growing seasons in the field under two different nutrient regimes (severe stress and optimal), and variation in early growth is described.

MATERIALS AND METHODS

The study site is located in Scotland County, North Carolina adjacent to the U.S. Forest Service/NC State University SETRES (Southeastern Tree Research and Experiment Site) study. The soil is a Wakulla series–sand to greater than 43 m, sandy, siliceous, thermic Psammentic Hapludult–very infertile, somewhat excessively drained with a total water holding capacity of 12-14 cm in a 2 meter profile. The site receives an average annual rainfall of 1200 mm. Temperatures average 17°C annually, 26°C summer, and 9°C winter. The existing 10-year-old loblolly pine stand was carefully removed and large block plots of different family-treatment combinations were established. Open-pollinated families from the North Carolina and South Carolina Coastal Plain and from the "Lost-Pines" area of Texas were included in the study. Five families from each provenance with average or slightly above average breeding values for volume production were used. Seeds were sown in containers (160 cc RL Super Cells) in the greenhouse in June 1993, and seedlings were field-planted in November 1993.

To facilitate the application of nutrients, a split-split-plot design was used with the two nutrient treatments as main plots, provenances as sub-plots, and families within provenances as sub-sub-plots. Each family plot consists of 100 measurement trees planted at 1.5 m by 2 m spacing. Buffer trees, 12 m around each treatment plot, were planted at the same spacing to eliminate the influence of one nutrient treatment on another. The study was replicated across 10 blocks, and a total of 19800 measurement trees (10 blocks × 2 nutrient treatments × 2 provenances × 5 families per provenance × 100 trees per family plot–less one LPT family in block 10 due to insufficient number of seedlings) were planted.

Foliar nutrient ratios (Hockman and Allen, 1990) have been used to guide annual fertilizer applications to maintain a balanced supply of all nutrients in the fertilized plots. Our goal has been to supply optimal levels of nutrients each year to stimulate rapid growth. Through the first four growing seasons the total nutrient additions (kg/ha) have been: 240 N, 45 P, 85 K, 3 Ca, 20 Mg, 35 S, 0.5 B, 2 Cu, 5 Fe, 5 Mn, and 2 Zn.

All trees were measured annually for height and starting in year 3 for breast height diameter. Analyses of variance were conducted on a family-plot-mean basis (Table 1). Means and within family-plot standard deviations and coefficients of variation were calculated for height for each 100-tree family plot. Within family-plot standard deviations and coefficients of variation were also subjected to analyses of variance to determine if sub-sub-plot uniformity varied.

TABLE 1. Significance levels for main effects and interactions tested in the analyses of variance for height over four years and DBH and stem volume at age four.

				Height		
DBH Vol.						
Source[1]	Yr1	Yr2	Yr3	Yr4	Yr4	Yr4
Treatment	***	***	***	***	***	***
Provenance	+	*	*	**	*	*
Trt × Prov		*				*
Family(Prov)	***	*	*	**	*	*
Trt × Fam(Prov)						*

[1] Treatment and provenance were considered fixed effects. Family within provenance and blocks were considered random effects.
+, *, **, *** Significant at P ≤ 0.10, 0.05, 0.01, 0.001, respectively.

Error terms for each main effect and interaction listed above are:
Treatment: [Block × Trt] + [Trt × Fam(Prov)] − [Block × Trt × Fam(Prov)]
Provenance: [Block × Prov] + [Fam(Prov)] − [Block × Fam(Prov)]
Trt × Prov: [Block × Trt × Prov] + [Trt × Fam(Prov)] − [Block × Trt × Fam(Prov)]
Family(Prov): [Block × Fam(Prov)] + [Trt × Fam(Prov)] − [Block × Trt × Fam(Prov)]
Trt × Fam(Prov): [Block × Trt × Fam(Prov)]

RESULTS AND DISCUSSION

Survival and growth of the trees has been excellent in the first four years. Survival averaged 93% after four growing seasons (no treatment or genetic effects), and height averaged 3.0 m. Deer browse and tipmoth caused some problems in the first two growing seasons, and 12.9% of the trees were damaged and not included in the analyses for growth traits.

Fertilizer Response

Growth responses to fertilization were very large and significant each year (Table 1). Height was 21%, 46%, 43%, and 43% greater in the fertilized plots for years one, two, three, and four, respectively (Figure 1). Volume differences at age 4 were even more dramatic (Figure 2), with the fertilized trees having 2.1 times greater stem volume than the controls. Although this is a well-drained site, from the results of the nutrition by irrigation study (SETRES) adjacent to this trial, we know that the primary limit to productivity is nutrition (Albaugh et al., 1998). The huge increase in productivity in the first four

FIGURE 1. Mean tree heights during the first four growing seasons in the field for trees from the Lost Pines Texas (LPT) and Atlantic Coastal Plain (ACP) provenances in the fertilized and control plots. Initial height of seedlings at planting was measured in 1994.

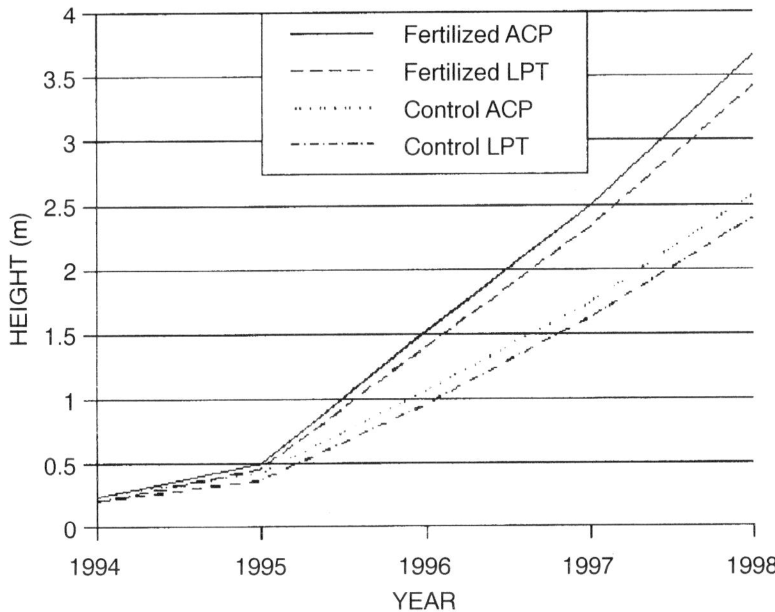

growing seasons is possible since all potential nutrient limitations (i.e., more than just N and P) were ameliorated.

One of the most dramatic effects of the nutrition amendments has been the increase in uniformity within the 100-tree family plots. The average within-plot coefficient of variation for fourth-year height was 22.6% for the control plots and 12.0% for the fertilized plots. The within plot standard deviations for height were also significantly different and were 0.56 m for the control plots and 0.42 m for the taller fertilized plots. While increased uniformity typically results from nutritional amendments on very poor sites, the dramatic differences in uniformity was surprising.

Provenance and Family Variation

As expected, the five families from the Atlantic Coastal Plain grew faster than the five Texas families (Figure 1). We anticipated that

FIGURE 2. Average individual-tree volume for the Lost Pines Texas (LPT) and Atlantic Coastal Plain (ACP) provenances in the control and fertilized plots at age four years.

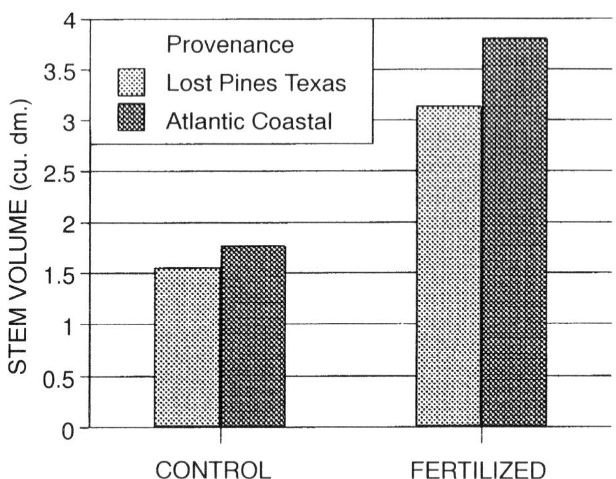

under the harsher environmental conditions in the control plots, the Texas families would perform relatively better. However, the ACP families were superior in both environments and the provenance by treatment interactions for height in all four years were not close to being significant.

The significant provenance by treatment interaction for stem volume at age four (Table 1) is indicative of the greater responsiveness to nutritional amendments of the Atlantic Coastal Plain provenance compared to the Lost Pines Texas provenance. Although there was no provenance rank change in the two environments, the difference in the magnitude of the provenance means (greater in the fertilized plots) is similar to previous trials (van Buijtenen, 1978).

The superior aboveground growth of the ACP families could be due in part to differential carbon allocation to above- and belowground tissues. Our work with one-year-old seedlings of these two provenances indicates that the ACP provenance may have a reduced cost for higher-order lateral root production compared to the LPT provenance especially under high fertility (Wu et al., 1998).

Families within provenances also differed for growth traits (Table 1). The family means at age four for the ACP families varied from 2.42 m to 2.67 m in the control plots and from 3.48 m to 3.78 m in the fertilized plots. The Texas families also differed in the control plots (2.29 m to 2.45 m) and in the fertilized plots (3.39 m to 3.53 m). The marked difference in productivity between the drought-hardy Lost Pines families and the ACP families is illustrated by the almost complete lack of overlap of the family means for height.

The lack of rank change across the treatments both at the provenance and family level was surprising for height and volume. Given the magnitude of the imposed environmental differences and the young age of the trees, differential performance of the families in the two treatments was expected. This result reinforces the tenet of the stability of open-pollinated families of loblolly pine as well as the better responsiveness of the ACP provenance compared to the Lost Pines provenance.

Future Work

This experiment will be a long-term (~20 years) field laboratory for ecologists, physiologists, and geneticists to study the bases for trees' responses to environmental stress. Productivity will continue to be assessed through rotation age to see if the early growth differences are maintained. We suspect that as the stand develops and competition for limited soil resources becomes more intensive, the Lost Pines Texas trees may be superior to the Atlantic Coastal Plain trees in the control plots. Future work will also emphasize both above- and belowground production and physiological processes and how they interact to affect productivity. Not only will traditional quantitative genetic analyses be conducted to evaluate genetic control for these traits, but genomic mapping to determine the significance of major gene control is also an integral part of the study. Megagametophytes for each of the 19,800 individuals in the trial are in cold storage (-80°C) and DNA will be extracted and genomic maps developed to determine marker–trait associations. Using the open-pollinated families in such a manner will allow us to determine if major genes with high breeding values (O'Malley and McKeand, 1994) are associated with adaptive response to environmental stress.

CONCLUSIONS

Height growth of families of loblolly pine from the "Lost Pines" provenance in Texas and the Atlantic Coastal Plains of NC and SC during the first four years has been evaluated as well as volume at age four. Response to fertilizer applications has been large with a 43% increase in height and a 109% increase in stem volume at age four years. The Atlantic Coastal families were significantly taller and had greater stem volume than the LPT families in both the fertilized and control plots. While no genotype by environment interactions were observed for height, there was significant provenance by treatment interaction for stem volume due to the greater responsiveness of the ACP provenance to nutrient amendments. Even given the tendency for low genotype by environment interaction for open-pollinated families of loblolly pine, the adaptability of the Atlantic Coastal families to such extreme environmental conditions was surprising. The long-term performance of the trees will be evaluated to see if this trend continues.

REFERENCES

Albaugh, T.J., H.L. Allen, P.M. Dougherty, L.W. Kress and J.S. King. 1998. Leaf-area and above- and belowground growth responses of loblolly pine to nutrient and water additions. For. Sci. 44:317-328.

Hockman J.N. and H.L. Allen. 1990. Nutritional diagnosis in young loblolly pine stands using a DRIS approach. Pp. 500-514, *In* S.P Gessel, D.S. Lacate, G.F. Weetman, and R.F. Powers (eds.). Sustained Productivity of Forest Soils. Univ. British Columbia, Faculty of Forestry Publication, Vancouver BC. 525 p.

Li, B., S.E. McKeand and R.J. Weir. 2000. Impact of forest genetics on sustainable forestry–results from two cycles of loblolly pine breeding in the U.S. Journal of Sustainable Forestry 10(1/2):79-85.

McKeand, S.E., R.J. Weir and A.V. Hatcher. 1989. Performance of diverse provenances of loblolly pine throughout the southeastern United States. South. J. Appl. For. 13:4650.

O'Malley, D.M. and S.E. McKeand. 1994. Marker assisted selection for breeding value in forest trees. Forest Genetics 1:231-242.

Schmidtling, R.C. 1994. Use of provenance tests to predict response to climatic change: loblolly pine and Norway spruce. Tree Phys. 14:805-817.

van Buijtenen, J.P. 1978. Response of "Lost Pines" seed sources to site quality. Pp. 228-234, *In* Proc. 5th N. Amer. For. Biol. Workshop. Gainesville, FL.

Wells, O.O. 1983. Southwide Pine Seed Source Study–loblolly pine at 25 years. South. J. Appl. For. 4:127-132.

Wells, O.O. and C.C. Lambeth. 1983. Loblolly pine provenance test in southern Arkansas–25th year results. South. J. Appl. For. 7:71-75.

Wu R.L., J.E. Grissom, D.M. O'Malley and S.E. McKeand. 2000. Root architectural plasticity to nutrient stress in two contrasting ecotypes of loblolly pine. J Sust For. 10(3/4): 307-317.

PART TWO: FRONTIERS OF PLANT PHYSIOLOGY

Influence of Procerum Root Disease on the Water Relations of Eastern White Pine (*Pinus strobus* L.)

J. R. Butnor
J. R. Seiler
J. A. Gray

INTRODUCTION

Procerum root disease (PRD) is caused by the deuteromycete fungus *Leptographium procerum* (Kendr.) Wingf, formerly *Verticicladiel-*

J. R. Butnor is affiliated with the Southern Research Station, USDA Forest Service, 3041 Cornwallis Road, Research Triangle Park, NC 27709 USA.

J. R. Seiler is affiliated with Virginia Polytechnic Institute and State University, Department of Forestry, Blacksburg, VA 24061 USA.

J. A. Gray is affiliated with the Virginia Polytechnic Institute and State University, Deparment of Plant Pathology, Physiology and Weed Science, Blacksburg, VA 24061 USA.

[Haworth co-indexing entry note]: "Influence of Procerum Root Disease on the Water Relations of Eastern White Pine (*Pinus strobus* L.)." Butnor, J. R., J. R. Seiler, and J. A. Gray. Co-published simultaneously in *Journal of Sustainable Forestry* (Food Products Press, an imprint of The Haworth Press, Inc.) Vol. 10, No. 1/2, 2000, pp. 95-105; and: *Frontiers of Forest Biology: Proceedings of the 1998 Joint Meeting of the North American Forest Biology Workshop and the Western Forest Genetics Association* (ed: Alan K. Mitchell et al.) Food Products Press, an imprint of The Haworth Press, Inc., 2000, pp. 95-105. Single or multiple copies of this article are available for a fee from The Haworth Document Delivery Service [1-800-342-9678, 9:00 a.m. - 5:00 p.m. (EST). E-mail address: getinfo@haworthpressinc.com].

la procera (Kendr.) and is most commonly isolated from *Pinus* sp. L., though the fungus has been isolated from other conifer species including Fraser fir (*Abies fraseri* [Pursh] Poir.), Douglas fir (*Pseudotsuga menziesii* [Mirb.] Franco) and Norway spruce (*Picea abies* [L.] Karst.) (Alexander et al., 1988). During the last two decades PRD has been responsible for significant economic losses in eastern white pine (*Pinus strobus* L.) Christmas tree plantations and trees in the urban landscape throughout the eastern United States (Lackner and Alexander, 1984).

Early studies reported rapid disease development in seedlings with mortality occurring 2 to 10 weeks after inoculation with *L. procerum* isolates (Lackner and Alexander, 1982). However, rapid disease development and seedling mortality have seldom been observed since. In Christmas tree plantations, white pines typically do not show any outward symptom of disease until they are near saleable maturity. The progression of PRD can vary greatly, but mortality usually occurs one or more years from the time the earliest symptoms are noted (Carlson, 1994). The initial symptoms of PRD in white pine include mild chlorosis, delayed bud break, and inhibited shoot elongation relative to healthy cohorts (Anderson and Alexander, 1979; Carlson, 1994). Resin exudation from the lower stem, basal stem irregularities due to localized cambium death, and basal stem swelling may accompany the initial foliar symptoms (Alexander et al., 1988; Carlson, 1994). As the disease progresses, wilting and marked chlorosis of the foliage are observed. Eventually, the foliage develops a reddish color associated with mortality (Carlson, 1994). Observations of severe foliar symptoms usually occur in association with substantial resin soaking of sapwood tissue (Horner, 1987; Carlson, 1994). Resin soaked tissue exhibits reduced moisture content and greatly reduced permeability to water, caused by resinous inclusions that can completely block the tracheids (Horner, 1985, 1987). When the basal stem is sectioned, wedge-shaped areas of vascular occlusion that originate in the outer cambium and taper to the pith are frequently observed (Carlson, 1994).

Carlson (1994) developed a disease severity scale using a staining technique to quantify percent-occluded sapwood cross-sectional area at the base of the main stem. Pre-dawn water potential, daily change in pre-dawn to midday water potential, leaf conductance, transpiration and photosynthetic rate all showed trends toward lower values with increasing disease severity (Carlson, 1994). However, in selected trees, physiological measures were not strongly correlated with disease severity. Some trees with relatively low disease severity ratings exhibited pre-dawn water potentials similar to severely diseased trees. Conversely, several trees with high disease ratings were able to maintain pre-dawn water potentials similar to uninfected trees. It is probable a disease severity scale based on percent occluded basal area would not accurately reflect changes related to hydraulic conductivity of the sapwood, a variable not measured by Carlson (1994). Subtle declines in sapwood permeability to water may not be detected by direct observation (Horner et al., 1987), or aided with stain. Horner (1985) demonstrated that resinous occlusion limits water transport in white pine using small samples cored with an increment borer, but did not attempt to measure or describe the impact of resin soaking on a whole stem basis, where resin soaked sapwood can be adjacent to functional tracheids.

The purpose of this study is to quantify the effects of PRD on hydraulic conductivity of white pine sapwood using a whole stem measurement technique. This technique allowed us to assess the function of the vascular system and therefore, quantify total stem hydraulic conductivity by accounting for resinous occlusion, physical blockage of tracheids by fungal hyphae and resistance to flow in cavitated tracheids. We specifically tested the hypothesis that PRD reduces stem conductivity, which leads to reduced leaf conductance and eventually to the death of the tree.

MATERIALS AND METHODS

Study Sites and Sample Material

Two Christmas tree plantations in Floyd County, Virginia were used for the study. Plantation 1 was under active cultivation and consisted of 8 ha of white pine and Scots pine (*P. sylvestris* L.) Christmas trees. Plantation 2 consisted of 20+ ha of white pine Christmas trees which had not been actively managed for 3-4 years. Heavy losses from PRD made this plantation economically nonviable. Four plots approximately 50 m in diameter were established at each plantation. Trees used for this study were near saleable maturity (7-9 years), were 2-2.5 m in height and measured 8-15 cm in diameter just above the root collar. In

June 1994, 12 visually symptomatic trees were selected for study in each of the eight plots. Every effort was made to choose 12 trees grading from barely to noticeably symptomatic. Mildly symptomatic trees displayed delayed bud break or retarded shoot elongation which otherwise appeared healthy. Trees exhibiting more severe symptoms including marked chlorosis, flattened areas on the lower stem, basal resinosis, and reduced shoot elongation for several seasons were also chosen. Six visually healthy control trees were located in the same plot interspersed with the diseased trees.

Gas Exchange

Leaf conductance was measured with a Li-Cor 1600 steady state porometer fitted with a 4 cm^2 closed system cuvette (Li-Cor Inc., Lincoln, NE) (Bingham and Coyne 1977). Measurements were made periodically from July 15 to October 15, 1994, when two consecutive days of clear weather was predicted. (Due to the number of trees being sampled and the distance between plots, it was necessary to sample the plantations on two consecutive days.) Trees were sampled between 09:00 and 12:00 EST on clear to partly cloudy days with P.AR. values ranging from 1500 to 2000, under ambient temperature and humidity. Measurements were made using one fascicle from each tree. Leaf conductance values were later corrected using the total surface area of each fascicle calculated using an equation presented by Ginn et al. (1991).

Hydraulic Conductivity

At the end of the growing season, in late October, the sample trees were harvested for hydraulic conductivity analysis. Trees were cut as close to the ground as possible and returned to the laboratory. The stems were re-cut to remove any soil or sap accumulation on the cut surface. A 5 cm long segment was removed from the base of the stem, debarked and weighed. The segment was soaked in clean tap water until hydraulic conductivity measures were made. To measure xylem hydraulic conductivity (Lp) an apparatus was developed using the basic principles of Sperry et al. (1988), but modified to accommodate debarked stem segments of up to 20 cm in diameter. Stem segments were fitted to the PVC pressurizing unit with a rubber gasket and the unit was filled with water from a reservoir elevated 4 m. Pressure (40 kPa) from the raised reservoir was applied, forcing water through the stem segment. After a period of stabilization, lasting 5-10 minutes the volume flow rate was measured by collecting the water expressed through the segment in a one-minute interval. This was repeated every other minute until the readings stabilized. Healthy trees stabilized in 5 to 10 minutes, while diseased trees required up to 45 minutes. Lp (cm s^{-1} kPa^{-1}) was calculated as:

$$Lp = (\Delta J_V/\Delta P)\ (1/A),$$

Where ΔJ_V ($cm^3\ s^{-1}$) is volume flow rate, ΔP (kPa) applied pressure difference and A (cm^2) is the surface area (Nobel et al., 1990). After conductivity analysis, each segment was oven-dried at 65°C and re-weighed to calculate initial percent moisture content (Panshin and de Zeeuw 1980).

Basal Occlusion Measures

After the 5 cm segments were removed for Lp analysis, the remaining bolt was soaked in 0.5 g/l Fast Green FCF solution (U.S. Biochemical Corporation, Cleveland, OH) for 24 hours. The following day a thin disk was cut from each bolt with a band saw, revealing a pattern of green staining on the surface. Unobstructed vessels stained blue/green, while resin soaked and other dysfunctional vessels remained unstained (Basham, 1970; Carlson, 1994). The segments were scanned with a high-resolution color scanner using a Macintosh Power PC and Adobe PhotoShop (Adobe Systems, Inc., Mountain View, CA). The unstained area was determined from the scan and percent basal occlusion was calculated.

Data Analysis

Data were analyzed using Proc GLM and Tukey's studentized range test (HSD) was used to determine significance of difference between the three disease classes (healthy, mildly symptomatic and diseased) and four physiological variables measured (leaf conductance, Lp, vascular occlusion and sapwood moisture content) (SAS Institute, 1991). A separate analysis was performed to describe the impact of disease class by date on leaf conductance. Proc ANOVA was used to perform a one-way analysis of variance to determine the relation between Lp, vascular occlusion (%) and sapwood moisture content (%) and leaf conductance, using leaf conductance values from the last sample period immediately prior to harvest (SAS Institute, 1991). Regression analysis (Proc REG) was used to define the relationships among the measures of sapwood function (Lp, occlusion, sapwood moisture content) (SAS Institute, 1991). A total of 27 diseased trees died during the study and could not be analyzed for sapwood functionality.

RESULTS AND DISCUSSION

Influence of Disease Class

Procerum root disease has a profound effect on foliar and vascular physiology of white pine. Leaf conductance was significantly lower in diseased trees than healthy trees on each sampling date (Figure 1). The magnitude of the difference in leaf conductance stayed about the same throughout the course of the study. This indicates that the trees may be in a type of steady-state condition for an extended period, where PRD

FIGURE 1. Leaf conductance of healthy, diseased and diseased trees presenting mild foliar symptoms over the duration of the 1994 field study. Each data point represents the mean leaf conductance from all eight plots located on two different plantations ± standard error of the mean data was collected on two consecutive days under similar weather conditions.

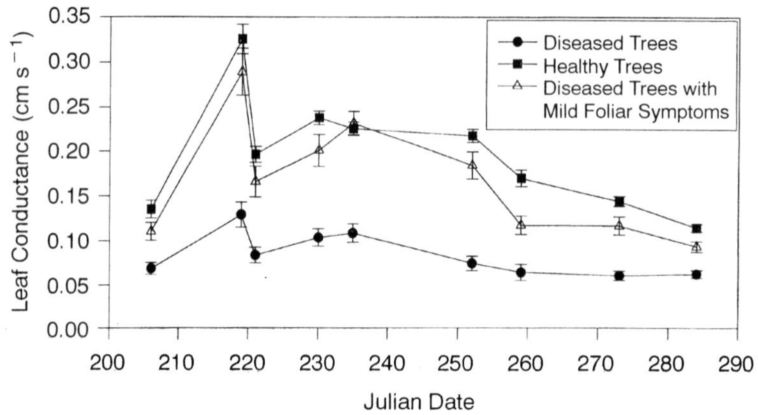

develops very slowly after reaching a developmental plateau. This is in line with Carlson (1994), who noted the development of PRD and PRD symptom expression usually preceded mortality by a year or more. Diseased trees with mild foliar symptoms were very similar to healthy trees in the first 30 days of measurement and the difference between the two classes was insignificant at two dates during this time (Figure 1). Later in the year, leaf conductance measured in the mildly symptomatic class remained consistently lower than measures in the healthy class. These mildly symptomatic trees are likely deviating from the healthy trees and exhibiting characteristics more consistent with diseased trees. Stem Lp and sapwood moisture content (%) were significantly higher in healthy trees, while mean vascular occlusion was greatest for the diseased trees (Table 1). Infection and colonization of host sapwood by *L. procerum* induces the host to fill affected regions with resins to limit fungal colonization (Horner, 1987). Under most circumstances this would be an effective strategy to slow or repel invading insects or fungal pathogens; however, *L. procerum* has the ability to survive and colonize resin-soaked woody tissue (Horner, 1985). Instead of limiting fungal colonization, the host is systematical-

TABLE 1. Results of GLM procedure defining relationship between disease class and water relations variables in white pine. Lp, occlusion (%), and sapwood moisture (%) were measured destructively over a two week period beginning in late October. Only leaf conductance measures from the final sample period (Day 284 and 285) just prior to harvest were used for this analysis.

Disease Class	n	Lp ($cm\ s^{-1}\ kPa^{-1}$)	Occlusion (%)	Sapwood Moisture (%)	Leaf Conductance ($cm\ s^{-1}$)
Healthy	48	3.20 a[2]	4.16 a[3]	208.72 a	0.01153 a
Mildly Symptomatic	14	2.27 b	15.32 a	145.39 b	0.0938 a
Diseased	55	0.51 c	47.77 b	72.32 c	0.0625 b
R^2		0.39	0.76	0.64	0.90
P-value		0.0001	0.0001	0.0001	0.0001

1. For date 284 and 285 only
2. Column means with the same letter are not significantly different using Tukey's HSD test ($\alpha = 0.05$)
3. Healthy trees do not truly have occluded sapwood. The technique used includes the small amount of heartwood these trees form as a portion of the stem that is not functional.

ly reducing its ability to conduct water. Sapwood moisture content had the best correlation with disease class ($R^2 = 0.90$, $p = 0.0001$) followed by Lp ($R^2 = 0.76$, $p = 0.0001$) (Table 1). Occlusion ($R^2 = 0.64$, $p = 0.0001$) and leaf conductance ($R^2 = 0.39$, $p = 0.0001$) were unable to delineate between mildly symptomatic and healthy trees.

Relationship of Leaf Conductance to Stem Hydraulic Properties

Leaf conductance was positively correlated with stem Lp and sapwood moisture content and inversely related to vascular occlusion, (Table 2), indicating that decreased leaf conductance is an expression of increasing disease severity. However, much of the variation in leaf conductance is not explained by either disease class or the vascular variables and is likely due to a variety of external variable that can influence stomatal function (Tables 1 and 2). The reduced leaf conductance is likely caused by the reduced capacity to conduct water, caused by resin infiltration of tracheids. As water-conducting vessels are filled with resin and rendered non-functional, moisture content of the sapwood is reduced (Figure 2 a, b). It was surprising that Lp had a lower correlation with leaf conductance than with occlusion given that Lp was more closely related to disease class. This is likely due to the inherently high degree of variability in the Lp measurements. Measurement of Lp is difficult, time consuming, destructive and explained less of the variation in leaf conductance than vascular occlusion, but did quantify the effect of occlusion on stem physiology in white pine. Regression analysis revealed a strong relationship between basal occlusion and stem Lp (Figure 3). As basal occlusion increased, making tracheids impermeable to water, stem Lp dropped markedly. When the

TABLE 2. ANOVA results defining the influences of Lp, occlusion (%), and sapwood moisture (%) content on leaf conductance in white pine. Only leaf conductance measures from the last sample date before harvest were used.

	Leaf Conductance				
Stem Measure	n	R^2	Intercept	Slope	P-value
Lp	117	0.36	0.0575	0.0166	0.0001
Occlusion (%)	117	0.48	0.1155	−0.011	0.0001
Sapwood Moisture (%)	117	0.46	0.0323	0.0004	0.0001

FIGURE 2. Relationship between Lp, occlusion (%), and sapwood moisture (%) defined using (a) 2nd order and (b) 3rd order linear regression analysis (n = 117).

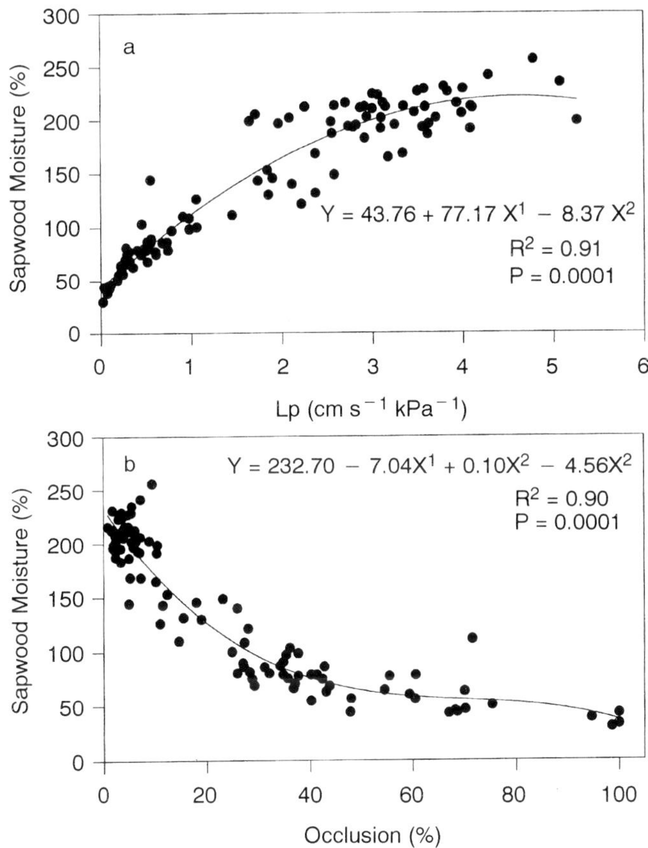

stem segments exhibited 20% occlusion, stem Lp was reduced over 50% and stem Lp approached zero at basal occlusion greater than 50%. From these results it appears that foliar disease expression is simply a matter of fluid dynamics, where water available to the leaves is lower and the leaves respond accordingly.

Summary

Diseased trees exhibited greater vascular occlusion, lower sapwood moisture content, and reduced hydraulic conductivity; however, these

FIGURE 3. Results of 3rd order linear regression analysis (n = 117) describing the relationship between measures of sapwood occlusion and Lp.

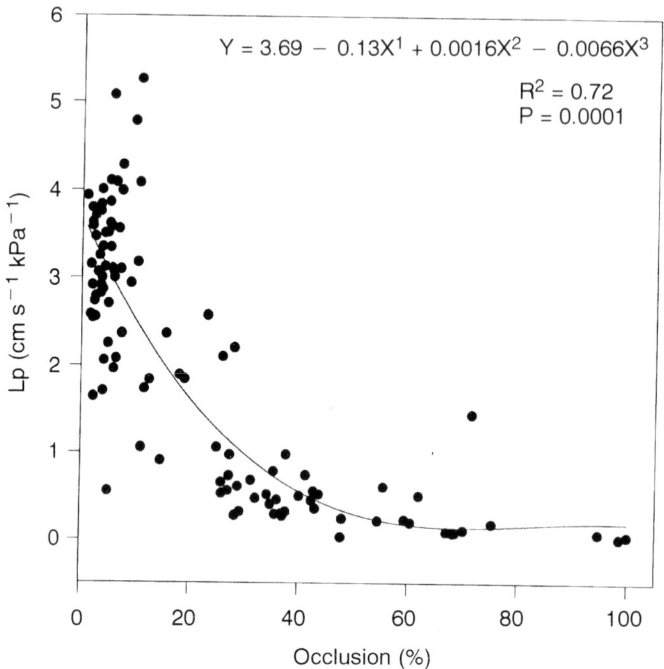

variables were only moderately correlated with leaf conductance. The blockage of tracheids with resin was previously shown to make sapwood less permeable to water using small sections. We have demonstrated using a whole stem measurement technique that vascular occlusion resulting from PRD reduces stem hydraulic conductivity leading to reduced leaf conductance. The amount of water conducted through the stem is reduced resulting in the desiccation of the sapwood and foliage leading to premature mortality in eastern white pine.

REFERENCES

Alexander, S.A., W.E. Horner, and K.J. Lewis. 1988. *Leptographium procerum* as a pathogen of pines, pp. 97-112, *In* Harrington, T.C. and F.W. Cobb, Jr. (eds.), Leptographium Root Diseases on Conifers. APS Press, St. Paul, MN. 149 pp.

Anderson, R.L. and S.A. Alexander. 1979. How to identify and control white pine root decline. USFS Southeastern area, State and Private Forestry. Forestry Bulletin SA-FR/P6 July1979, 4 pp.

Basham, H.G. 1970. Wilt of loblolly pine inoculated with blue-stain fungi of the genus *Ceratocystis*. Phytopathology 60:750-754.

Bingham, G.E. and P.I. Coyne. 1977. A portable temperature controlled steady-state porometer for the field measurements of transpiration and photosynthesis. Photosynthetica 11:148-160.

Carlson, J.A. 1994. Procerum Root Disease Physiology and Disease Interactions with Ozone. Ph.D. Dissertation, Virginia Polytechnic Institute and State University, Blacksburg, VA. 95 pp.

Ginn, S.E., J.R. Seiler, B.H. Cazell, and R.E. Kreh. 1991. Physiological and growth responses of 8 year old loblolly pine stands to thinning. For. Sci. 37:1030-1040.

Horner, W.E. 1985. Etiologic Studies of *Verticicladiella procera* Kendr. in Pine Christmas Trees. Ph.D. Dissertation. Virginia Polytechnic Institute and State University, Blacksburg, VA. 168 pp.

Horner, W.E., S.A. Alexander, and K.J. Lewis. 1987. Colonization patterns of *Verticicladiella procera* in Scots and eastern white pine and associated resin-soaking, reduced sapwood moisture content and reduced needle water potential. Phytopathology 77:557-560.

Lackner, A.L. and S.A. Alexander. 1982. Occurrence and pathogenicity of *Verticicladiella procera* in Christmas tree plantations in Virginia. Plant Dis. 66:211-212.

Lackner, A.L. and S.A. Alexander. 1984. Incidence and development of *Verticicladiella procera* in Virginia Christmas tree plantations. Plant Dis. 68:210-212.

Nobel, P.S., P.J. Schulte, and G.B. North. 1990. Water influx characteristics and hydraulic conductivity for roots of *Agave deserti*. J. Exp. Bot. 41:409-419.

Panshin, A.J. and C. de Zeeuw. 1980. Textbook of Wood Technology: Structure, Identification, Properties, and Uses of the Commercial Woods of the United States and Canada. McGraw- Hill, New York. 722 pp.

SAS Institute Inc. 1991. SAS System for Linear Models 3rd edition. SAS Institute, Cary, NC.

Sperry, J.S., J.R. Donnelly, and M.T. Tyree. 1988. A method for measuring hydraulic conductivity and embolism in xylem. Plant, Cell and Environment 11:35-40.

CO_2 Diffusion in Douglas Fir Bark: Implications for Measuring Woody-Tissue Respiration with Removable Cuvettes

Lucas A. Cernusak
Nate G. McDowell
John D. Marshall

INTRODUCTION

Physiological growth analyses can make significant contributions to sustainable forestry by mechanistically investigating the controls over net primary production in forest ecosystems (Landsberg and Gower, 1997). The success of these analyses relies on accurate measurement of physiological processes. Respiration in the woody tissue of forest trees can consume up to 33% of annual net daytime carbon assimilation (Ryan et al., 1994). Woody-tissue respiration can be measured *in situ* using removable cuvettes (Ryan, 1990; Sprugel, 1990; Sprugel and Benecke, 1991). However, questions regarding the use of such cuvettes have arisen. For example, it is important that the amount

---

Lucas A. Cernusak, Nate G. McDowell, and John D. Marshall are affiliated with the Department of Forest Resources, University of Idaho, Moscow, ID 83844-1133.

Address correspondence to Lucas A. Cernusak at the above address (E-mail: cern3317@novell.uidaho.edu).

The authors gratefully acknowledge the support of the McIntyre-Stennis Program. The authors thank Mike Ryan for providing the cuvettes used in this study.

[Haworth co-indexing entry note]: "CO_2 Diffusion in Douglas Fir Bark: Implications for Measuring Woody-Tissue Respiration with Removable Cuvettes." Cernusak, Lucas A., Nate G. McDowell, and John D. Marshall. Co-published simultaneously in *Journal of Sustainable Forestry* (Food Products Press, an imprint of The Haworth Press, Inc.) Vol. 10, No. 1/2, 2000, pp. 107-113; and: *Frontiers of Forest Biology: Proceedings of the 1998 Joint Meeting of the North American Forest Biology Workshop and the Western Forest Genetics Association* (ed: Alan K. Mitchell et al.) Food Products Press, an imprint of The Haworth Press, Inc., 2000, pp. 107-113. Single or multiple copies of this article are available for a fee from The Haworth Document Delivery Service [1-800-342-9678, 9:00 a.m. - 5:00 p.m. (EST). E-mail address: getinfo@haworthpressinc.com].

of woody-tissue contributing to a measurement is correctly determined, especially when one considers that stand level estimates of woody-tissue respiration are often based on measurements made over just a few square decimeters. The difficulty arises from the gasket material used to seal the cuvettes to tissue sections. Is the respiring tissue volume: (a) the portion between the interior edges of the gaskets, or (b) the portion between the gasket midpoints? The latter volume may exceed the former by more than 15%.

Measuring the volume of respiring tissue between the gasket interior edges implicitly assumes that CO_2 evolved beneath the gaskets does not diffuse laterally around the gaskets over the course of the measurement. On the other hand, measuring the tissue volume to the gasket midpoints assumes that lateral diffusion occurs and that it occurs equally in each direction. This last assumption of directionality should be valid so long as the CO_2 concentration inside the chamber is approximately equal to that outside the chamber.

Our first objective in this study was to determine the location on the gasket (interior edge or midpoint) that best describes the true volume of woody tissue contributing to respiration measurements. We designed a simple experiment to test the hypothesis that CO_2 will diffuse laterally beneath a gasket in the time course of a typical measurement. Our second objective was to examine a previously observed CO_2 efflux pulse immediately following cuvette closure in branches, which subsequently declined over the course of several minutes (L. Cernusak, unpublished data). We wished to investigate the distribution of this apparent respiratory pulse among woody-tissue types and to determine its approximate duration.

METHODS

Determining the Correct Gasket Positions

We excised 10 branches from the bases of three open-grown Douglas fir trees (*Pseudotsuga menziesii* [Mirb.] Franco var. *glauca*) on the University of Idaho Experimental Forest in January 1998. Upon returning to the laboratory, we coated the cut surfaces of the branches with paraffin to minimize disruption of the normal CO_2 diffusion pathway. While this technique may cast some doubt on the applicability of measured rates to undisturbed trees (Sprugel et al., 1995), it conveniently allowed us to make our measurements at constant temperature.

Our experiment involved two sets of measurements on each branch. In the first set we wrapped two strips of closed cell foam (Li-Cor, Inc., Lincoln, NE) 1.5 cm in width around each tissue sample and tightened split Plexiglas cuvettes over them. We leak-tested each cuvette with a flow meter to ensure that chamber efflux was greater than 95% of influx. We measured CO_2 efflux from each sample section with a closed-system, portable Li-6200 photosynthesis system (Li-Cor, Inc., Lincoln, NE). We then removed the cuvettes and applied closed cell foam to approximately half of the woody-tissue surface enclosed in the first measurements (see Figure 1). The new foam was wrapped tightly with electrical tape to approximate the pressure of the clamps on the original gaskets and to minimize lateral leakage of CO_2 between the foam and the bark surface. We then repeated the respiration measurements. To ensure that we captured the steady-state flux rate for each branch and to further investigate the previously observed CO_2 efflux pulse, we monitored apparent respiration continuously at 5-10 minute intervals following each cuvette closure. To prevent build up of CO_2 in the system between observations, we disengaged the cuvettes by opening quick-connect joints near the Plexiglas. Care was taken to bracket the ambient CO_2 concentration during all gas ex-

FIGURE 1. A schematic representation of the surface area coverage experiment used to determine the correct gasket position for calculating woody-tissue volumes. The first set of measurements was made without the closed cell foam barrier on the tissue inside the chamber; the second set with it.

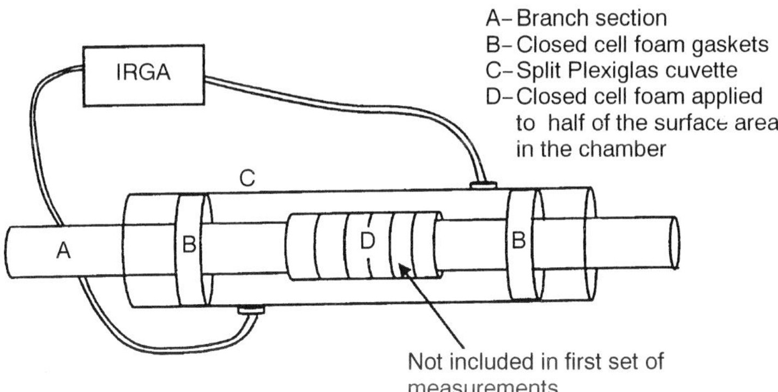

change measurements. At the conclusion of the experiment, we measured the sapwood volume of each sample section (Sprugel, 1990). The branch sections measured for respiration ranged from 0.7-2.8 cm in diameter, and 5.9-20.0 cm in length between gasket interior edges. The total system volume during gas exchange measurements varied from 220-360 cm^3, depending on the size of the branch and cuvette.

Stem Respiration Measurements

We made respiration measurements on large diameter Douglas fir stems (22-43 cm at breast height) to see if CO_2 efflux was influenced by cuvette tightening in these tissues. We made these measurements in northeastern Washington (48°56′N, 118°36′W) in November 1997. Stem chamber plates for fitting cuvettes and thermocouples for measuring stem temperature were previously installed (N.G. McDowell, in preparation). We secured cuvettes on five trees with a nylon strap and ratchet and leak-tested each chamber until a 95% seal was obtained. We made respiration measurements as described for the branches. The stem temperature and the time after cuvette closure were recorded for each observation. Respiration rates were normalized to 15°C to eliminate temperature-dependent variation as follows:

$$R = R_{15} Q_{10}^{([T-15]/10)} \tag{1}$$

where R is the respiration rate per unit sapwood volume at the time of measurement, R_{15} is the estimated respiration rate at 15°C, Q_{10} is the thermal coefficient of respiration, and T is the stem temperature at the time of measurement. We developed a site specific Q_{10} of 2.0 concurrently with our measurements (N.G. McDowell, in preparation). Sapwood volume beneath each chamber was determined after Ryan (1990).

Analysis

We determined steady-state branch respiration rates by fitting a nonlinear, least-squares regression equation to each branch as follows:

$$R(t) = R_s + \beta_1 t^{\beta_2} \tag{2}$$

where *R(t)* is the respiration rate at time t, t is the time after cuvette closure, R_s is the steady-state respiration rate, and β_1 and β_2 are regression parameters. As a check on this method, we conducted a second analysis in which we determined steady-state branch respiration rates by averaging measurements made *more* than 15 minutes after cuvette closure. We tested for differences between pre- and post-coverage steady-state branch respiration rates for the two data sets using paired *t* tests. We used analysis of covariance to test for a significant effect of time after cuvette closure on CO_2 efflux in the large diameter stems. Analyses were performed in SYSTAT 7.0 (SPSS Inc., Chicago, IL).

RESULTS AND DISCUSSION

Pre- and post-coverage branch respiration rates along with paired *t* tests are presented in Table 1. Whether steady-state branch respiration rates were solved for by nonlinear regression or by averaging measurements after 15 minutes, pre-coverage values did not differ significantly from post-coverage values (P = 0.49 and 0.39, respectively). Our steady-state respiration rates for Douglas fir branches at 21°C are consistent with rates observed for branches of other conifer species (Sprugel, 1990; Maier et al., 1998). The similarity in respiration rates before and after surface area coverage suggests that resistance to lateral diffusion beneath the bark surface of Douglas fir branches is relatively small. A concentration gradient steep enough to overcome this resistance was established within approximately 15 minutes.

The magnitude and duration of the apparent respiratory pulse following cuvette closure was quite variable (see Figure 2). We believe the observed increase in CO_2 efflux results from compression of spongy bark beneath the gaskets upon cuvette tightening. In our mea-

TABLE 1. Mean respiration rate before and after the surface area coverage treatment, mean difference between the two sets of measurements, standard deviation of the difference, and *P* value for the two analytical methods.

		Respiration Rates[1]				
Analytical Method	n	Mean Before Coverage	Mean After Coverage	Mean Difference	SD Difference	P
Averaging	10	170.8	176.6	−5.76	20.15	0.39

[1] Expressed as μmol CO_2 m^{-3} sapwood s^{-1}

FIGURE 2. Regression lines showing the decrease in CO_2 efflux following cuvette closure in three representative Douglas fir branches. Elevation of the CO_2 efflux rate due to the sampling technique varied from almost 200% to none at all.

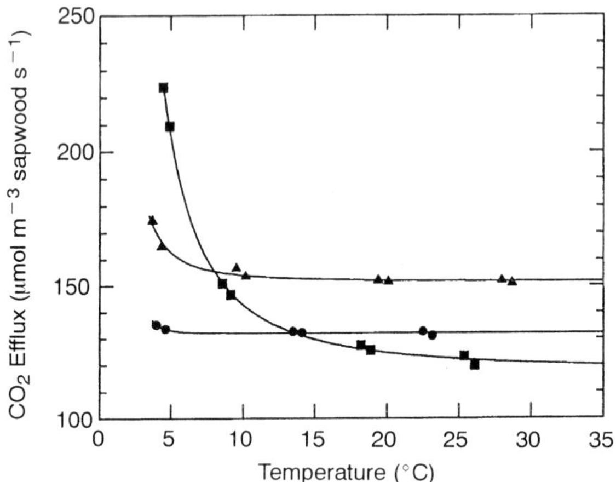

surements we used C-clamps to secure the cuvettes and applied pressure until the desired seal was obtained. The CO_2 concentration within the bark of Douglas fir branches at 21°C is on the order of 2500 μmol mol^{-1} (L. Cernusak, unpublished data). Compressing the bark would decrease pore volume beneath the bark surface, forcing CO_2-rich air into the chamber as the cortex collapsed. CO_2 efflux would therefore increase until a new steady-state concentration gradient developed.

Analysis of covariance showed no significant effect of cuvette closure on CO_2 efflux in the large diameter stems ($P = 0.16$). The mean respiration rate at 15°C was 17.9 μmol CO_2 m^{-3} sapwood s^{-1}. This is similar to maintenance respiration rates reported for stems of other North American conifers (Ryan, 1990; Ryan et al., 1995). The extent of increase in CO_2 efflux upon cuvette closure likely depends on the compressibility of the bark over which the cuvette is fitted, the amount of force applied to the cuvette, and the rigidity of the gasket material. The large diameter stems generally had thick, rigid bark. There was also a layer of non-hardening putty beneath each chamber plate that may have absorbed most of the pressure during cuvette tightening. In contrast, the branches had a flexible cortex, covered by a smooth periderm; the two together could be compressed between one's fingers. The gasket material was of a rigid nature relative to other closed cell foams, and had no putty beneath it. Thus, it is unlikely that stems and branches will necessarily differ in their responses to cuvette closure; rather the combination of the tissue structure and the method of cuvette attachment should be of most importance.

CONCLUSIONS

We conclude that calculations of woody-tissue volume in removable cuvette respiration measurements should be a function of all tissue between the gasket midpoints. In addition, when making respiration measurements on branches and small diameter stems, care should be taken to avoid artificially elevated flux rates due to compression of the bark beneath the gaskets. Our data for Douglas fir branches suggest that 15 minutes is a reasonable wait before making an observation; however, this may vary among species and with measurement conditions.

REFERENCES

Landsberg, J.J. and S.T. Gower. 1997. Applications of physiological ecology to forest management. Academic Press, San Diego, CA.

Maier, C.A., S.J. Zarnoch and P.M. Dougherty. 1998. Effects of temperature and tissue nitrogen on dormant season stem and branch maintenance respiration in a young loblolly pine (*Pinus taeda*) plantation. Tree Physiol. 18: 11-20.

Ryan, M.G. 1990. Growth and maintenance respiration in stems of *Pinus contorta* and *Picea engelmannii*. Can. J. For. Res. 20: 48-57.

Ryan, M.G., S. Linder, J.M. Vose and R.M. Hubbard. 1994. Dark respiration in pines. In: pp. 50-63. H.L. Gholz, S. Linder, and R.E. McMurtrie (eds.). Environmental constraints on the structure and productivity of pine forest ecosystems: a comparative analysis. Ecol. Bull. 43, Copenhagen.

Ryan, M.G., S.T. Gower, R.M. Hubbard, R.H. Waring, H.L. Gholz, W.P. Cropper and S.W. Running. 1995. Woody tissue maintenance respiration of four conifers in contrasting climates. Oecologia 101: 133-140.

Sprugel, D.G. 1990. Components of woody-tissue respiration in young *Abies amabalis* (Dougl.) Forbes trees. Trees 4: 88-98.

Sprugel, D.G. and U. Benecke. 1991. Measuring woody-tissue respiration and photosynthesis. In: pp. 329-355. J.P. Lassoie and T.M. Hinckley (eds.). Techniques and approaches in forest tree ecophysiology. CRC Press, Boca Raton, FL.

Sprugel, D.G., M.G. Ryan, J.R. Brooks, K.A. Vogt and T.A. Martin. 1995. Respiration from the organ level to the stand. In: pp. 255-299. W.K. Smith and T.M. Hinckley (eds.). Resource physiology of conifers: acquisition, allocation and utilization. Academic Press, San Diego, CA.

Respiratory Parameters Define Growth Rate, Species Distribution and Adaptation to Temperature

R. S. Criddle
J. N. Church
D. Hansen

INTRODUCTION

Development of procedures for short-term measurements of plant physiological properties to define plant growth rate responses to temperature could allow rapid identification of plants best suited for growth in a given temperature environment. Plants grown at one site could be evaluated for their potential growth properties at distant sites without the extensive field trials that are currently required. Previous attempts to identify parameters that define plant growth rate responses to temperature have considered temperature dependencies of photosynthesis and of various enzyme reactions (e.g., Chiariello et al., 1989; Burke et al., 1988). None have been successful in developing a theoretical basis employing quantitative predictive parameters to iden-

R. S. Criddle and J. N. Church are affiliated with the Section of Molecular and Cellular Biology, University of California, Davis, CA 95616.

D. Hansen is affiliated with the Department of Chemistry, Brigham Young University, Provo, UT 84602.

Address correspondence to R. S. Criddle at the above address (E-mail: rscriddle@ucdavis.edu).

[Haworth co-indexing entry note]: "Respiratory Parameters Define Growth Rate, Species Distribution and Adaptation to Temperature." Criddle, R. S., J. N. Church, and D. Hansen. Co-published simultaneously in *Journal of Sustainable Forestry* (Food Products Press, an imprint of The Haworth Press, Inc.) Vol. 10, No. 1/2, 2000, pp. 115-123; and: *Frontiers of Forest Biology: Proceedings of the 1998 Joint Meeting of the North American Forest Biology Workshop and the Western Forest Genetics Association* (ed: Alan K. Mitchell et al.) Food Products Press, an imprint of The Haworth Press, Inc., 2000, pp. 115-123. Single or multiple copies of this article are available for a fee from The Haworth Document Delivery Service [1-800-342-9678, 9:00 a.m. - 5:00 p.m. (EST). E-mail address: getinfo@haworthpressinc.com].

tify high and low temperature limits for growth and growth rates within the allowed temperature range. This communication demonstrates the use of respiration-based measurements to describe *Eucalyptus* growth as a function of temperature. The data produce predicted growth rate responses to temperature change for trees from three species of *Eucalyptus*. The growth-temperature patterns obtained are consistent with temperature ranges in the native growth climates of the three species.

The ability to compete favorably and grow over a range of temperatures is determined by a complex mix of physiological factors. Genetic differences allow for many dissimilar solutions to the growth and survival problems plants face. All these solutions, however, require energy. No matter what the combination of mechanisms used by a particular plant to grow amid an array of stresses and competition from other organisms, all have energy costs. Plants that use energy more efficiently for the acquisition of carbon, nitrogen, etc., for growth processes are more likely to survive. Because temperature is always changing in most native habitats, survival and effective competition requires efficient energy use over the range of temperatures encountered. Major decreases in energy-use efficiency are expected at the temperature limits for growth of a species.

The energy-use efficiency of plants during aerobic metabolism can be described in terms of the substrate carbon conversion efficiency (ε), i.e., the fraction of substrate carbon incorporated into plant structural biomass (Hansen et al., 1994). The relation between ε (unitless), specific growth rate R_{SG} (moles C time^{-1} mass^{-1}), and specific respiration rate R_{CO_2} (moles CO_2 time^{-1} mass^{-1}) is given by equation 1:

$$R_{SG} = R_{CO_2}\,[\varepsilon/(1 - \varepsilon)] \qquad (1)$$

The temperature dependence of growth rate is a function of the temperature dependencies of metabolic rate and metabolic efficiency. The ratio of the rate of metabolic heat loss from a plant (q, μW) to the rate of CO_2 production by the plant provides a convenient measure of ε. This relation is intuitively apparent in that the more heat lost from the plant per mole of CO_2 produced through respiration, the lower the efficiency of storage of respiratory energy in structural biomass. The precise relation is given in equation 2 (Hansen et al., 1994):

$$q/R_{CO_2} = 455(I - \gamma_p/4) - [\varepsilon/(I - \varepsilon)]\Delta H_B \qquad (2)$$

In this equation, γ_p is the average chemical oxidation state of the substrate carbon for respiration/biosynthesis, ΔH_B is the enthalpy change for conversion of substrate carbon to biomass carbon, but including all other elements, and the constant 455 kJ mol^{-1} is from Thornton's rule (Erickson, 1987). Thus, if γ_p and ΔH_B do not change, changes in the ratio q/R_{CO_2} measure the changes in ε with temperature.

If growth rate is described as the rate of storage of chemical energy in structural biomass, with substrate as the reference energy state (Criddle et al., 1997), then growth rate can be expressed as the difference between the rate of energy produced by respiration and the rate of energy lost to the surroundings. This relation is expressed in equation (3), which is a combination of equations 1 and 2:

$$R_{SG}\Delta H_B = 455R_{CO_2} - q \quad (3)$$

$R_{SG}\Delta H_B$ gives growth rate as the rate of storage of chemical energy in units of kJ sec^{-1} mg^{-1} tissue. When $455R_{CO_2} > q$, R_{SG} is positive (assuming a small, constant, positive value for ΔH_B). When $q > 455R_{CO_2}$, growth stops, the rate of accumulation of energy into structural biomass is less than zero, and a plant must rely on its reserves to survive. The instantaneous response of growth rate to temperature can thus be calculated from measurements of the temperature dependencies of R_{CO_2} and q. In all plants studied to date, the temperature dependencies of q and R_{CO_2} differ (Criddle et al., 1997). Thus, the ratio q/R_{CO_2}, and therefore the efficiency, must change with temperature (see equation 2). Measurements of R_{CO_2} and q over the entire growth temperature range of a plant therefore allow examination of questions of how efficiencies and growth temperature range are linked and how this relates to climates for optimal growth. Compartment models for plant respiration attempting to accomplish this goal by division of respiratory energy into growth and maintenance components are conceptually flawed and have lead to wrong conclusions about the relations between plant metabolism and temperature responses to growth (Hansen et al., 1998).

METHODS

Trees from three *Eucalyptus* species, adapted for growth in different native environments but grown in common gardens, were studied to investigate the relation between the temperature responses of growth and climates of origin. Measurements of q and R_{CO_2} values were made on rapidly expanding, small, whole leaves (approximately 150 mg wet weight) at several temperatures (Criddle et al., 1997). The species studied were *E. saligna* (Smith), *E. grandis* (W. Hill ex Maiden), and *E. globulus* (Labill. subsp. *maidenii* [F. Muell] Kirkpatr., commonly known as *E. maidenii*) (Boland et al., 1984). *E. saligna* ranges from the south coast of New South Wales into Queensland to the north, at elevations from near sea level to about 1,100 m. Seed for the tree shown in Figure 1 was collected from the middle of the range at an elevation of 500 m. The observed temperature range for *E. saligna* is from about −2°C mean low for the coldest month (mean low) to 33°C mean high for the warmest month (mean high) with a yearly mean near 16.5°C. *E. grandis* occurs from the central coast of New South Wales into northern Queensland from sea level to 600 m, except in the far north where it exists from 500 to 1,100 m. Seed for the *E. grandis* tree with data shown in Figure 1 came from about 100 km south of the Queensland border where 3°C is the mean low, 30°C is the mean, and 16.5°C is the mean annual temperature. *E. globulus* has a narrow native growth range along the SE coast of New South Wales at altitudes from 200 to 900 m. Seed for the *E. globulus* tree with data shown in Figure 1 was from an altitude of 500 m and mean low, mean high, and annual mean temperatures of 4°C, 25°C and 15°C, respectively. All trees studied to obtain Figure 1 were approximately 12 week-old seedlings grown hydroponically in a controlled environment chamber with a 16 h, 25°C day and 8 h, 15°C night. Additional *E. globulus* and *E. grandis* trees were examined using approximately 10 month-old seedlings grown in the field at Corning, CA with controlled additions of water and fertilizer. The seed origins of the field-grown trees are unknown.

RESULTS

Figure 1 presents calculated values of R_{SG} ΔH_B and substrate carbon conversion efficiency (ε) for one tree from each of the three species, based on measurements of q and R_{CO_2} from 5 to 35°C. The ranges of temperature encountered at the native growth site for each tree are included in Figure 1. Because γ_p and ΔH_B may differ among

FIGURE 1a. Effect of temperature on the rate of biomass accumulation by three *Eucalyptus* species with different native temperature ranges.

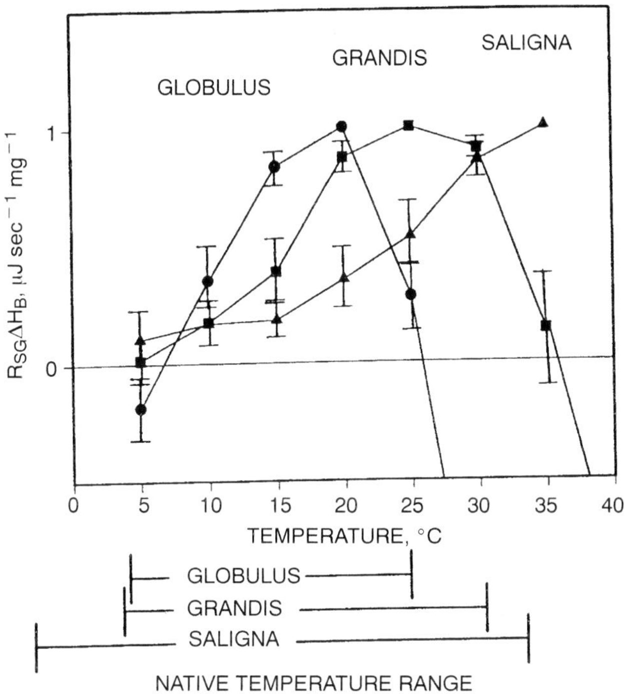

species, the curves of Figure 1a were normalized to the maximum for *E. grandis.* Normalizing to obtain identical maximum values has no effect on the shapes of the curves or the temperature range predicted for viable growth. The reproducibility of repeat measurements of $R_{SG}\Delta H_B$ vs. temperature for each tree is indicated by the error bars in Figure 1a. The error bars represent standard deviations for three separate measurements on each tree.

$R_{SG}\Delta H_B$ for *E. saligna* has a maximum at 35°C and is positive over the entire temperature range studied. $R_{SG}\Delta H_B$ for *E. grandis* passes through a maximum near 25°C. Values are positive over a relatively narrow temperature range and decrease to indicate lack of growth below 5 and above 35°C. These predicted limits for growth may be compared with the 3 and 30°C mean low and mean high temperatures in the *E. grandis* native growth habitat. $R_{SG}\Delta H_B$ for *E. globulus* has a maximum near 20°C and remains positive over a narrow temperature range from about 7 to 26°C. This range should be compared with the 4 to 25°C range of mean low and mean high temperatures at the native growth site.

Figure 1b shows values of substrate carbon conversion efficiency calculated (without normalization) as a function of temperature using equations 1 and 3, measured values of q and R_{CO2}, and assuming $\gamma_p = 0$ and $\Delta H_B = 20$. Measurements of ε vs. temperature for additional *E. grandis, E. saligna*, and *E. globulus* trees yielded overall patterns similar to those shown in Figure 1, but with significant intra-species variability. The error bars in Figure 1b indicate standard deviations of the within-species differences among 5 *E. grandis* and 5 *E. globulus*; one grown in a controlled environment chamber and four grown in the field. The *E. saligna* standard deviations are from only two chamber-grown trees.

Note that ε has a larger maximum value (≅0.66) for the narrowly

FIGURE 1b. Effect of temperature on substrate carbon conversion efficiency of three *Eucalyptus* species.

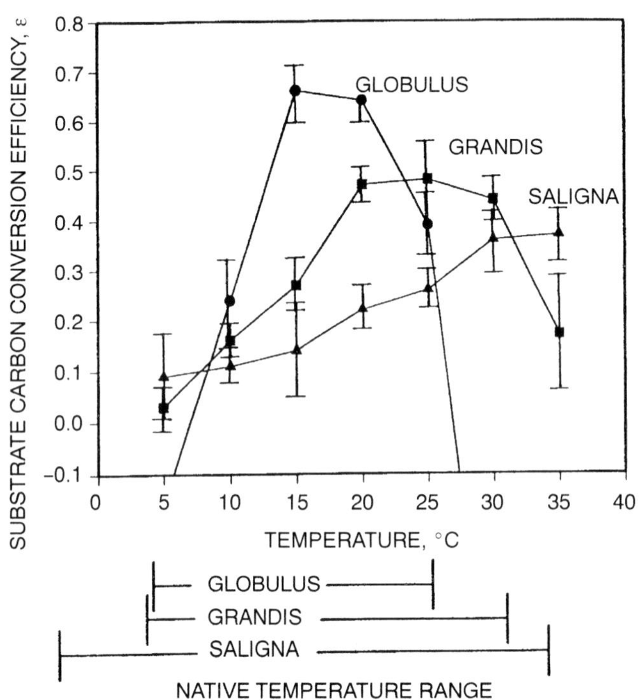

adapted *E. globulus* than for *E. grandis* (≅0.48) and *E. saligna* (≅ 0.39), which are adapted to broader climatic temperature ranges. However, ε for *E. globulus* is large over only a narrow range of temperature. Apparently, a trade-off exists between the maximum obtainable efficiency and the breadth of the temperature range to which the trees are adapted (see also Clark, 1993).

The curves of $R_{SG}\Delta H_B$ vs. temperature in Figure 1 make it possible to examine the relative growth rates of the species under different patterns of temperature fluctuation. Assuming a sinusoidal daily variation in temperature and known mean high and mean low values, the time each tree spends at a given temperature can be calculated. For each unit of time during the day, an average temperature can be calculated and used to produce values of time-at-temperature. Determination of the product (time-at-temperature) × (growth rate at each temperature) and summation over all values yields daily total growth. (Note that plants in their native habitat will not experience the entire seasonal range of temperatures on any given day. However, for initial comparison of species responses to temperature fluctuations it does not matter whether we consider sinusoidal variations over a day or a year.)

Figure 2 shows calculated values of growth for each of the three trees from Figure 1, for three assumed mean temperatures, and for

FIGURE 2. Effect of mean temperature and temperature range on growth of trees from three *Eucalyptus* species.

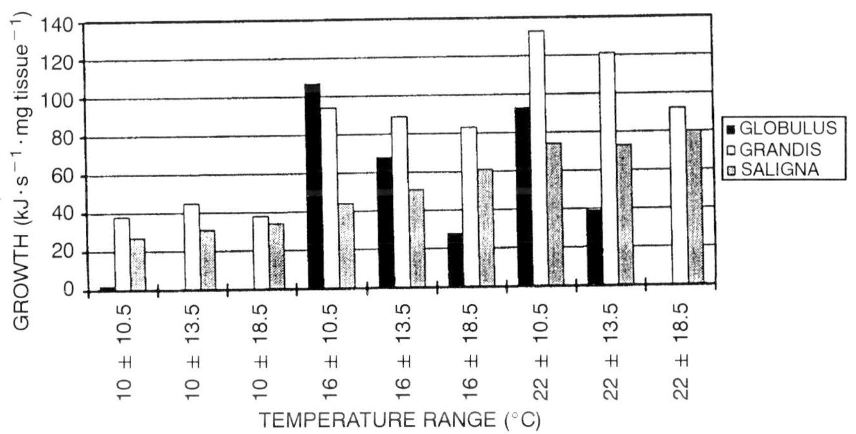

three temperature fluctuation ranges. One set of calculations uses a mean of 16°C and fluctuations of ± 10.5, 13.5, and 18.5°C. The 16°C temperature is near the mean temperature for the native range of all three species. The values of ± 10.5, 13.5, and 18.5°C correspond to the native temperature ranges for the *E. globulus*, *E. grandis*, and *E. saligna* trees studied, respectively. With 16°C as the average temperature, the *E. globulus* tree grows best when the temperature range is narrow. Growth of the *E. grandis* tree decreases with increasing range of temperature. The *E. saligna* tree growth increased as the temperature range was increased. With 22°C as the average, the *E. globulus* tree again grows best in a narrow temperature range and fails to grow in the 18.5°C temperature range. The *E. grandis* tree is calculated to be the best grower in all three temperature ranges considered. Relative to trees from the other species, the *E. saligna* tree growth is improved by larger temperature ranges. When 10°C is used as the average temperature, calculations show that growth for the trees from all species is diminished, and the *E. globulus* tree survives only with the narrow temperature range. With increasing growth temperature range, the *E. grandis* tree growth increases then decreases slightly and the *E. saligna* tree growth increases regularly. Because specific growth rate responses to temperature for all *Eucalyptus* trees examined in this study had species-specific patterns, we propose that the relative growth rate differences in Figure 2 will, within limits of intra-species variability, apply to growth rate differences among the three species.

CONCLUSIONS

The results in Figures 1 and 2 show that measured values of respiratory parameters of rapidly expanding tissues from individual *Eucalyptus* trees can be used to calculate the relative growth rates of each tree over the entire growth temperature range. Calculated values of relative growth rates at different temperatures are reproducible for repeat measurements on a single plant, show some variability within a species, and show much wider inter-species variability. Interspecies differences in response to temperature among the measured plants reflect the known differences in native growth climates of the three species examined. Relative growth rates of the trees measured differ with climatic temperature conditions. Measurements on individual *E. globulus, E. grandis*, and *E. saligna* trees suggest that species-specific differences in metabolic rate and growth rate responses to temperature are tightly correlated with native growth range. Calculated specific growth rates show that the range of temperatures tolerated for growth of the trees from each species is inversely related to their maximum substrate carbon conversion efficiencies. A trade-off exists between maximal efficiency and growth temperature range. In addition to the marked influence on growth and survival within a climate, the trade-off between efficiency and range also defines the relative growth rates and therefore the relative distribution of trees within any range of common growth.

REFERENCES

Boland, D.J. M.I.H. Brooker, G.M. Chippendale, N. Hall, B.M.P. Hyland, R.D. Johnston, D.A. Kiening and J.D. Turner. 1984. Forest Trees of Australia. Nelson, CSIRO, Melbourne, Australia.

Burke, J.J., J.R. Mahan and J.L. Hatfield. 1988. Crop-specific thermal kinetic windows in relation to wheat and cotton biomass production. Agron. J. 80: 553-556.

Chiariello, N.R., H.A. Mooney and K. Williams. 1989. Growth, carbon allocation and cost of plant tissues. In: R.W. Pearch, J.R. Ehleringer, H.A. Mooney and P.W. Rundel, eds. Plant Physiological Ecology, Chapman and Hall, New York. pp. 327-365.

Clark, A. 1993. Temperature and extinction in the sea: a physiologist's view. Paleobiology 19: 499-518.

Criddle, R.S., B.N. Smith and L.D. Hansen. 1997. A respiration based description of plant growth rate responses to temperature. Planta 201: 441-445.

Erickson, L.E. 1987. Energy requirements in biological systems. In: A.M. James, ed. Thermal and Energetic Studies of Cellular Biological Systems. IOP Publishing Ltd., Bristol. pp. 14-33.

Hansen, L.D., M.S. Hopkin, D.R. Rank, T.S. Anekonda, R.W. Breidenbach and R.S. Criddle. 1994. The relation between plant growth and respiration: a thermodynamic model. Planta 194: 77-85.

Hansen, L.D., R.W. Breidenbach, B.N. Smith, J.R. Hansen and R.S. Criddle. 1998. Misconceptions about the relation between plant growth and respiration. Botanica Acta 111:1- 6.

Carbon Dioxide Efflux Rates from Stems of Mature *Quercus prinus* L. and *Acer rubrum* L. Trees Do Not Appear to Be Affected by Sapflow Rates

Nelson T. Edwards
Stan D. Wullschleger

INTRODUCTION

The goal of the long-term Throughfall Displacement Experiment on the Walker Branch Watershed in eastern Tennessee is to develop a mechanistic understanding of how ecosystems adjust to changes in precipitation inputs that result from a warming global climate (Hanson

---

Nelson T. Edwards and Stan D. Wullschleger are Research Staff Members, Environmental Sciences Division, Oak Ridge National Laboratory, P.O. Box 2008, Oak Ridge, TN 37831-6422 USA.

The research was sponsored by the Program for Ecosystem Research, Environmental Sciences Division, Office of Health and Environmental Research, U.S. Department of Energy under contract No. DE-AC05-96OR22464 Lockheed Martin Energy Research Corp. Research was conducted on the Oak Ridge National Environmental Research Park.

Publication No. 4805, Environmental Sciences Division, Oak Ridge National Laboratory.

[Haworth co-indexing entry note]: "Carbon Dioxide Efflux Rates From Stems of Mature *Quercus prinus* L. and *Acer rubrum* L. Trees Do Not Appear to Be Affected by Sapflow Rates." Edwards, Nelson T., and Stan D. Wullschleger. Co-published simultaneously in *Journal of Sustainable Forestry* (Food Products Press, an imprint of The Haworth Press, Inc.) Vol. 10, No. 1/2, 2000, pp. 125-131; and: *Frontiers of Forest Biology: Proceedings of the 1998 Joint Meeting of the North American Forest Biology Workshop and the Western Forest Genetics Association* (ed: Alan K. Mitchell et al.) Food Products Press, an imprint of The Haworth Press, Inc., 2000, pp. 125-131. Single or multiple copies of this article are available for a fee from The Haworth Document Delivery Service [1-800-342-9678, 9:00 a.m. - 5:00 p.m. (EST). E-mail address: getinfo@haworthpressinc.com].

et al., 1995; Hanson et al., 1998). A key objective of the research is to evaluate the carbon cycling responses of the organisms to changes in precipitation. One requirement for accomplishing this objective is an accurate assessment of stem respiration because of the large amount of carbon cycled via this pathway. However, a major concern over the interpretation of stem respiration measurements is the suggestion that water moving upward through trees dissolves respired CO_2 and transports it to the leaves where it may be used in photosynthesis or otherwise released to the atmosphere. This suggestion has grown primarily out of the observation that a depression in measured stem respiration often occurs between midday and late-afternoon when transpiration rates are high (Edwards and McLaughlin, 1978; Lavigne, 1987; Kakubari, 1988). The objective of the experiment presented here was to determine whether measured stem CO_2 efflux rates in mature deciduous trees on Walker Branch Watershed are affected by transpiration-driven sapflow rates.

METHODS

The study site is located on Walker Branch Watershed (35°58′ N and 84°17′ W), a part of the U.S. Department of Energy's National Environmental Research Park near Oak Ridge, TN. Mean air temperature is 13.3°C and mean annual precipitation is 140 cm. The forest is an uneven-aged stand, with trees generally ranging from 40 to 75 years old and a few trees about 150 years old. The predominant overstory species are white oak (*Quercus alba* L.), chestnut oak (*Quercus prinus* L.), and red maple (*Acer rubrum* L.). The predominant understory species are flowering dogwood (*Cornus florida* L.) and red maple (Edwards and Hanson, 1996).

Three randomly selected overstory chestnut oak trees (37 to 69 cm diameter, 23 to 29 m height) and one randomly selected overstory red maple tree (35 cm diameter, 27 m height) were chosen for simultaneous measurements of sapflow rates and stem respiration rates. Measurements in the chestnut oaks were taken between 0430 h and 2200 h on 12 July, 1996 while those in the red maple were taken between 0800 h and 2300 h on 29 August, 1996. Stem respiration was measured with a closed infrared gas analysis system (Edwards and Hanson, 1996). A flat, rectangular frame (17.5 cm wide × 35 cm long) made of 3 mm thick and 3 cm wide aluminum was attached to the north side of each tree approximately 1.5 m above the ground. Spaces between the frame and the tree were filled with insulating foam and sealed. The frames provided a uniform base for attaching an aluminum cuvette for isolating the interior from the outside air and creating a volume from which stem CO_2 exchange rates could be measured. The cuvette was fastened to the tree frame using two nylon straps that extended around the tree. Ratchets were used to pull the cuvette against a closed-cell foam gasket sandwiched between the frame and the cuvette forming an airtight seal. During each measurement, air was pumped across the stem surface from a manifold on one side of the cuvette to a manifold on the opposite side and then to the gas analyzer. Sapwood temperature was measured during each respiration measurement with a thermocouple inserted into the outer portions of the sapwood to a depth of 2 cm. Sapwood volume under respiration chambers was calculated by multiplying sapwood thickness (determined from increment cores) times the area covered by the respiration chamber.

Estimates of sapflow density (kg H_2O dm^{-2} h^{-1}) were obtained for the four trees with a constant energy-input method (Granier, 1987). A pair of 2-cm probes, separated vertically by 12 to 15 cm, was inserted into the sapwood on the north side of each tree. The upper probe was heated at a constant rate (200 mW), whereas the lower probe served as an unheated reference. Each probe contained a copper-constantan thermocouple and between the probes, these thermocouples were connected in opposition so as to produce a temperature differential. Measured temperature differentials are influenced by the rate at which water flows past the probes. Temperature differentials were monitored every 60 s and 30-min averages were stored to a datalogger.

RESULTS

Red maple stem CO_2 efflux rates were greatest during periods of greatest sapflow rates (Figure 1). Sapflow rates ranged from 0 kg dm^{-2} h^{-1} at 0800h to 1.5 kg dm^{-2} h^{-1} at about 1600 h. Stem respiration in the red maple ranged from a low of 66 μmol CO_2 m^{-3} sapwood s^{-1} at 0800h to 92 μmol CO_2 m^{-3} sapwood s^{-1} at 1600 h and were almost identical to predicted rates based on stem temperature and a previously determined Q_{10} of 2.0. In the chestnut oaks, stem respiration remained relatively constant and no distinct diurnal pat-

FIGURE 1. Diurnal patterns of stem respiration, stem temperature, and sapflow rates in four overstory trees (three chestnut oaks and one red maple) in eastern Tennessee.

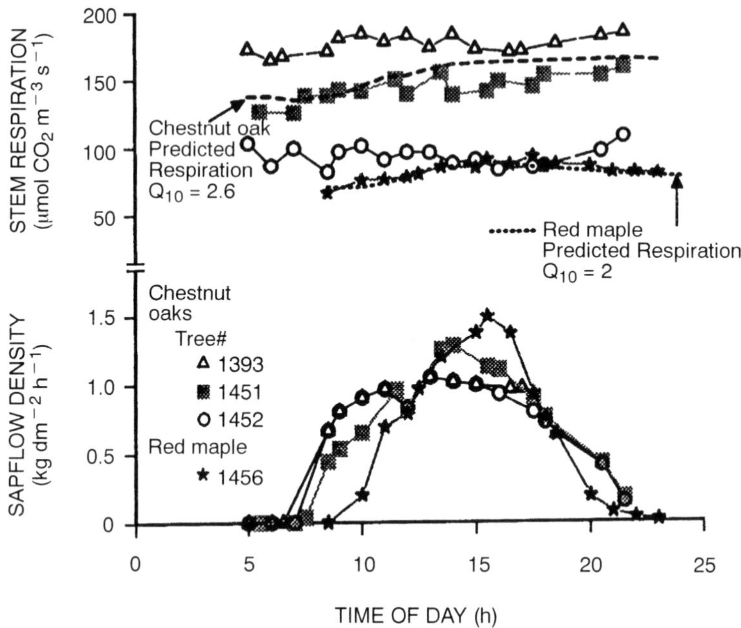

terns were observed. The average rates for all three trees deviated little from predicted rates (Q_{10} = 2.6). Average respiration was greater in tree #1393 (175 μmol CO_2 m^{-3} sapwood s^{-1}) than in trees #1451 and #1452 (143 and 93 μmol CO_2 m^{-3} sapwood s^{-1}, respectively). Similarly, average sapflow rate was greatest (0.74 kg dm^{-2} h^{-1}) in tree #1393 and least in trees #1451 and #1452 (0.69 and 0.60 kg dm^{-2} h^{-1}, respectively). The respiration rates and sapflow rates were in reverse order of tree size. Regression analysis revealed no correlation between respiration and sapflow in chestnut oak, and a weak but positive correlation (r^2 = 0.34) between respiration and sapflow in red maple.

DISCUSSION

Studies with seedlings have shown that the transpiration stream does transport CO_2 thereby reducing stem CO_2 efflux (e.g., Martin et al. 1994), but detectable reductions in stem CO_2 efflux by this phenomenon have not yet been demonstrated in large deciduous trees. The amount of CO_2 that can dissolve in xylem sap can be calculated and the amount is relatively large (7.98×10^{-7} mol CO_2 ml^{-1} H_2O at 25°C) if xylem sap is assumed to be pure water (Martin et al., 1994). However, given that the water moving through the trees in this study entered the roots from a naturally high CO_2 soil environment, it is possible that the water was already near CO_2 saturation before entering the trees. For example, Yavitt et al. (1995) reported soil atmospheric CO_2 concentrations as high as 19,000 $\mu L\ L^{-1}$ in a northern hardwood forest in New York and Castelle and Galloway (1990) reported concentrations as high as 15,000 $\mu L\ L^{-1}$ in a deciduous forest in Virginia. Our sapflow and stem respiration measurements were made when both transpiration and respiration rates were near their annual maximum (both below- and aboveground), yet there was no apparent diurnal reduction in CO_2 efflux from the stems during periods of greatest sapflow. Furthermore, the measurements were made on the large portions of the stems where sapflow would be expected to have its greatest impact on CO_2 efflux (Negisi, 1975). In red maple, CO_2 efflux was greatest during periods of greatest sapflow rates. In chestnut oak, we observed a relatively constant rate of total stem CO_2 efflux. We would have expected a decrease in CO_2 efflux during peak sapflow if stem respired CO_2 were transported in the transpiration stream. An argument could be made, however, that while the CO_2 efflux rates did not decrease during the highest sapflow rates, perhaps the CO_2 efflux rates were less than would have been expected based on temperature predictions of respiration rates. In chestnut oak, stem temperature did increase from mid- to late afternoon without a corresponding increase in stem respiration, but the temperature range was very narrow. Thus, there was only a slight increase in Q_{10} predicted CO_2 efflux rates in the afternoon. In red maple, measurements were made when the temperature range was greater and a distinct peak in stem temperature was matched by a peak in stem respiration. Regression analysis showed that temperature and stem respiration were positively correlated ($r^2 = 0.79$). The question still remains as to whether the relatively constant CO_2 efflux rate in chestnut oak was due to a relatively constant or perhaps a reduced rate of growth respiration (given that maintenance respiration would have increased even with the small temperature increase), or was it due to net CO_2 transport in the transpiration steam? If we accept the latter argument then we would expect that the chestnut oak tree with the greatest sapflow rates to have the lowest CO_2 efflux rates. However, the reverse was true. The tree with the lowest sapflow rates (tree #1452) also had the lowest CO_2 efflux rates and the tree with the greatest sapflow rates (#1393) also had the greatest CO_2 efflux rates. Previously reported mid-day depressions in large tree stem respiration may be due to reduced substrate availability (Edwards and McLaughlin, 1978) or to mid-day tissue water stress (Negisi, 1975; Kakubari, 1988; Lavigne, 1987). Either of these findings support the possibility of decreased growth activity and thus decreased growth respiration at mid-day.

CONCLUSION

The results of this study indicate that stem CO_2 efflux rates in mature red maple and chestnut oak trees are not altered by diurnal patterns of sapflow, thus suggesting very little, if any, net movement of CO_2 in the transpiration stream.

REFERENCES

Castelle, A.J. and J.N. Galloway. 1990. Carbon dioxide dynamics in acid forest soils in Shenandoah National Park, Virginia. Soil Sci. Soc. Am. J. 54:252-257.

Edwards, N.T. and S.B. McLaughlin. 1978. Temperature-independent diel variations of respiration rates in *Quercus alba* and *Liriodendron tulipifera*. Oikos 31:201-206.

Edwards, N.T. and P.J. Hanson. 1996. Stem respiration in a closed-canopy upland oak forest. Tree Physiology 16:433-439.

Granier, A. 1987. Evaluation of transpiration in a Douglas fir stand by means of sap flow measurements. Tree Physiology 3:309-320.

Hanson, P.J., D.E. Todd, N.T. Edwards, and M.A. Huston. 1995. Field performance of the Walker Branch Throughfall Displacement Experiment. In: pp. 307-313. A. Jenkins, R.C. Ferrier, and C. Kirby (eds.). Ecosystem Manipulation International Symposium. Ecosystems Research Report No 20. ECSC-EC-EAEC, Brussels–Luxembourg.

Hanson, P.J., D.E. Todd, M.A. Huston, J.D. Joslin, J.L. Croker, and R.M. Auge. 1998. Description and field performance of the Walker Branch Throughfall Displacement Experiment: 1993-1996. ORNL/TM-13586. Oak Ridge National Laboratory, Oak Ridge, TN 37831, USA.

Kakubari, Y. 1988. Diurnal and seasonal fluctuations in the bark respiration of standing *Fagus sylvatica* trees at Solling, West Germany. J. Jpn. For. Soc. 70 (2): 64-70.

Lavigne, M.B. 1987. Differences in stem respiration responses to temperature between balsam fir trees in thinned and unthinned stands. Tree Physiology 3: 225-233.

Martin, T.A., R.O. Tesky, and P.M. Dougherty. 1994. Movement of respiratory CO_2 in stems of loblolly pine (*Pinus taeda* L.) seedlings. Tree Physiology 14:481-495.

Negisi, K. 1975. Diurnal fluctuation of CO_2 release from the bark of standing young *Pinus densiflora* trees. J. Jap. For. Soc. 57(11):375-383.

Yavitt, J.B., T.J. Fahey, and J.A. Simmons. 1995. Methane and carbon dioxide dynamics in a northern hardwood ecosystem. Soil Sci. Soc. Am. J. 59:796-804.

Root Growth Plasticity of Hybrid Poplar in Response to Soil Nutrient Gradients

Alexander L. Friend
Juanita A. Mobley
Elizabeth A. Ryan
H. D. Bradshaw, Jr.

INTRODUCTION

Forest trees are often intensively selected and bred for yield improvement, yet most of this effort has been devoted to shoot rather than root traits. Genetic variation exists, but has not been exploited, in a variety of tree root traits such as specific root length (length of root per dry weight, Eissenstat 1991), vertical distribution (Heilman et al., 1994), horizontal extent of roots (Friend et al., 1991), and root growth plasticity in response to nutrient-enriched soil patches (George et al., 1997). Root growth plasticity, the increase in root growth associated with enriched zones (or patches) of soil, is of particular relevance

---

Alexander L. Friend, Juanita A. Mobley, and Elizabeth A. Ryan are affiliated with the Department of Forestry, Box 9681, Mississippi State University, MS 39762-9681 (E-mail: afriend@cfr.msstate.edu).

H. D. Bradshaw, Jr. is affiliated with the College of Forest Resources, University of Washington, Seattle, WA 98195.

Journal Article No. FO-091 of the Forest and Wildlife Research Center, Mississippi State University.

[Haworth co-indexing entry note]: "Root Growth Plasticity of Hybrid Poplar in Response to Soil Nutrient Gradients." Friend, Alexander L. et al. Co-published simultaneously in *Journal of Sustainable Forestry* (Food Products Press, an imprint of The Haworth Press, Inc.) Vol. 10, No. 1/2, 2000, pp. 133-140; and: *Frontiers of Forest Biology: Proceedings of the 1998 Joint Meeting of the North American Forest Biology Workshop and the Western Forest Genetics Association* (ed: Alan K. Mitchell et al.) Food Products Press, an imprint of The Haworth Press, Inc., 2000, pp. 133-140. Single or multiple copies of this article are available for a fee from The Haworth Document Delivery Service [1-800-342-9678, 9:00 a.m. - 5:00 p.m. (EST). E-mail address: getinfo@haworthpressinc.com].

because it may enhance the efficiency of root investment into nutrient acquisition in heterogeneous soils. We explored the variation in root growth plasticity in *Populus* hybrids. *P. trichocarpa* Torr. & Gray × *P. deltoides* Marsh. hybrids are some of the highest-yielding trees in the world, with total biomass productivity of up to 35 Mg ha^{-1} y^{-1} (Scarascia-Mugnozza et al., 1997) and they have been the focus of intensive genetics and physiological studies. A three-generation pedigree of parentals, F_1 hybrids, and F_2 offspring have been particularly useful for quantifying gene sequences responsible for physiological traits (Bradshaw, 1996). We studied this pedigree because of its demonstrated variability and because of the potential to contribute root traits to existing selection and breeding schemes. The objective of this study was to quantify the variation in fine root growth of hybrid poplar in response to nutrient-enriched patches of soil and make some informed speculation about the application of this variation to yield physiology and poplar ideotypes.

METHODS

Hardwood cuttings (approx. 15 cm len. and 1 cm dia.) were produced in Boardman, Oregon, USA from a three-generation pedigree of *P. trichocarpa* × *P. deltoides* hybrids (Bradshaw, 1996), shipped to Mississippi in February 1996, and stored for one month. The experiment was conducted outdoors near Starkville, MS, USA (33° 28′ N 88° 49′ W) between April and June 1996. Cuttings were planted in 3 L (15 cm. dia) pots of sand, grown for five weeks, and then exposed to patches of high and low nutrients over the next 20 days. Nineteen clones from the pedigree were studied: parents (*P. trichocarpa* female 93-968, *P. deltoides* male 14-129), F_1 hybrids (female 53-246, male 53-242) and 15 F_2 hybrid offspring (331–see Table 1), with nine individuals of each clone (n = 9). Pots were watered daily for five days and three times per week thereafter. Plants received a background supply of N after planting to sustain growth but allow N deficiency to develop. Each pot received 100 ml of nutrient solution 3, 4, and 5 weeks after planting. The nutrient solution contained 0.5 mM NH_4NO_3, 0.18 mM KCl, 0.06 mM KH_2PO_4, 0.05 mM $MgSO_4$, 0.03 mM $CaCl_2$, and micronutrients in balanced proportions (Ingestad and Lund, 1986). High and low nutrient patches were maintained in each pot for the next 20 days by daily nutrient additions (150 mL of nutrient solution) to a

PVC discharge tube (+N patch) on one side of the pot and an equal volume of distilled water to a comparable tube (−N patch) on the opposite side of the pot. Tubes (3 cm inside dia.) were inserted to a 5 cm depth and covered to prevent contamination. During treatment, pots were watered three times per week with tap water (before nutrient addition to the tubes). Plants were harvested on 16 June 1996.

At harvest, roots were recovered from patches with coring tubes (4.4 cm inside dia.) which encompassed treated patches. After core removal, sand was gently washed from the remaining roots, and the plant was divided into foliage, stems (including the cutting below the soil surface), and roots. Tissues were dried at 70°C and weighed. Fine roots (< 2 mm dia.) were removed from cores by handsorting with tweezers, and stored in 20% methanol. A subsample of five clones (53-242, 331-1114, -1127, -1128, -1149) which varied in growth response, based on the fresh weight of roots in +N: −N patches, was analyzed for specific root length (m g^{-1}) by measuring length with an image analysis system (Harris and Campbell, 1989) before drying. Roots from patches were dried and weighed as for the other tissue types.

Whole plant dry mass (foliage, stems, and roots) was analyzed in a completely random design, with clone as the treatment. Five clones were dropped from the analysis due to poor survival, but 14 well represented ($n \geq 4$) clones were used. Root fraction was determined by regression analysis of root mass vs. foliage mass. Roots from patches were analyzed in a completely random design as a 14 × 2 factorial, with 14 clones and 2 patch N treatments (+N or −N); specific root length was a 5 × 2 factorial. Root plasticity, as determined from the +N: −N mass ratio, was analyzed for each clone in a completely random design.

RESULTS

Survival varied from 0-100% among the clones, with five of the 19 clones dropped from the study due to poor survival (Table 1a). Growth performance of the remaining material varied by clone for foliage ($p = 0.06$), stem ($p = 0.001$), root ($p = 0.03$), and total mass ($p = 0.002$). Values for the three tissue types were positively correlated (Table 2). Total root mass (including roots inside and outside of patches) varied over a three-fold range (Table 1b), was strongly correlated with foliage mass (Figure 1), but was weakly correlated with stem mass

TABLE 1. Ranking of clones for survival (a), total root growth (b), and root growth plasticity to patch N (c), listed in descending order. Clones are numbered for F_1 (53–series) and F_2 (331–series) material. Clonal ranking for root growth plasticity (c) is according to the +N patch: – N patch biomass ratio (+N: – N). Values presented are means ± standard error. Similar letters indicate no significant differences (α = 0.05) within cells of the column according to Duncan's Multiple Range test.

a. Survival[1]		b. Total root growth		c. Root growth plasticity to patch N			
Clone	%	Clone	Root mass (mg)	Clone	+N patch (mg)	– N patch (mg)	+N: – N
331-1086	100	331-1151	608 ± 137 a	331-1114	56 ± 16	12 ± 4	4.6
331-1151	89	331-1075	585 ± 164 ab	331-1122	51 ± 10	12 ± 3	4.2
53-246	78	331-1086	504 ± 51 ab	331-1149	48 ± 14	15 ± 5	3.2
331-1061	78	53-242	436 ± 167 abc	331-1061	24 ± 6	9 ± 3	2.7
331-1126	78	331-1126	409 ± 59 abc	331-1068	38 ± 14	15 ± 5	2.5
331-1075	67	331-1149	356 ± 80 abc	331-1126	58 ± 8	24 ± 6	2.4
331-1122	67	331-1122	346 ± 49 abc	53-242	51 ± 14	21 ± 8	2.4
331-1127	67	331-1114	322 ± 44 abc	53-246	39 ± 8	17 ± 4	2.3
331-1128	67	331-1128	313 ± 71 abc	331-1580	47 ± 18	21 ± 9	2.2
331-1149	67	53-246	285 ± 53 bc	331-1151	64 ± 14	32 ± 6	2.0
53-242	56	331-1127	285 ± 44 bc	331-1086	50 ± 5	29 ± 5	1.7
331-1068	56	331-1068	280 ± 116 bc	331-1075	47 ± 12	29 ± 9	1.6
331-1114	44	331-1580	273 ± 98 bc	331-1127	23 ± 9	19 ± 5	1.2
331-1580	44	331-1061	181 ± 43 c	331-1128	22 ± 3	26 ± 2	0.9

[1] Clones with less than 33% survival (14-129, 93-968, 331-1063, 31-1069, 331-1169) were eliminated from this analysis.

(Table 2). Root growth in patches varied with clone (p = 0.014) and patch N (p = 0.0001), but the two did not interact (clone × N, p = 0.40). When averaged over all clones, twice as much root mass accumulated in +N patches (44 ± 3 mg) as – N patches (21 ± 2 mg) (mean ± s.e.m.). Roots from +N patches had larger specific root lengths (172 ± 18 m g^{-1}) than roots from – N patches (129 ± 8 m g^{-1}) (p = 0.04), but there were no clonal effects (p = 0.70) or clone × N interactions (p = 0.69). Clonal effects were observed for ratios of +N patch: – N patch root mass (p = 0.06), with a maximum of 8.1 ± 4.8 in clone 53-242, a minimum of 0.9 ± 0.1 in clone 331-1128, and an overall average of 3.3. This mean of ratios from individual plants was both volatile (9-fold range) and noisy (average within-clone coefficient of variation 67%), particularly for the top clone (53-242) which had a CV of 132%. A different method of calculating the ratio (+N mean: – N mean, by clone) was less volatile, with a maximum of 4.6, a minimum of 0.9, and an average of 2.4. This measure (ratio of means)

TABLE 2. Correlations between dry mass of tissue types. Values presented are correlation coefficients (r) analyzed for raw data (n = 86) and clonal means (n = 14). Superscripts indicate statistical significance; ($p < 0.01$)**, ($p < 0.05$)*, ($p \leq 0.05$)$^{n.s.}$.

	Root mass (g)	Stem mass (g)
Leaf mass		
(raw data)	0.88**	0.66**
(clonal means)	0.90**	0.70**
Root mass		
(raw data)		0.70**
(clonal means)		0.51$^{n.s.}$

FIGURE 1. Total root mass accumulation (y) as a function of foliage mass accumulation (x) eight weeks after planting hardwood cuttings. Points represent clonal means for 14 F_1 and F_2 clones. The relationship ($y = a + bx$) is $y = -0.13 + 0.54x$, $r^2 = 0.81$. Dashed lines indicate the 95% confidence interval for the regression, and numbers identify the clones that were outside of the confidence bands.

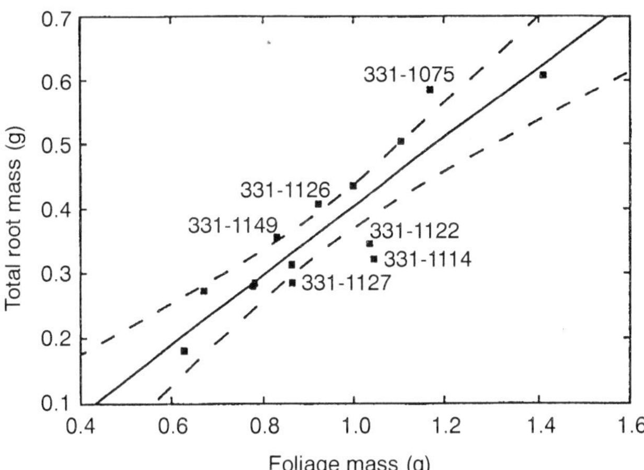

was used to rank the clones for root plasticity (Table 1c). The two measures of plasticity agreed for the top- and bottom-ranked clones. The most plastic clones were 331-1114 and 331-1122 which had a +N: – N ratio of > 4.0 by either measure. The least plastic clones were 331-1127 and 331-1128 which had a +N: – N ratio of < 1.3 by either measure.

DISCUSSION

We attributed cases of low survival to heat and evaporative load in excess of normal adaptation, as the site was more than 5° south of either parent source, and the cuttings were rooted outdoors in sand. Nevertheless, cuttings accumulated nearly 2 g of foliage over 8 weeks and root growth varied considerably by clone (Table 1b). A portion of the clonal variation seen in absolute growth might be attributable to non-genetic factors such as variation in initial cutting size, which was not quantified. However, clonal variation in root growth from our study (Table 1b), was positively correlated ($r = 0.78$, $p = 0.002$) with variation in root growth of the same clones from a similar study (Riemenschneider, personal communication). The high correlation observed between root and leaf mass (Figure 1, Table 2) points to the importance of leaf development and current assimilation in supporting vigorous root system development, even though other factors are involved in adventitious root initiation (Friend et al., 1994). Outliers from this root:foliage relationship (Figure 1) indicated that some clones had a relatively large root investment (331-1149, -1126, -1075) while others had a relatively small root investment (331-1127, -1114, -1122).

Root growth plasticity varied over a large range, and was independent of whole-plant performance in that the most and least plastic clones were intermediate in total root growth and survival (Table 1). This independence was also supported by non-significant ($p > 0.19$) correlations between root growth plasticity and foliage biomass, total root biomass, or total plant biomass. It is interesting to note that the most plastic clones (331-1114 and 331-1122) were the farthest outliers on the low end of the root:foliage allometric relationship (Figure 1). This suggests that greater discrimination for favorable soil environments may have enabled these clones to carry a relatively low root mass due to more effective root deployment. The variation seen in relative root investment (outliers from Figure 1) and root plasticity (Table 1c) may have application to tree improvement if the efficiency of carbon allocated to roots could be improved. Using the terminology of Grime (1994), our results illustrate root systems that show various combinations of scale (relative root investment) and precision (root growth plasticity), e.g., large scale-low precision: 331-1075, small scale-high precision: 331-1114 and 331-1122, and low scale-low precision: 331-1127. It is hypothesized that root systems of small scale and high precision (e.g., 331-1114) would benefit yield of hybrid poplar growing in plantation culture, especially when water and nutrients are supplied in patches from drip irrigation. Such a combination could permit overall root investment to be small and yet effective at resource acquisition through vigorous root proliferation in resource-rich patches of soil. This hypothesis has an appealing potential to contribute to a *Populus* ideotype and the associated debate over whether phenotypic plasticity can be used in tree selection (Dickmann and Keathley 1996). To reach this potential, the phenomenon of genetic variation in root growth plasticity should be explored under representative field conditions. Specifically, the sensitivity of whole tree growth or nutrient accumulation to genetic variation in root growth plasticity should be evaluated under stand conditions for more than one growing season.

CONCLUSIONS

- Clonal variation was observed in the degree to which root growth was stimulated by nutrient-enriched patches of soil (root growth plasticity).
- It was hypothesized that high root growth plasticity would be a yield-enhancing trait.
- Root growth plasticity should be studied under representative field conditions to determine its relevance to tree selection and yield physiology.

REFERENCES

Bradshaw, H.D., Jr., 1996. Molecular genetics of *Populus*. In: pp. 183-189. R.F. Stettler, H.D. Bradshaw, Jr., P.E. Heilman, and T.M. Hinckley (eds) Biology of *Populus* and its implications for management and conservation. NRC Research Press, National Research Council of Canada, Ottawa.

Dickmann, D.I., and D.E. Keathley. 1996. Linking physiology, molecular genetics, and the *Populus* ideotype. In: pp. 491-514. R.F. Stettler, H.D. Bradshaw, Jr., P.E. Heilman, and T.M. Hinckley (eds) Biology of *Populus* and its implications for management and conservation. NRC Research Press, National Research Council of Canada, Ottawa.

Eissenstat, D.M. 1991. On the relationship between specific root length and the rate of root proliferation: a field study using citrus rootstocks. New Phytol. 118: 63-68.

Friend, A.L., M.D. Coleman and J.G. Isebrands. 1994. Carbon allocation to root and shoot systems of woody plants. In: pp. 245-273. T.D. Davis and B.E. Haissig (eds). Biology of adventitious root formation. Plenum Press, New York.

Friend, A.L., G. Scarascia-Mugnozza, J.G. Isebrands, and P.E. Heilman. 1991. Quantification of two-year-old hybrid poplar root systems: morphology, biomass, and ^{14}C distribution. Tree Physiol. 8: 109-119.

George, E., B. Seith, C. Schaeffer, and H. Marschner. 1997. Responses of *Picea*, *Pinus* and *Pseudotsuga* roots to heterogeneous nutrient distribution in soil. Tree Physiol. 17: 39-45.

Grime, J.P. 1994. The role of plasticity in exploiting environmental heterogeneity. In: pp. 1-19. M.M. Caldwell and R.W. Pearcy (eds.) Exploitation of environmental heterogeneity by plants: ecophysiological processes above- and belowground. Academic Press, New York.

Harris, G.A. and G.S. Campbell 1989. Automated quantification of roots using a simple image analyzer. Agron. J. 81: 935-938.

Heilman, P.E., G. Ekuan, and D. Fogle. 1994. Above- and belowground biomass and fine roots of 4-year-old hybrids of *Populus trichocarpa* × *Populus deltoides* and parental species in short-rotation culture. Can. J. For. Res. 24: 1186-1192.

Ingestad, T. and Ann-B. Lund. 1986. Theory and technique for steady state mineral nutrition and growth of plants. Scand. J. For. Res. 1: 439-453.

Riemenschneider, D.E., Forest Geneticist, USDA Forest Service, Rhinelander, WI. Personal communication, 13 March 1998.

Scarascia-Mugnozza, G.E., R. Ceulemans, P.E. Heilman, J.G. Isebrands, R.F. Stettler, and T.M. Hinckley. 1997. Production physiology and morphology of *Populus* species and their hybrids grown under short rotation. II. Biomass components and harvest index of hybrid and parental species clones. Can. J. For. Res. 27: 285-294.

Effects of Ultraviolet-B Radiation on Needle Anatomy and Morphology of Western Larch, Interior Spruce and Lodgepole Pine

P. M. Krol
D. P. Ormrod
W. D. Binder
S. J. L'Hirondelle

INTRODUCTION

Ultraviolet-B (UV-B) radiation (280-320 nm) is increasing in the biosphere because anthropogenic ozone scavengers diffusing to the stratosphere are depleting the protective stratospheric ozone layer (Kerr and McElroy, 1993). For every 1% decrease in stratospheric

P. M. Krol and D. P. Ormrod are affiliated with the Centre for Forest Biology, University of Victoria, Biology Department, P.O. Box 3020, Station CSC, Victoria, B.C., Canada V8W 3N5 (E-mail: pkrol@uvic.ca).

W. D. Binder and S. J. L'Hirondelle are affiliated with the B.C. Ministry of Forests, Research Branch, Glyn Road Research Station, P.O. Box 936, Station Prov. Gov., Victoria, B.C., Canada V8W 3E7.

The photographs and measurements were made in the University of Victoria Advanced Imaging Laboratory with the assistance of T. Gore and H. Down. J. N. Owens, G. Catalano and D. Fernando provided microtechnique suggestions.

This research was supported by the British Columbia Ministry of Forests through a FRBC grant to W. D. Binder (Project HQ96383-RE).

[Haworth co-indexing entry note]: "Effects of Ultraviolet-B Radiation on Needle Anatomy and Morphology of Western Larch, Interior Spruce and Lodgepole Pine." Krol, P. M. et al. Co-published simultaneously in *Journal of Sustainable Forestry* (Food Products Press, an imprint of The Haworth Press, Inc.) Vol. 10, No. 1/2, 2000, pp. 141-148; and: *Frontiers of Forest Biology: Proceedings of the 1998 Joint Meeting of the North American Forest Biology Workshop and the Western Forest Genetics Association* (ed: Alan K. Mitchell et al.) Food Products Press, an imprint of The Haworth Press, Inc., 2000, pp. 141-148. Single or multiple copies of this article are available for a fee from The Haworth Document Delivery Service [1-800-342-9678, 9:00 a.m. - 5:00 p.m. (EST). E-mail address: getinfo@haworthpressinc.com].

ozone there is a 2% increase in UV-B reaching the earth's surface (Caldwell, 1979). This study describes morphological and anatomical changes in needles of ecologically and economically important conifers exposed to differential UV-B doses.

MATERIALS AND METHODS

Plant Material and UV-B Treatments

Twenty-one-year-old seedlings of western larch (*Larix occidentalis* Nutt.), interior spruce (*Picea glauca* × *Picea engelmannii*) and lodgepole pine (*Pinus contorta* Dougl. ex. Loud var *latifolia* Engelm.) were placed in each of 12 1.2 m × 1.2 m UV-B exposure chambers in a greenhouse. The daily 6h UV-B doses consisted of 0, 4 (summer ambient at Victoria, B.C.), 8 and 12 kJ m^{-2} day^{-1} (biologically effective UV-B be, normalized to 300 nm) from UV-B-313 fluorescent lamps (Q-Panel, Cleveland, Ohio) with three chambers per dose.

Needle Curling

After 6 weeks of UV-B exposure, about ten needles per species per UV-B dose (3 seedlings per chamber) were excised from the upper two-fifths of the terminal leaders. Needle outlines were traced on transparent film and the curled length determined by laying the needle tracings on graph paper. The farthest point away from X = 0 was recorded as the curled length of the needle (the greater the curling, the shorter the curled length). Dental floss was used to follow the curling of each needle tracing and then placed along a ruler to determine the uncurled length of each needle. The curled/uncurled length ratio was calculated for each needle and plotted against UV-B dose.

Needle Sections

After 14 weeks of UV-B exposure, approximately 15 needles per species per UV-B dose (5 seedlings per chamber) were excised from the upper two-fifths of the terminal leaders; sliced into base (basal), mid and tip (distal) segments; and immediately placed in FAA fixative (Johansen, 1940). Segments were dehydrated, using a modified Johansen series, embedded edgewise in Tissueprep 2 and softened for four weeks in Gifford's Softening solution at 37°C, and microtomed to 8 μm thick needle sections.

Staining and Measurements

The needle sections were stained with safranin (1% aqueous), which indicates the presence of cutin, suberin, lignin, and phenolic compounds, and fast green as a counterstain (in 95% alcohol) (Johansen, 1940). The needle thickness and adaxial epidermal (epidermis plus hypodermis) thickness were measured at four to six locations on each section using a calibrated 6.3× and 16× objective lens, respectively. A large drop of 1% fluoroglucinol in 20% HCl (Johansen, 1940), which stains lignin, was placed on four deparaffinated slides per species for 45 seconds and then poured off. The slides were immediately viewed under a light microscope.

Statistical Analyses

At $\alpha = 0.05$, t-tests of differences for each species between 0 UV-B and each dose were calculated with Minitab (1995) for needle curling, needle thickness and adaxial epidermal thickness.

RESULTS

Needle Curling

An increase in needle curling, indicated by decreasing curled/uncurled length ratio, was significant for larch needles beyond the 4 kJ m^{-2} day^{-1} UV-B dose (Figure 1). Increased curling of spruce needles was evident only at 12 kJ m^{-2} day^{-1}. There was no significant effect of UV-B on needle curling in pine.

Needle Sections (Staining and Measurements)

At 12 kJ m^{-2} day^{-1}, safranin indicated increased phenolic compounds associated with an apparent increase in mesophyll cell density at and near the adaxial surface of the larch needles compared to those of the 0 UV-B treatments (Figure 2a). Mid segments exposed to all

FIGURE 1. Larch, spruce and pine needle curling (curled/uncurled length ratio) with standard error bars vs. UV-B dose (kJ m^{-2} day^{-1})

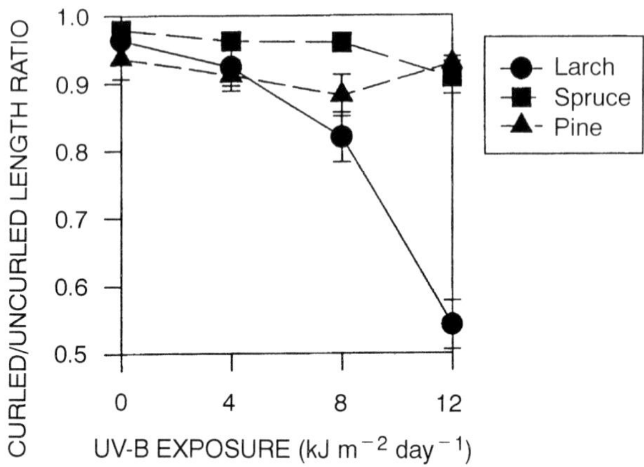

UV-B doses and tip segments exposed to 4 kJ m^{-2} day^{-1} indicated a significant decrease in needle thickness (Table 1). The significant decrease in adaxial epidermal thickness of larch tip segments exposed to 8 kJ m^{-2} day^{-1} (Table 2) is a result of the loss of the outer epidermis on many segments during the microtechnique process. Increased UV-B radiation resulted in increased phenolic compounds in spruce needles, most of which were located near the mesophyll cell walls (Figure 2b). It should be noted that, although the spruce needles in Figure 2b indicate differences in thickness, overall differences were not significant as shown in Table 1. Base and tip needle thickness of spruce increased significantly at 4 and 8 kJ m^{-2} day^{-1} and, 4 kJ m^{-2} day^{-1}, respectively (Table 1). The adaxial epidermal thickness of base segments increased significantly while mid segments decreased significantly when exposed to 4 and 12 kJ m^{-2} day^{-1} UV-B doses (Table 2). Spruce tip adaxial epidermal thickness values were low because the microtechnique process had removed the outer epidermis on most segments. Pine needles appeared to contain constitutive phenolic compounds without UV-B (Figure 2c). Pine base segments were significantly thicker when exposed to all UV-B doses while the thickness of pine mid segments decreased significantly when exposed to 4 kJ m^{-2} day^{-1} (Table 1). Base and mid segments exhibited significant increases in adaxial epidermal thickness when exposed to UV-B doses over 8 kJ m^{-2} day^{-1} and, all UV-B doses, respectively (Table 2).

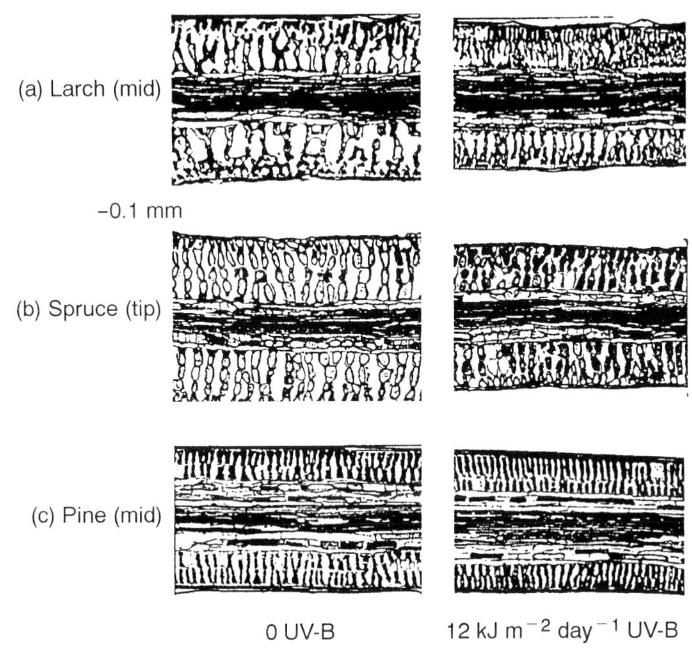

FIGURE 2. Larch, spruce and pine needle sections exposed to 0 and 12 kJ m^{-2} day^{-1} UV-B (adaxial side is uppermost)

TABLE 1. Larch, spruce and pine needle thickness (mm) when exposed to 0, 4, 8 and 12 kJ m^{-2} day^{-1} UV-B$_{be}$.

Species Segment	Larch base	Larch mid	Larch tip	Spruce base	Spruce mid	Spruce tip	Pine base	Pine mid	Pine tip
0	0.638	0.619	0.541	0.730	0.720	0.605	0.559	0.589	0.472
4 kJ	0.637	0.521*	0.488*	0.781*	0.747	0.686*	0.633*	0.544*	0.490
8 kJ	0.662	0.576*	0.517	0.637*	0.703	0.616	0.691*	0.598	0.510
12 kJ	0.660	0.595*	0.527	0.750	0.750	0.613	0.637*	0.613	0.532

*Significantly different from 0 UV-B

Lignin

Larch base segments were slightly lignified on the outer periclinal wall of the abaxial epidermal cells in all controls and UV-B doses. Spruce needles were not lignified at all while pine base and mid

TABLE 2. Larch, spruce and pine adaxial epidermal thickness (mm) when exposed to 0, 4, 8 and 12 kJ m^{-2} day^{-1} UV-B_{be}.

Species Segment	Larch base	Larch mid	Larch tip	Spruce base	Spruce mid	Spruce tip	Pine base	Pine mid	Pine tip
0	0.024	0.025	0.024	0.023	0.025	0.014	0.020	0.018	0.021
4 kJ	0.023	0.022	0.022	0.025*	0.021*	0.017*	0.021	0.020*	0.021
8 kJ	0.026	0.022	0.017*	0.024	0.022	0.017*	0.027*	0.021*	0.020
12 kJ	0.025	0.026	0.022	0.026*	0.021*	0.013	0.023*	0.022*	0.020

*Significantly different from 0 UV-B

portions displayed lignification of the adaxial and abaxial hypodermal cell walls in all controls and UV-B doses (not shown).

DISCUSSION

Severe needle curling occurred in larch at 8 and 12 kJ m^{-2} day^{-1}. Wilson and Greenberg (1993) also observed curling of *Brassica napus* seedling leaves exposed to enhanced UV-B. Curling may be due to the effects on needle development of: (1) an UV-B photoreceptor; (2) degradation of auxin in the adaxial epidermal layer; and/or, (3) free radical injury of proteins, DNA or photosystems.

Lignin does not play a major role as an UV-B attenuator (Schnitzler et al., 1996) but does provide mechanical support for tissue (Taiz and Zeiger, 1991). The presence of lignin on all sides of pine needles could contribute to the prevention of leaf curling. This may not explain less curling in spruce compared to larch needles, however, as lignin was not detected in spruce needles using the fluoroglucinol staining technique.

Pine epidermal and mesophyll cells contained phenolic compounds, even in the absence of UV-B. Constitutive phenolic compounds are present in the epidermal and mesophyll tissue of *Pinus taeda* (Sullivan et al., 1996) and absorb UV-B radiation; this is a mechanism responsible for UV-B resistance.

Increases in needle and adaxial epidermal thickness may be another protective response to reduce UV-B penetration into the needle (Murali et al., 1988). In general, such increases in thickness occurred in pine needles while larch and spruce needles demonstrated minimum thickness values at 4 and 8 kJ m^{-2} day^{-1}, respectively. Doses of 12 kJ m^{-2} day^{-1} resulted in higher leaf thickness values although such conditions would be detrimental for these two species in view of the needle distortion that takes place. It is difficult to understand why some needles are thinner at moderate UV-B doses (4 or 8 kJ m^{-2} day^{-1}) and thicken to control levels at higher doses. Both the absence of UV-B and the higher doses may represent stressful conditions that result in thicker needles. Uncertainties associated with the small sample sizes used in this preliminary study could also account for such differences.

CONCLUSIONS

Lodgepole pine can be considered an UV-B resistant conifer species, interior spruce intermediate, and, western larch sensitive based on morphological and anatomical changes that occur when exposed to increasing doses. The effects of UV-B doses on the anatomy, morphology and photosynthetic rates of these and other conifer species need broader study. Such studies will provide information on conifer species sensitivity to future global increases in UV-B in the biosphere as well as insights for further physiological and biochemical research to determine the mechanisms of UV-B sensitivity.

REFERENCES

Caldwell, M.M. 1979. Plant life and ultraviolet radiation: Some perspective in the history of the earth's UV climate. Bioscience 29: 520-525.

Johansen, D.L. 1940. Plant Microtechnique. McGraw-Hill, New York.

Kerr, J.B., and C.T. McElroy. 1993. Evidence for large upward trends of ultraviolet-B radiation linked to ozone depletion. Science 262: 1032-1034.

Minitab Statistical Software for Windows (Student edition) 1995.

Murali, N.S., Teramura A.H., and S.K. Randall. 1988. Response differences between two soybean cultivars with contrasting UV-B radiation sensitivities. Photochem. Photobiol. 48: 653-657.

Schnitzler, J.P., Jungblut, T.P., Heller, W., Kofferlein, M., Hutzler, P., Heinzmann, U., Schmelzer, E., Ernst, D., Langebartels, C., and H. Sandermann Jr. 1996. Tissue localization of UV-B-screening pigments and of chalcone synthase mRNA in needles of Scots pine seedlings. New Phytol. 132: 247-258.

Sullivan, J.H., Howells, B.W., Ruhland, C.T., and T.A. Day. 1996. Changes in leaf expansion and epidermal screening effectiveness in *Liquidambar styraciflua* and *Pinus taeda* in response to UV-B radiation. Physiol. Plant. 98: 349-357.

Taiz, L., and E. Zeiger. 1991. Plant Physiology. The Benjamin/Cummings Publishing Company Inc. Redwood City, CA.

Wilson, M.I., and B.M. Greenberg. 1993. Specificity and photomorphogenic nature of ultraviolet-B-induced cotyledon curling in *Brassica napus* L. Plant Physiol. 102: 671-677.

Wintertime Patterns of Chlorophyll Fluorescence in Red Spruce (*Picea rubens* Sarg.) Foliage

S. T. Lawson
T. D. Perkins
G. T. Adams

INTRODUCTION

Chlorophyll fluorescence (CF) is a measure of photosynthetic function in plants and can be used to examine numerous aspects of physiology in conifers (Mohammed et al., 1995; Vindaver et al., 1991). CF has been used to assess frost hardiness (Adams and Perkins, 1993; Binder and Fielder, 1996) and early detection of winter injury (Lindgren and Hallgren, 1993). Adams (1996) and Lindgren and Hallgren (1993) showed that wintertime CF in coniferous foliage varied in response to air temperature and light conditions prior to sampling. These studies demonstrate the value of CF as a rapid, non-invasive, and quantifiable indicator of photosynthetic integrity and health of foliage.

S. T. Lawson and G. T. Adams are Research Assistants, and T. D. Perkins is Director of Research, University of Vermont, Proctor Maple Research Center, P.O. Box 233, Underhill Center, VT 05490 USA.

Address correspondence to S. T. Lawson at the above address (E-mail: slawson@zoo.uvm.edu).

Research was funded by an USDA NRI grant to T. D. Perkins.

[Haworth co-indexing entry note]: "Wintertime Patterns of Chlorophyll Fluorescence in Red Spruce (*Picea rubens* Sarg.) Foliage." Lawson, S. T., T. D. Perkins, and G. T. Adams. Co-published simultaneously in *Journal of Sustainable Forestry* (Food Products Press, an imprint of The Haworth Press, Inc.) Vol. 10, No. 1/2, 2000, pp. 149-153; and: *Frontiers of Forest Biology: Proceedings of the 1998 Joint Meeting of the North American Forest Biology Workshop and the Western Forest Genetics Association* (ed: Alan K. Mitchell et al.) Food Products Press, an imprint of The Haworth Press, Inc., 2000, pp. 149-153. Single or multiple copies of this article are available for a fee from The Haworth Document Delivery Service [1-800-342-9678, 9:00 a.m. - 5:00 p.m. (EST). E-mail address: getinfo@haworthpressinc.com].

Winter injury to current-year foliage has been suggested as a primary cause of the decline of red spruce (*Picea rubens* Sarg.) in northeastern North America (DeHayes, 1992). The exact mechanism by which injury occurs is not certain. Possible inciting factors include winter desiccation (Herrick and Friedland, 1991), insufficient cold tolerance (DeHayes et al., 1990), precocious dehardening (Strimbeck et al., 1995), or rapid freezing (Perkins and Adams, 1995). Measures of physiological status in red spruce foliage during the winter can be useful in determining the stress response of trees to changing environmental conditions. In this study, we monitored CF and cold tolerance in current-year red spruce foliage over a winter season to determine long-term (seasonal) and short-term (thaw periods) variations.

METHODS

Current-year foliage from six red spruce saplings (2-5 m tall) growing at 400 m elevation at the Proctor Maple Research Center, Underhill Center, Vermont, was sampled from 21 November 1995 to 5 April 1996. Samples were collected several times weekly during stable weather and up to several times daily during thaw periods and periods of rapid air temperature changes.

One shoot from each tree was collected then returned to the lab and immediately prepared for measurement. Needles were removed from the stem and cut into small pieces using a razor blade. These needle segments were loaded into a CF measurement tray with five replicates per shoot. Samples were dark-equilibrated for 15 minutes before CF measurement. On selected sub-freezing days, additional shoots were collected at 15 minute intervals to assess CF response over time, at controlled laboratory temperatures of +4°C and +20°C for up to one week.

A PK Morgan CF1000 fluorometer modified to measure low wintertime fluorescence levels was used for all CF measurements. A saturating light pulse of 400 $\mu mol \cdot m^{-2} \cdot s^{-1}$ and a measurement interval of 5 s were used. Data collected includes initial fluorescence (F_o), maximum fluorescence (F_m), quenching fluorescence (F_q), and terminal fluorescence (F_t). Variable fluorescence (F_v) is calculated as the difference between F_m and F_o.

Cold tolerance was assessed using a laboratory freezing chamber and CF analysis using the methods of Adams and Perkins (1993).

Twelve shoots were collected from each tree and placed in foam trays on moist toweling. Trays were subjected to temperatures from 0°C to −55°C (freeze rate of $-10°C \cdot hr^{-1}$), with one shoot from each tree removed at each 5°C increment. Samples were then placed in the cold room for 4 days to allow recovery before CF measurement. Damaged foliage is known to exhibit a sharp decrease in F_v/F_m. Cold tolerance was determined as the temperature at which F_v/F_m decreased to 80% of the control value (0°C).

RESULTS

A steady decrease in chlorophyll fluorescence (F_v/F_m) was observed from the initial measurements on 21 November to relatively stable winter levels in late-December (Figure 1, top). Cold tolerance of first-year foliage also followed this pattern. F_v/F_m was found to close-

FIGURE 1

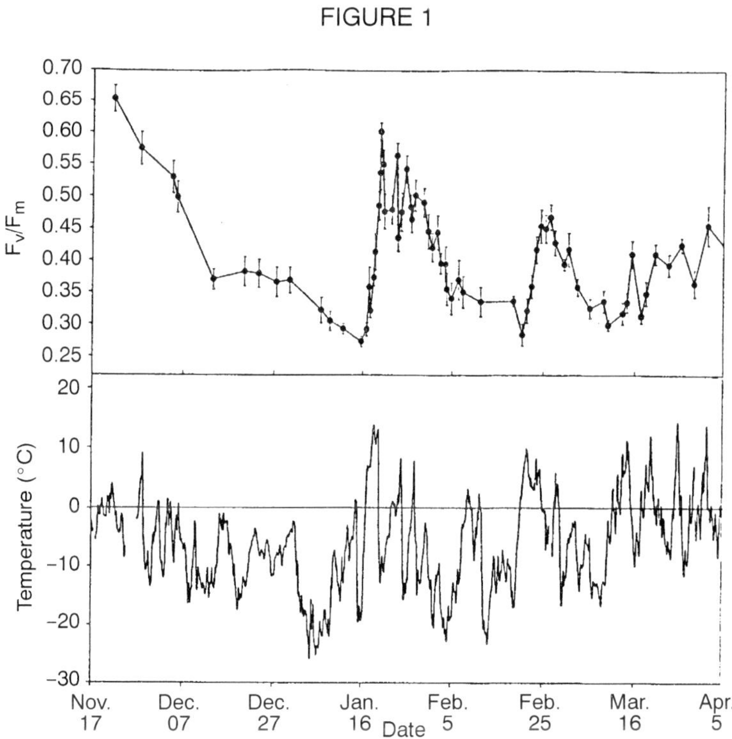

ly follow air temperature and was significantly correlated to mean hourly temperatures over the preceding 72 hours ($r^2 = 0.537$). CF also exhibited a steady increase with temperature when shoots were brought from sub-freezing field conditions and continuously measured in the lab. F_v/F_m increased approximately 5-8% per day when foliage was stored at 4°C and 10% per day at 20°C.

Chlorophyll fluorescence remained stable at a low level until a significant thaw on 17-20 January resulted in a large and rapid increase in mean F_v/F_m from 0.27 ± .02 to 0.60 ± .04 (Figure 1, top). Upon a return to subfreezing temperatures, CF and cold tolerance regained pre-thaw values over a period of several days. Estimated cold tolerance varied from a pre-thaw value of −39.5°C (Jan. 16, 1996) to −37.0°C during the thaw (Jan. 19, 1996) to −43.0°C, post-thaw (Jan. 20, 1996). A thaw on 23-25 January (Figure 1, bottom) resulted in cold tolerance decreasing to −36°C (Jan. 23, 1996). Much colder temperatures returned on 27 January, with cold tolerance increasing to −50°C (Jan. 29, 1996).

DISCUSSION

These results demonstrate that CF generally exhibits a pattern similar to air temperature and cold tolerance throughout the winter season. During mid-winter thaws, large and rapid increases of chlorophyll fluorescence (F_v/F_m) can occur. Adams (1996) also documented increases in CF during a mid-winter thaw, with a strong correlation to preceding ambient air temperature ($r^2 = 0.73$). Recent evidence has shown that red spruce foliage responds to transient winter thaws by dehardening (Strimbeck et al., 1995) and the resumption of net photosynthesis (Schaberg, 1996). We observed dehardening of foliage that was proportional to maximum temperature and duration of thaw periods. Cold tolerance followed the same pattern as CF, but with a short lag period. Thus, current-year red spruce shoots can experience a loss of winter hardiness during thaw periods and regain cold tolerance when sub-freezing temperatures return.

CONCLUSIONS

Chlorophyll fluorescence monitoring can provide a valuable assessment of the physiological status in red spruce foliage during winter.

This tool may offer a relatively simple and efficient method for determining cold tolerance. Most importantly, changes in wintertime CF may serve as an early indicator of dehardening in current-year spruce shoots. With continued monitoring, we may more clearly understand the role of thaw-induced physiological changes in the winter injury phenomenon.

REFERENCES

Adams, G.T. and T.D. Perkins. 1993. Assessing cold tolerance in *Picea* using chlorophyll fluorescence. Env. Exp. Bot. *33*: 377-382.

Adams, G.T. 1996. Wintertime photostress in red spruce foliage. Master's thesis, Department of Botany, University of Vermont.

Binder, W.D. and P. Fielder. 1996. Chlorophyll fluorescence as an indicator of frost hardiness in white spruce seedlings from different latitudes. New Forests *11*: 233-253.

DeHayes, D.H., C.E. Waite, and M.A. Ingle. 1990. Storage temperature and duration influence cold tolerance of red spruce foliage. For. Sci. *36*: 1153-1158.

DeHayes, D.H. 1992. Winter injury and developmental cold tolerance of red spruce. pp. 295-337. *In:* C. Egar, and M.B. Adams eds. *Ecology and Decline of Red Spruce in the Eastern United States*. Springer-Verlag. New York.

Herrick, G.T. and A.J. Friedland. 1991. Winter desiccation and injury of subalpine red spruce. Tree Physiol. *8*: 23-26.

Lindgren, K. and J. Hallgren. 1993. Cold acclimation of *Pinus contorta* and *Pinus sylvestris* assessed by chlorophyll fluorescence. Tree Physiol. *13*: 97-106

Mohammed, G.H., W.D. Binder, and S.L. Gillies. 1995. The role of chlorophyll fluorescence: A review of its practical forestry applications and instrumentation. Scand. J. For. Res. *10*: 383-410.

Perkins, T.D. and G.T. Adams. 1995. Rapid freezing induces winter injury symptomatology in red spruce foliage. Tree Physiol. *15*: 259-266.

Schaberg, P.G. 1996. Cold-season photosynthesis of red spruce in Vermont. Ph.D. dissertation, Department of Botany, University of Vermont.

Strimbeck, G.R., P.G. Schaberg, D.H. DeHayes, J.B. Shane, and G.J. Hawley. 1995. Midwinter dehardening of montane red spruce during a natural thaw. Can. J. For. Res. *25*: 2040- 2044.

Vindaver, W.E., G.R. Lister, R.C. Brooke, and W.D. Binder. 1991. A manual for the use of variable chlorophyll fluorescence in the assessment of the ecophysiology of conifer seedlings. B.C. Ministry of Forests, Victoria, B.C. FRDA Report 163, 60 pp.

Winter Season Tree Sap Flow and Stand Transpiration in an Intensively-Managed Loblolly and Slash Pine Plantation

Timothy A. Martin

INTRODUCTION

On the lower gulf coastal plain of the southeastern United States, meteorological conditions during the winter months of November to February remain conducive to aboveground physiological activity, with frequent periods of high radiation, and temperatures seldom falling below freezing. While a number of studies have investigated southern pine physiology during the winter months (Drew and Ledig,

---

Timothy A. Martin is affiliated with the School of Forest Resources and Conservation, University of Florida, Box 110410, Gainesville, FL 32611-0410 (E-mail: tmartin@harm.sfrc.ufl.edu).

The author thanks Tom Hinckley for loaning the sap flow systems, and the University of Florida Department of Agricultural and Biological Engineering for providing the meteorological data. The author acknowledges the help of Jeff English, Robert McGarvey and Duncan Wilson with fieldwork, and the helpful discussions and reviews provided by Eric Jokela and Duncan Wilson.

This study was funded by the Forest Biology Research Cooperative at the University of Florida. The research was conducted on land managed by Jefferson Smurfit Corporation.

Florida Agricultural Experiment Station Journal Series Number XXXX.

[Haworth co-indexing entry note]: "Winter Season Tree Sap Flow and Stand Transpiration in an Intensively-Managed Loblolly and Slash Pine Plantation." Martin, Timothy A. Co-published simultaneously in *Journal of Sustainable Forestry* (Food Products Press, an imprint of The Haworth Press, Inc.) Vol. 10, No. 1/2, 2000, pp. 155-163; and: *Frontiers of Forest Biology: Proceedings of the 1998 Joint Meeting of the North American Forest Biology Workshop and the Western Forest Genetics Association* (ed: Alan K. Mitchell et al.) Food Products Press, an imprint of The Haworth Press, Inc., 2000, pp. 155-163. Single or multiple copies of this article are available for a fee from The Haworth Document Delivery Service [1-800-342-9678, 9:00 a.m. - 5:00 p.m. (EST). E-mail address: getinfo@haworthpressinc.com].

1981; Boltz et al., 1986; Day et al., 1991; Teskey et al., 1994; Murthy et al., 1997), most of these studies have focused on organ-level responses. Few, if any, have examined physiological processes at the whole-tree or stand level. The primary objective of this study was to quantify winter season transpiration rates of trees and stands of loblolly (*Pinus taeda* L.) and slash (*Pinus elliottii* Engelm.) pine. Transpiration is an important process, both in terms of hydrology (Whitehead and Kelliher, 1991; McNulty et al., 1996) as well as tree and stand physiology (e.g., Monteith 1995), and is closely linked to productivity at various scales (Leuning, 1995; Le Maitre and Versfeld, 1997). As such, measurements of winter transpiration rates will be useful for understanding the environmental limits to southern pine productivity. In addition, by conducting measurements in plantations that have undergone intensive cultural treatments, we will gain insight into the water balance of stands under management scenarios that are becoming increasingly common (Sedjo and Botkin, 1997).

METHODS

This research was conducted in one stand of *P. taeda* and one stand of *P. elliottii* planted in 1983, 20 km northeast of Gainesville, Florida as part of a larger experiment investigating the effects of intensive weed control and fertilization on growth of southern pines (Swindel et al., 1988). The sap flow measurements took place on 468 m^2 study plots that contained 56 *P. taeda* or 52 *P. elliottii* trees (approximately 1150 trees ha^{-1}). After planting, understory vegetation was permanently excluded with herbicide and mechanical treatments, and fertilizer containing micro- and macronutrients was applied one to three times per year from establishment until 1993. Basal areas of the *P. taeda* and *P. elliottii* stands were 40.5 and 34.8 m^2 ha^{-1}, respectively. Average tree height in the *P. taeda* and *P. elliottii* plots were 18.8 and 17.1 m, respectively (range 14.3-20.6 m and 14.6-19.3 m, respectively) at the time of sap flow measurements. Average *P. taeda* and *P. elliottii* diameter at breast height (DBH) were 20.5 and 19.8 cm, respectively (range 10-29.7 cm and 15.2-25.7 cm, respectively).

Tree sap flow was measured using the tissue heat balance technique described in detail by Cermák et al. (1973, 1982), Kucera et al. (1977), and Cermák and Kucera (1981). Gauges were installed on four sample trees in each stand. Sample trees were chosen so that each tree represented one fourth of the total basal area of the stand (sample trees were at the 16th, 59th, 78th and 93rd percentile of the cumulative stand basal area for *Pinus taeda,* and at the 30th, 50th, 70th and 94th percentile for *Pinus elliottii*). Gauges were installed and insulated as described in Martin et al. (1997). On the two largest sample trees in each stand, two gauges were installed on opposite sides of the stem to account for possible circumferential variation in sap flow rates (Cermák and Kucera, 1985) and sap flow was calculated as the average of the two sides. Sap flow was logged each second and saved as 15 minute averages. Measurements continued from early November 1997 until late February 1998.

Tree-level sap flow data were integrated to the stand level as described by Cermák and Kucera (1990) and Martin et al. (1997). This was accomplished by quantifying the relationship between size of the sampled sap flow trees (in this case, individual tree basal area) and daily total tree sap flow. This relationship was then applied to the basal area distribution of the stand to calculate transpiration for each day.

To evaluate the environmental evaporative demand (i.e., the maximum transpiration rate possible given the meteorological conditions and no limitations due to reductions in stomatal conductance), a physiologically-based potential transpiration (T_{pot}) was calculated from weather station data using the Penman-Monteith equation:

$$T_{pot} = \frac{sR + \rho_a C_p D g_a}{\lambda\left[s + \gamma\left(1 + g_a/g_{crown}\right)\right]} \qquad (1)$$

where:

T_{pot} is in mm s^{-1}

s is the slope of the saturated vapor pressure versus temperature curve (kPa K^{-1})

R is incoming radiation (W m^{-2}), (ρ_a is the density of dry air (kg m^{-3})

C_p represents the specific heat capacity of air (J kg^{-1} K^{-1})

λ is the latent heat of evaporation of water (J kg^{-1})

γ is the psychrometer constant (kPa K^{-1}).

Because there is little or no information in the literature on the micrometeorology of southern pines, canopy and aerodynamic conductances

(g_{crown} and g_a) were set to 25 and 200 mm s^{-1}, respectively, the maximum values for conifers given in Kelliher et al. (1993). This had the effect of simulating transpiration from a tree canopy with high leaf area with stomata fully open during all daylight hours. Potential transpiration was calculated hourly during daylight hours, and summed for each day.

RESULTS AND DISCUSSION

Tree Sap Flow

Sap flow rate throughout the day was very responsive to meteorological changes. For example, on January 3, patterns of sap flow closely mirrored changes in solar radiation (Figure 1).

These sensitive responses to radiation are most likely the result of radiation-mediated changes in crown-level stomatal conductance, rather than the direct effects of radiation on transpiration (the R term in Equation 1). This is because, in conifers, boundary layer conductance is much larger than stomatal conductance, making vapor pressure deficit and stomatal conductance the primary controllers of transpiration rate; changes in radiation have little direct effect on transpiration in these "well coupled" systems (Meinzer, 1993).

Daily water loss from individual trees had a strong linear correlation with tree basal area (Figure 2), with R^2 values of this relationship usually exceeding 0.80. The use of a linear scaling relationship with these data produces a positive x-intercept, which implies that trees with basal areas smaller than this intercept have zero transpiration. In this study, the x-intercept almost always fell below the basal area of the smallest tree in the study plots. Only on days with very low transpiration rates (cool, moist, cloudy days) did the x-intercept slightly exceed the basal area of the smallest tree in the plots (i.e., December 27, 1997, Figure 2). In these cases, for the purposes of scaling water loss to the stand scale, the one or two trees with basal areas smaller than the intercept were assumed to have no water loss. Stand transpiration estimated with this method differed on average by less than 3% from an independent non-regression method, which scaled transpiration upward, based on four tree basal area size classes (data not shown).

Regressions between tree daily water loss and tree basal area were not significantly different between the species on any day measured.

FIGURE 1. Diurnal patterns of *Pinus taeda* and *Pinus elliottii* tree sap flow and environmental variables on January 3, 1998.

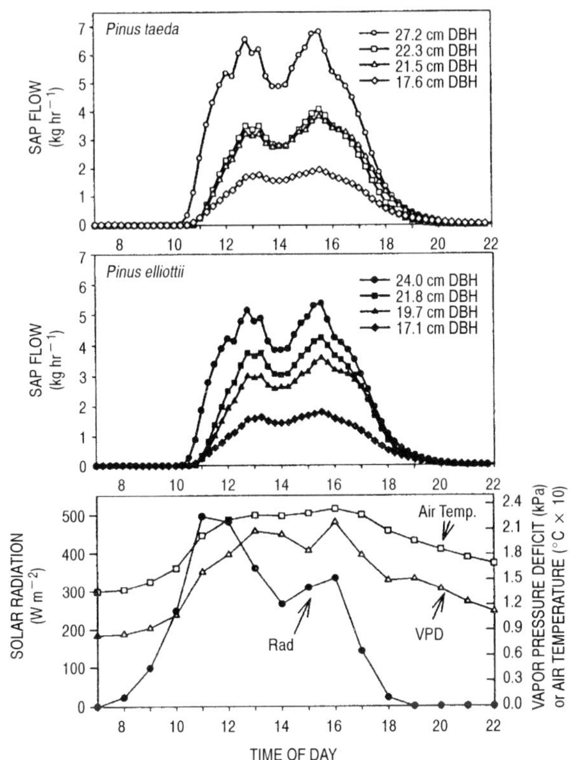

Average daily transpiration ranged from 8.1 kg for the smallest tree measured (17.1 cm DBH) to 35.3 kg for the largest tree (27.2 cm DBH). Maximum rates of water loss for the largest trees measured exceeded 75 kg (Table 1). These winter rates of daily tree water loss are comparable to spring and summer tree water loss measured in *Pinus halapensis* Mill., *Pinus pinaster* Ait. and *Pinus sylvestris* L. trees of similar size (Granier et al., 1990; Schiller and Cohen, 1995; Granier et al., 1996).

Stand Transpiration

Stand transpiration during the 108 day measurement period ranged from effectively zero on rainy days to over 4 mm day^{-1} on warm, dry

FIGURE 2. Relationship between tree size and tree water loss for *Pinus taeda* and *Pinus elliottii* trees on days with differing potential transpiration (T_{pot}, Equation 1).

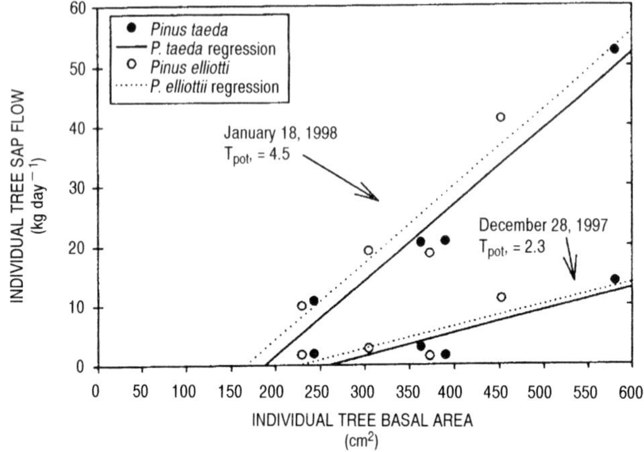

TABLE 1. Biometric parameters and sap flow statistics for eight *Pinus taeda* and *Pinus elliottii* trees in north-central Florida, U.S.A.

Species	Diameter at 1.37 m (cm)	Basal Area (cm^2)	Projected Crown Area (m^2)[a]	Maximum Daily Transpiration (kg)	Average Daily Transpiration (kg)
Pinus taeda					
	27.2	581	20.0	75.1	35.3
	22.3	391	13.1	50.9	19.4
	21.5	363	6.5	45.3	18.6
	17.6	243	4.2	17.0	8.1
Pinus elliottii					
	24.0	452	12.9	71.6	28.3
	21.8	373	8.5	42.3	14.3
	19.7	305	5.5	38.4	14.7
	17.1	230	3.6	28.4	8.1

[a] Area of an ellipse calculated from measurements of crown diameters in two directions

days (Figure 3). As was the case with tree-scale transpiration, these winter stand transpiration rates were comparable to spring and summer season stand transpiration measured with sap flow methods in European pine forests (Granier et al., 1990; Granier et al., 1996) and in a less intensively managed loblolly pine forest in North Carolina

FIGURE 3. Daily and cumulative stand transpiration in *Pinus taeda* and *Pinus elliottii* stands, and potential transpiration and precipitation during the winter of 1997-1998.

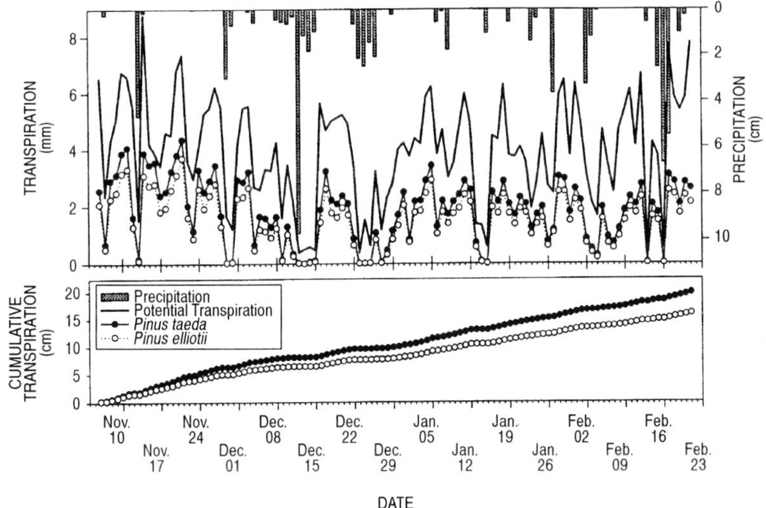

(Oren et al., 1998; tree age = 12, stand basal area = 15.9 m^2 ha^{-1}). It should be noted that throughout the measurement period, the water table remained within 1.3 m of the soil surface, suggesting that soil water deficits were not an important influence on tree transpiration.

While the relationship between tree size and tree water loss was identical for the two species (Figure 2), *P. taeda* stand transpiration was consistently 20 to 25% higher than transpiration of the *P. elliottii* stand. Mean *P. taeda* stand transpiration was 1.8 mm day^{-1}, compared to the *P. elliottii* average of 1.5 mm day^{-1}. Over the three months of measurement, this translated into a 3.8 cm difference in cumulative water loss between the stands (19.8 and 16.0 cm cumulative *P. taeda* and *P. elliottii* stand transpiration, respectively, Figure 3). The contrasts in stand transpiration were the result of differences in tree size distributions between the two species, and are consistent with trends in canopy leaf area for the two stands. Litterfall and destructive sampling data show that all-sided leaf area index (LAI) in the *P. taeda* stand ranged from 7.9 m^2 m^{-2} in November to 5.9 m^2 m^{-2} in February, while *P. elliottii* LAI ranged from 6.2 to 5.0 m^2 m^{-2} over the same period, demonstrating between-species LAI differences on the order of 20% (E. Jokela and T. Martin, unpublished data).

CONCLUSIONS

Due to the mild winter climate conditions in northern Florida, intensively-managed *P. taeda* and *P. elliottii* stands transpire during the winter at rates observed in more northern pine forests during the spring and summer. Because transpiration and productivity are linked, this suggests that considerable carbon gain takes place during the winter months in this region.

The relationship between tree basal area and daily water loss was not statistically different bet veen *P. taeda* and *P. elliottii,* indicating similar rates of water loss per unit stem or sapwood area. More research needs to be done to determine whether transpiration rates per unit leaf area (and therefore canopy conductance) remains the same between species.

The *P. taeda* stand transpired approximately 20% more water than the *P. elliottii* stand under identical climate conditions, partially due to species differences in stand structure (diameter distributions and leaf area index). However, species differences in canopy conductance could also partially explain these results. Further work is needed to investigate potential canopy conductance differences between the species.

REFERENCES

Boltz, B. A., B. C. Bongarten, and R. O. Teskey. 1986. Seasonal patterns of net photosynthesis of loblolly pine from diverse origins. Canadian Journal of Forest Research 16:1063-1068.

Bunce, J. A. 1996. Does transpiration control stomatal responses to water vapour pressure deficit? Plant, Cell and Environment 19:131-135.

Cermák, J., M. Deml, and M. Penka. 1973. A new method of sap flow rate determination in trees. Biol. Plant. 23:469-471.

Cermák, J. and J. Kucera. 1981. The compensation of natural temperature gradient at the measuring point during the sap flow rate determination in trees. Biol. Plant. 23:469-471.

Cermák, J., J. Ulehla, J. Kucera, and M. Penka. 1982. Sap flow rate and transpiration dynamics in the full-grown oak (*Quercus robus* L.) in floodplain forest exposed to seasonal floods as related to potential evapotranspiration and tree dimensions. Biologia Plantarum 24:446-460.

Cermák, J. and J. Kucera. 1990. Scaling up transpiration data between trees, stands and watersheds. Silva Carelica 15:101-120.

Day, T. A., S. A. Heckathorn, and E. H. DeLucia. 1991. Limitations of photosynthesis in *Pinus taeda* L. (loblolly pine) at low soil temperatures. Plant Physiology 96:1246-1254.

Drew, A. P. and T. Ledig. 1981. Seasonal patterns of carbon dioxide exchange in the shoot and root of loblolly pine (*Pinus taeda*) seedlings. Botanical Gazette 142:200-205.

Granier, A., V. Bobay, J. H. C. Gash, J. Gelpe, B. Saugier, and W. J. Shuttleworth. 1990. Vapour flux density and transpiration rate comparisons in a stand of Maritime pine (*Pinus pinaster* Ait.) in Les Landes forest. Agricultural and Forest Meteorology 51:309-319.

Granier, A., P. Biron, B. M. M. Kostner, L. W. Gay, and G. Najjar. 1996. Comparisons of xylem sap flow and water vapour flux at the stand level and derivation of canopy conductance for Scots pine. Theoretical and Applied Climatology 53:115-122.

Kelliher, F. M., R. Leuning, and E. -D. Schulze. 1993. Evaporation and canopy characteristics of coniferous forests and grasslands. Oecologia 95:153-163.

Kucera, J., J. Cermák, and M. Penka. 1977. Improved thermal method of continual recording the transpiration flow rate dynamics. Biol. Plant. 19:413-420.

Leuning, R. 1995. A critical appraisal of a combined stomatal-photosynthesis model for C_3 plants. Plant, Cell and Environment 18:339-355.

Le Maitre, D. C. and D.B. Versfeld. 1997. Forest evaporation models: relationships between stand growth and evaporation. Journal of Hydrology 193:240-257.

Martin, T. A., K. J. Brown, J. Cermák, R. Ceulemans, J. Kucera, F. C. Meinzer, J. S. Rombold, D. G. Sprugel, and T. M. Hinckley. 1997. Crown conductance and tree and stand transpiration in a second-growth *Abies amabilis* forest. Canadian Journal of Forest Research 27:797-808.

Meinzer, F. C. 1993. Stomatal control of transpiration trends in ecology and evolution 8:289-294.

McNulty, S. G., J. M. Vose, and W. T. Swank. 1996. Loblolly pine hydrology and productivity across the southern United States. Forest Ecology and Management 86:241-251.

Monteith, J. L. 1995. A reinterpretation of stomatal responses to humidity. Plant, Cell and Environment 18:357-364.

Murthy, R., S. J. Zarnoch, and P. M. Dougherty. 1997. Seasonal trends of light-saturated net photosynthesis and stomatal conductance of loblolly pine trees grown in contrasting environments of nutrition, water and carbon dioxide. Plant, Cell and Environment 20:558-568.

Oren, R., N. Phillips, G. Katul, B. E. Ewers, and D. E. Pataki. 1998. Scaling xylem sap flux and soil water balance and calculating variance: a method for partitioning water flux in forests. Annales des Sciences Forestieres 55:191-216.

Schiller, G. and Y. Cohen. 1995. Water regime of a pine forest under a Mediterranean climate. Agricultural and Forest Meteorology 74:181-193.

Sedjo, R. A. and D. Botkin. 1997. Using forests to spare natural forests. Environment 39(10):14-20, 30.

Swindel, B. F., D. G. Neary, N. B. Comerford, D. L. Rockwood, and G. M. Blakeslee. 1988. Fertilization and competition control accelerate early southern pine growth on flatwoods. Southern Journal of Applied Forestry 12:116-121.

Teskey, R. O., H. L. Gholz, and W. P. Cropper, Jr. 1994. Influence of climate and fertilization on net photosynthesis of mature slash pine. Tree Physiology 14:1215-1227.

Teskey, R. O. and D. W. Sheriff. 1996. Water use by *Pinus radiata* trees in a plantation. Tree Physiology 16:273-279.

Whitehead, D. and F. M. Kelliher. 1991. Modeling the water balance of a small *Pinus radiata* catchment. Tree Physiology 9:17-33.

Long-Term Nitrogen Fertilization Increases Winter Injury in Montane Red Spruce (*Picea rubens*) Foliage

T. D. Perkins
G. T. Adams
S. T. Lawson
P. G. Schaberg
S. G. McNulty

INTRODUCTION

Current-year red spruce (*Picea rubens* Sarg.) foliage is predisposed to winter injury by one or more types of anthropogenic pollutants, particularly acidic deposition (DeHayes, 1992). The resultant defoli-

T. D. Perkins is Director of Research, and G. T. Adams and S. T. Lawson are Research Assistants, Proctor Maple Research Center, The University of Vermont, P. O. Box 233, Harvey Road, Underhill Center, VT 05490 USA.

P. G. Schaberg is Research Plant Physiologist, Northeastern Experiment Station, USDA Forest Service, Burlington, VT 05402 USA.

S. G. McNulty is Program Manager, Southern Global Change Program, USDA Forest Service, Raleigh, NC 27705 USA.

Address correspondence to T. D. Perkins at the above address.

This research was supported by a USDA NRI grant to T. D. Perkins and a Cooperative Agreement with Dr. Mel Tyree of the U.S. Forest Service.

[Haworth co-indexing entry note]: "Long-Term Nitrogen Fertilization Increases Winter Injury in Montane Red Spruce (*Picea rubens*) Foliage." Perkins, T. D. et al. Co-published simultaneously in *Journal of Sustainable Forestry* (Food Products Press, an imprint of The Haworth Press, Inc.) Vol. 10, No. 1/2, 2000, pp. 165-172; and: *Frontiers of Forest Biology: Proceedings of the 1998 Joint Meeting of the North American Forest Biology Workshop and the Western Forest Genetics Association* (ed: Alan K. Mitchell et al.) Food Products Press, an imprint of The Haworth Press, Inc., 2000, pp. 165-172. Single or multiple copies of this article are available for a fee from The Haworth Document Delivery Service [1-800-342-9678, 9:00 a.m. - 5:00 p.m. (EST). E-mail address: getinfo@haworthpressinc.com].

ation, when severe and repeated, leads to dieback and eventual mortality of affected red spruce individuals.

The role of soil-deposited pollutants in predisposing red spruce to winter injury is less well understood. Two early studies showed that large amounts of soil-applied nitrogen (N) had a positive effect on the cold tolerance of potted red spruce seedlings (Klein et al., 1989; DeHayes et al., 1989). In several field experiments on mature trees, some up to three years in duration, N fertilization had no effect, or even a slight positive influence, on the ability of red spruce foliage to survive winter conditions (Perkins and Adams, unpublished). However, given the long-term nature of forest stands, and the increasing concern over nitrogen saturation in the northeast (Abed et al., 1989), we sought to establish the relationships among soil-applied N, foliar N, winter physiology, and winter damage in high-elevation red spruce.

METHODS

Study Area and Nitrogen Treatments

This work was conducted in a series of fertilization plots established on Mt. Ascent, in southeast Vermont (42° 26′N, 72° 27′W, 762 m elevation) in 1988 (McNulty et al., 1996). Two replicate 15 m × 15 m plots of five fertilization types and levels were established in 1988, in stands dominated by red spruce. These plots were fertilized in three equal amounts applied in June, July, and August of each year to yield total annual additions of 0, 15.7, 19.8, 25.6, and 31.4 kg $N \cdot ha^{-1} \cdot yr^{-1}$ (McNulty et al., 1996). Ambient bulk precipitation added an additional 5.4 kg $N \cdot ha^{-1} \cdot yr^{-1}$ (McNulty and Aber, 1993). Monitoring within these plots has showed evidence of N-induced disruptions in foliar cations and carbon relations (Schaberg et al., 1997) and reductions in growth and increases in mortality of red spruce trees (McNulty et al., 1996).

Foliar Collections

Current-year foliage was collected in August 1995. Foliar chemistry was determined according to procedures given by Schaberg et al. (1997). Monthly visits to the site were made from December 1995 through March 1996 to collect foliage for winter physiological measurements. Foliage was collected from mid-upper crown branches of three dominant trees near the center of each plot. Foliage was transferred to the laboratory in an insulated container packed with snow. The following morning, the samples were examined for presence of winter injury using a scale of 0-9 (10% injury classes). Chlorophyll fluorescence (F_v/F_m) was measured the day following collection to detect non-visual winter injury. Fluorescence of a replicate set of foliage was made after four days in the cold room (4°C) to assess potential recovery. Chlorophyll fluorescence has been demonstrated to be a rapid and quantitative measure of foliar winter damage (Adams and Perkins, 1993). Cold tolerance was established using a controlled temperature chamber using methods of Adams and Perkins (1993). Red spruce shoots were frozen at a rate of $10°C \cdot hr^{-1}$ and samples removed at various temperatures to bracket the range of expected foliar cold tolerance. Chlorophyll fluorescence was used to determine the temperature at which cold tolerance was exceeded. Dehardening capacity (the cold tolerance foliage would deharden to) and extent (the amount of dehardening) were determined by remeasuring cold tolerance of duplicate foliage samples after four days exposure to above freezing conditions in a cold room (4°C) under low light (50 $\mu mol \cdot m^{-2} \cdot s^{-1}$). Rapid freezing tolerance was assessed using a Cryomed liquid-nitrogen controlled rate freezer at freeze rates of 0.0, -10.0, -12.5, and $-15.0°C \cdot min^{-1}$ after Perkins and Adams (1995). Foliar water content and desiccation rates were established gravimetrically (Perkins et al., 1993). Chlorophyll fluorescence measurements were also made as an indicator of the degree of photostress (Adams, 1996).

Slight winter injury was observed on a few trees within the plots during January 1996. Damage became visually pronounced in February, reaching a maximum in March. In both February and May (just prior to extensive abscission of affected needles), visual injury to current-year foliage on crowns of test trees within each plot was estimated on a scale of 0-9 (representing 10% damage classes of current-year foliage) from several sides, using binoculars.

Differences among physiological response and levels of winter injury were assessed using ANOVA in PC SAS. Statistical power was low, as there were only two replicate plots per N treatment. Differences were considered significant at $p \leq 0.05$. Relationships among

FIGURE 1. Air temperatures on Mt. Ascutney during the winter of 1995-96. Measurements were recorded hourly.

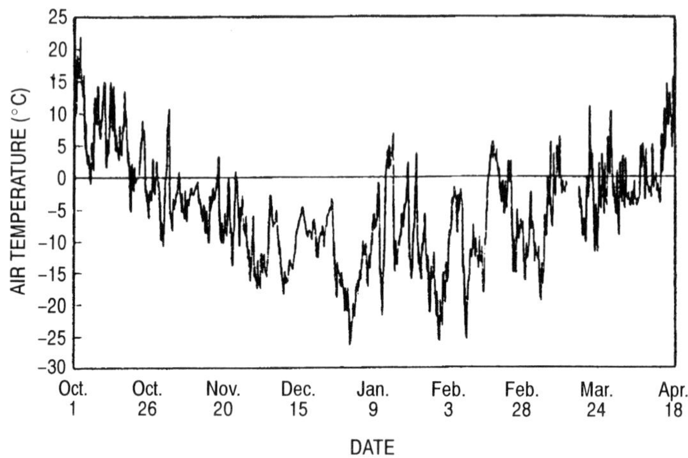

nutrient concentrations and measures of winter damage were established using correlation analysis in PC SAS.

RESULTS

Physiological Measurements

Since most of the physiological testing relied upon visual or chlorophyll fluorescence detection of injury induced by the test conditions, the early appearance of winter injury (January 1996) rendered most measurements after that time unreliable. Thus it is not possible to fully accredit the injury which occurred in the winter of 1995-1996 on Mt. Ascutney, Vermont to any one particular stressor. The minimum winter temperatures on Mt. Ascutney in the period of time between the two site visits did not approach the normal expected levels of cold tolerance for this species (minimum temperature −26.5°C, Figure 1). Cold tolerance of current-year foliage in December did not vary among N-treatments ($p = 0.78$), averaging 49.4°C for sampled trees overall (Table 1). Increasing N-treatment decreased both the temperature to which red spruce foliage dehardened ($p = 0.03$) as well as the amount of dehardening ($p = 0.03$). Trees that received the highest

TABLE 1. Winter physiological parameters of current year red spruce foliage on Mt. Ascutney during the winter of 1995-1996. Since water injury was observed beginning with the January 1996 sample, values are given for only the December 1995 sampler period. Numbers represent means ± standard error of the two replicate treatment plots (three trees per plot). Significance of the ANOVA is given.

	Nitrogen Addition (kg-ha^{-1}-yr^{-1})					
Physiological Variable	0.0	15.7	19.8	25.6	31.4	p value
Cold Temperature (°C)	−43.4 ± 1.2	−43.5 ± 1.6	−42.9 ± 0.5	−45.1 ± 0.5	−44.5 ± 0.5	0.78
Dehardened Tolerance (°C)	−35.5 ± 1.1	−37.4 ± 2.1	−33.0 ± 1.5	−42.4 ± 0.3	−44.0 ± 0.1	0.03
Dehardened Extent (°C)	8.0 ± 2.3	6.1 ± 0.5	9.9 ± 1.1	2.8 ± 0.2	0.6 ± 0.4	0.05
Rapid Freeze Survival (%)	16.7 ± 11.8	22.2 ± 7.9	22.2 ± 7.9	50.0 ± 3.9	22.2 ± 7.9	0.23
Max Fluorescence (F_m)	153.6 ± 0.6	101.0 ± 7.6	186.5 ± 22.6	127.6 ± 31.9	138.4 ± 8.4	0.54
Water Content (% dw)	124.2 ± 1.6	115.0 ± 4.5	119.3 ± 0.7	112.0 ± 1.8	116.0 ± 2.6	0.31
Dessication Rate (% dw)	5.5 ± 0.6	8.2 ± 1.9	6.4 ± 0.7	8.1 ± 1.0	12.1 ± 4.8	0.01

amount of N-fertilization dehardened the least. No significant difference in susceptibility to rapid freezing was observed ($p = 0.23$). Similarly, water content was not significantly affected by N-addition ($p = 0.31$). The rate at which foliage lost water was significantly different among treatments ($p = 0.01$).

Winter Injury

Winter injury was observed on trees in treatment plots slightly earlier than in trees outside plots. Injury generally increased with increasing N-addition regardless of the method used to measure or express injury (Figure 2), with correlation coefficients ranging from 0.43 to 0.93. Correlations from visual estimates of winter injury were lower than those based upon fluorescence measurements. This likely arises from the fact that winter injury tended to be concentrated on southern sides of exposed tree crowns, and visual estimates were made on all sides, whereas foliage collected for physiological and damage assessments were always collected from the upper-third of the south side of the trees. It is also possible that the different forms of N-fertilization within the treatment plots contributed to the somewhat elevated levels of winter injury observed in the 15 kg N · ha^{-1} · yr^{-1}

FIGURE 2. Relationships between nitrogen treatment and winter injury on Mt. Ascutney, Vermont. Visual winter injury to current-year foliage was assessed for the crowns of three trees within each plot in February and again in May 1996. Shoot fluorescence (F_v/F_m) was measured on sampled branches in March 1996 and expressed as a percentage of fluorescence loss from the control N-treatment. Shoot mortality was calculated by placing samples of collected foliage under low light in a cold room (4°C). Foliage that did not regain typical levels of wintertime fluorescence (Fv/Fm > 0.45) was considered injured. Numbers next to lines indicate the correlation coefficient for the relationship between nutrient addition and the various measures of winter injury.

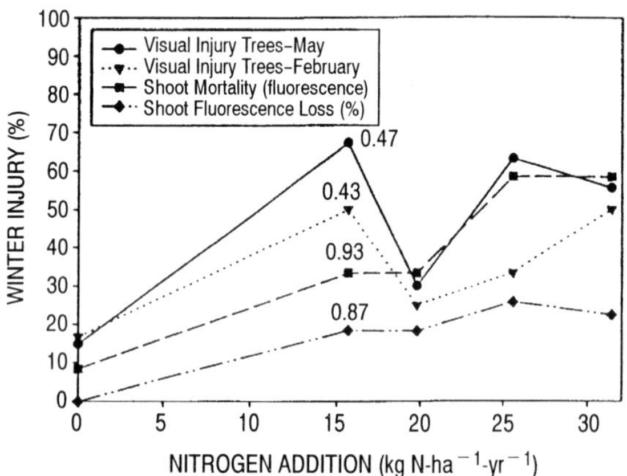

treatment. Nitrogen addition was correlated with several foliar nutrients (Table 2). The various measures of winter injury were also significantly correlated with foliar nutrients.

DISCUSSION

While several earlier studies have shown that acidic deposition can reduce the cold tolerance of current-year red spruce foliage and pointed out the importance of foliar nutrition in winter injury (see review by DeHayes, 1992), no work to date has shown that soil-applied N can increase the susceptibility of mature red spruce trees to winter injury. Many prior studies have clearly demonstrated that winter injury may be exacerbated by foliar-mediated processes. The strong relationships among N-addition, foliar nutrition and winter injury in this study show that the winter physiology of red spruce trees

TABLE 2. Correlation coefficients (r^2) among nitrogen additions or measures of injury during the winter of 1995-1996 and foliar nutrient content of current-year red spruce foliage on Mt. Ascutney (see Schaberg et al., 1997). Negative values indicate inverse relationships. Visual injury was assessed on trees in the field. Fluorescence was measured on sampled branches and is expressed as a percentage of the control N-treatment. Survival is based upon measurements of chlorophyll fluorescence after exposure of foliage to conditions that promote recovery. All relationships are highly significant ($p \leq 0.01$).

	Foliar Element			
	N	Ca	Mg	Al
N Addition	0.71	−0.93	−0.89	−0.49
Visual Injury (May 1996)	0.86	−0.85	−0.81	−0.66
Visual Injury (Feb 1996)	0.52	−0.84	−0.73	−0.41
Fluorescence (F_v/F_m)	0.90	−0.94	−0.97	−0.74
Foliar Survival	0.80	−0.88	−0.84	−0.46

may also be impacted via soil-mediated processes, although it is unknown whether nitrogen itself or another nutrient affected by the N-fertilization, that is the important nutrient regulating winter injury.

Because of the unexpected early appearance of winter injury, the physiological measurements done in the course of this work provide few clues as to the mechanism(s) that caused the damage. Seedling studies have shown that cold tolerance is enhanced by N-fertilization (Klein et al., 1989; DeHayes et al., 1989). This study showed no clear effect of N on cold tolerance. However, dehardening was clearly positively affected (trees receiving higher N-fertilization dehardened less). Interestingly, although winter injury does not seem to be caused by desiccation, increasing N tended increase the rate at which foliage loses water.

CONCLUSIONS

Although foliar-mediated processes that cause winter injury are clearly recognized and reasonably well understood, soil-mediated processes have been largely overlooked. These results indicate that, in addition to the foliar effects of acidic deposition, nitrogen deposition to soils can also influence winter physiological processes and winter injury susceptibility in current-year red spruce foliage. With continued deposition of N-containing compounds to northern forests, it is likely that we will observe further changes in summer and winter physiology and growth of trees in high-elevation sites.

REFERENCES

Aber, J.D., Nadelhoffer, K.J., Steudler, P., and Melillo, J.M. 1989. Nitrogen saturation in northern forest ecosystems. Bioscience 39: 378-386.

Adams, G.T. 1996. Wintertime Photostress in Red Spruce Foliage. M.S. Thesis, University of Vermont, Burlington, VT.

Adams, G.T. and T.D. Perkins. 1993. Assessing cold tolerance in *Picea* using chlorophyll fluorescence. Environ. Exp. Bot. 33: 377-382.

DeHayes, D.H. 1992. Winter injury and developmental cold tolerance of red spruce. In: C. Eagar and M.B. Adams (Eds.). The Ecology and Decline of Red Spruce in the Eastern United States. Springer-Verlag, New York.

DeHayes, D.H., M.A. Ingle, and C.E. Waite. 1989. Nitrogen fertilization enhances cold tolerance of red spruce seedlings. Can. J. For. Res. 19: 1037-1043.

Klein, R.M., T.D. Perkins, and H.L. Myers. 1989. Nutrient status and winter hardiness in red spruce foliage. Can. J. For. Res. 19: 754-758.

McNulty, S.G. and J.D. Aber. 1993. Effects of chronic nitrogen additions on nitrogen cycling in a high elevation spruce-fir stand. Can. J. For. Res. 23:1252-1263.

McNulty, S.G., J.D. Aber, and S.D. Newman. 1996. Nitrogen saturation in a high elevation New England spruce-fir stand. For. Ecol. Mgmt. 84: 109-121.

Perkins, T.D. and G.T. Adams. 1995. Rapid freezing induces winter injury symptomatology in red spruce foliage. Tree Physiol. 15: 259-266.

Perkins, T.D., G.T. Adams, S. Lawson, and M.T. Hemmerlein. 1993. Cold tolerance and water content of current-year red spruce (*Picea rubens* Sarg.) foliage over two winter seasons. Tree Physiol. 13: 119-144.

Schaberg, P.G., T.D. Perkins, and S.G. McNulty. 1997. Effects of chronic low-level N additions on gas exchange, shoot growth and foliar elemental concentrations of mature montane red spruce. Can. J. For. Res. 27: 1622-1629.

Cold Tolerance and Photosystem Function in a Montane Red Spruce Population: Physiological Relationships with Foliar Carbohydrates

P. G. Schaberg
G. R. Strimbeck
G. J. Hawley
D. H. DeHayes
J. B. Shane
P. F. Murakami
T. D. Perkins
J. R. Donnelly
B. L. Wong

P. G. Schaberg is Research Plant Physiologist, P. F. Murakami is Research Associate and B. L. Wong is Plant Physiologist, Northeastern Research Station, USDA Forest Service, Burlington, VT 05402 USA.

G. R. Strimbeck is Research Assistant Professor, G. J. Hawley is Senior Researcher, D. H. DeHayes is Professor, J. B. Shane is Senior Researcher and J. R. Donnelly is Professor, School of Natural Resources, The University of Vermont, Burlington, VT 05405 USA.

T. D. Perkins is Director of Research, Proctor Research Center, Underhill Center, VT 05490 USA.

This research was supported by Northeastern Research Station Project 4103 and the McIntire-Stennis Research Program.

[Haworth co-indexing entry note]: "Cold Tolerance and Photosystem Function in a Montane Red Spruce Population: Physiological Relationships with Foliar Carbohydrates." Schaberg, P. G. et al. Co-published simultaneously in *Journal of Sustainable Forestry* (Food Products Press, an imprint of The Haworth Press, Inc.) Vol. 10, No. 1/2, 2000, pp. 173-180; and: *Frontiers of Forest Biology: Proceedings of the 1998 Joint Meeting of the North American Forest Biology Workshop and the Western Forest Genetics Association* (ed: Alan K. Mitchell et al.) Food Products Press, an imprint of The Haworth Press, Inc., 2000, pp. 173-180. Single or multiple copies of this article are available for a fee from The Haworth Document Delivery Service [1-800-342-9678, 9:00 a.m. - 5:00 p.m. (EST). E-mail address: getinfo@haworthpressinc.com].

INTRODUCTION

Red spruce (*Picea rubens* Sarg.) growing in northern montane forests of eastern North America appears to be distinctive with respect to at least two aspects of winter physiology. First, red spruce attains only a modest level of midwinter cold tolerance compared to other north temperate conifers and appears barely capable of avoiding freezing injury at commonly occurring ambient winter temperatures (DeHayes, 1992). Second, red spruce is capable of net photosynthesis during winter in response to relatively brief exposure to mild temperatures (Schaberg et al., 1995). Despite considerable study, most of our understanding of red spruce cold tolerance and winter photosynthesis is derived from seedling studies in modified environments or examination of only a few mature trees in forests. As such, actual levels of attainment and the range among native trees are not well understood.

Foliar carbohydrate concentrations, especially sugars, may influence cold tolerance development (Sakai and Larcher, 1987), could limit photosynthesis via feedback inhibition (Krapp and Stitt, 1995) and enhance foliar respiration levels (Farrar, 1985). Although numerous studies have shown that foliar sugar and cold tolerance levels can be related, the extent of a causal association between these parameters has been complicated by the confounding influence of season (Parker, 1959; Hinesley et al., 1992) or experimental treatment (Ögren, 1997; Ögren et al., 1997). Cold tolerance and foliar sugar concentrations change dramatically and somewhat synchronously with season in north temperate tree species. It is not possible to effectively examine the extent of a potential functional relationship between these parameters by sampling across season. Furthermore, our current understanding of gas exchange-sugar relationships is based largely on growing season assessments of non-woody plants. In this paper, we document the range in midwinter cold tolerance and photosystem capacity (dark and light reactions) among 60 native red spruce trees in north central Vermont and examine the extent and nature of their physiological relationship with foliar sugar concentrations. We chose to focus measurement efforts across many individuals on a single date in midwinter when trees have reached their maximum depth of cold tolerance so that functional associations could be examined on an individual basis without the confounding influence of seasonal changes. The information reported provides a comprehensive assessment of red spruce freezing injury vulnerability and the role of winter carbon relations in the survival of the species.

MATERIALS AND METHODS

On January 16, 1996, over 20 current-year shoots per tree were collected from each of 60 mature red spruce on a southeast-facing slope of Mt. Mansfield, Vermont. Ambient air temperatures on the mountain were continually below 0°C for 48 days prior to sampling (NOAA records). To better determine the physiological range of parameters measured, 20 dominant or codominant trees were sampled from each of three elevations (650, 850 and 1050 m). Shoots were sealed in plastic bags, packed in snow, transported to the laboratory, and immediately prepared for assessment of cold tolerance, gas exchange and chlorophyll fluorescence. Foliage used in carbohydrate assessments was stored at −60°C prior to analysis. Cold tolerance was determined by controlled freezing tests using electrolyte leakage as a tissue injury assay and was estimated as T_m, the temperature at the midpoint of a sigmoid curve fitted to electrolyte leakage data (Strimbeck, 1997). Net photosynthetic capacity (P_{max}) was measured with a LI-6262 CO_2/H_2O IRGA (Li-Cor Inc., Lincoln NE, USA) using the methods of Schaberg et al. (1998). Shoots were taken directly from snow-packed storage, recut under water, attached to a supply of distilled water, and were placed in cuvettes maintained at near-optimal growing season temperature (16.7 ± 0.2°C) and light (582 ± 16 μmol m^{-2} s^{-1}) conditions (Alexander et al., 1995). Shoots processed in this manner reached stable photosynthetic maxima comparable to rates observed in freshly excised shoots within 20-40 minutes of initial illumination (Schaberg et al., 1998). Temperature-induced increases in P_{max} can occur within 3 h (Schaberg et al., 1998), but were avoided by timely photosynthetic assessment (within 45 min of removal from storage). The ratio of variable to maximum chlorophyll fluorescence (F_v/F_m) was used as an indicator of photosystem II function. Chlorophyll fluorescence was measured with a PK Morgan CF1000 fluorometer using the methods of Adams (1996). For carbohydrate samples, cuticular waxes were extracted (Hinesley et al., 1992), then sugars were extracted and sucrose, glucose, fructose and starch concentrations were enzymatically determined (Hendrix, 1993). Raffinose concentrations were evaluated using a Waters HPLC and Sugar-Pak column (Millipore Corp., Milford MA, USA) with a 50 mM Ca EDTA solvent.

Correlation analyses were used to evaluate relationships among measurement parameters. Elevational data were used in partial correlations with glucose and total sugar concentrations because these parameters increased slightly with increasing elevation ($P \leq 0.01$ for each as determined using analyses of variance). No other differences attributable to elevation were found. Differences were considered statistically significant when $P \leq 0.05$.

RESULTS AND DISCUSSION

Cold Tolerance

Mean cold tolerance was −41.7°C and ranged from −30.1 to −53.7°C (Table 1). Approximately 36% of trees evaluated had T_m values higher than −40°C. Because wintertime minimum temperatures on Mt. Mansfield are typically between −30 and −40°C (91% of the last 44 winters), our data support the conclusion of DeHayes (1992) that red spruce current-year foliage attains cold tolerance levels

TABLE 1. Means, standard errors and ranges of cold tolerance (T_m), photosynthetic capacity (P_{max}), variable to maximum fluorescence (F_v/F_m), and sugar and starch concentrations for current-year foliage of red spruce on Mt. Mansfield, Vermont during midwinter.

Parameter	Mean ± se	Range among trees
Cold Tolerance		
T_m (°C)	−41.7 ± 0.6	−30.1- −53.7
Photosystem Function		
P_{max} (μmol $m^{-2}s^{-1}$)	−1.10 ± 0.10	0.62- −3.28
F_v/F_m	0.28 ± 0.01	0.22-0.40
Carbohydrate Concentrations (mg/g)		
Sucros	28.71 ± 0.93	15.83-45.99
Glucose	4.37 ± 0.28	1.24- 9.65
Fructose	12.89 ± 0.38	2.33-17.66
Raffinose	19.35 ± 0.43	14.42-31.21
Total Sugar	65.37 ± 1.50	38.91-93.94
Starch	33.07 ± 0.06	2.12- 4.12

barely sufficient to protect it from minimum temperatures encountered in northern montane habitats. Given that acid deposition and winter thaws reduce red spruce cold tolerance by 4 to 10°C (DeHayes, 1992), the unique susceptibility and documented increase in freezing injury of red spruce over the past 50 years appears explainable.

Photosystem Function

Average P_{max} was -1.24 μmol m^{-2} s^{-1} (Table 1), and photosynthetic capacities for individual trees ranged from -3.28 (net respiration) to 0.63 μmol m^{-2} s^{-1} (net photosynthesis). Fewer than ten percent of trees showed any capacity for photosynthesis.

Average F_v/F_m was 0.28 and the range was 0.22 to 0.40. All values were well below the 0.832 level typical of foliage with a well functioning photosynthetic apparatus and indicate that a significant inhibition of photosystem II existed (Mohammed et al., 1995).

Carbohydrate Concentrations

Sucrose and raffinose are the dominant foliar carbohydrates (Table 1) as has been found in other spruces during winter (Neish, 1958; Parker, 1959), and are the primary sugars produced following experimentally induced winter photosynthesis (Neish, 1958).

Associations Among Parameters

Few correlations involving T_m and P_{max} were significant. Cold tolerance (T_m) was negatively correlated with raffinose (Figure 1A), sucrose (Figure 1B) and total sugar ($r = -0.38$, $P \leq 0.01$, not depicted) concentrations, indicating that foliage with higher sugar levels tended to be more cold tolerant. Sucrose, raffinose and other sugars protect both protein (Carpenter and Crowe, 1988) and membrane (Anchorduguy et al., 1987) systems from freezing injury *in vitro*. Sugars may also limit cellular freeze dehydration by their bulk colligative effects (Levitt, 1980), or by promoting vitrification of the partially dehydrated cell solution (Hirsh, 1987). The preferential accumulation of sugars in the cytoplasm (Koster and Lynch, 1992) may impart special protection to this region.

Interestingly, F_v/F_m, often used as a measure of plant stress, was not

FIGURE 1. Data plots, correlation coefficients and P-values for associations of physiological parameters for red spruce during winter that were significantly correlated ($P \leq 0.05$) with cold tolerance (T_m) and photosynthetic capacity (P_{max}): T_m and raffinose concentration (A), T_m and sucrose concentration (B). P_{max} and sucrose concentration (C), and P_{max} and F_v/F_m (D).

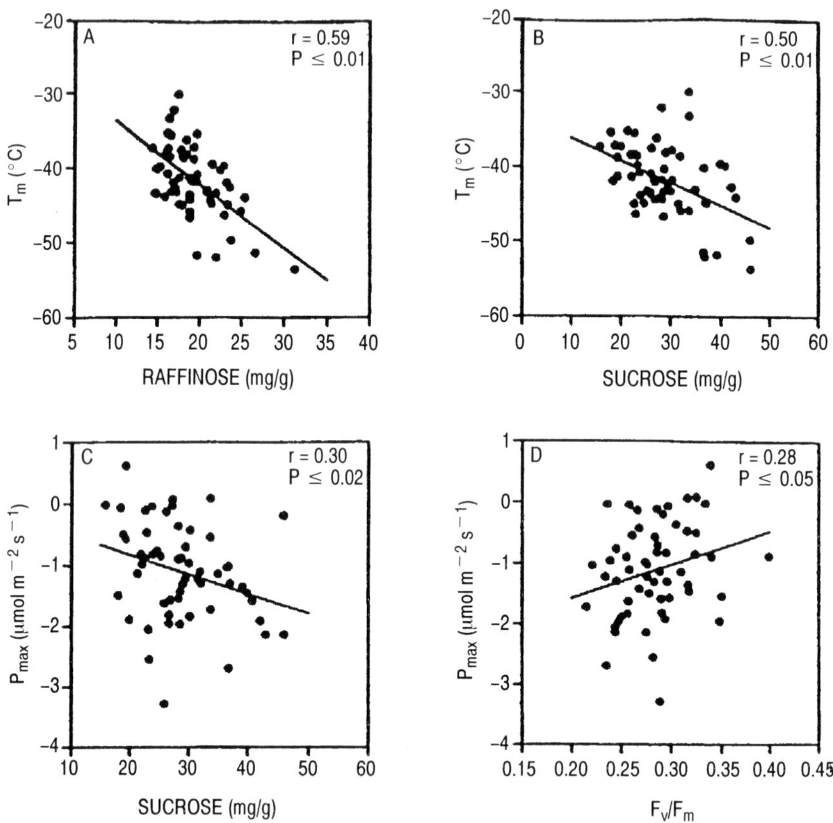

correlated with individual tree cold tolerance ($r = -0.031$, $P = 0.994$). Given the energy intensive nature of cold tolerance development, one might expect that plants exhibiting greater levels of stress would achieve a lesser depth of cold tolerance.

P_{max} was negatively correlated with sucrose concentration (Figure 1C) and positively correlated with F_v/F_m (Figure 1D). Possible physiological links among these parameters have been proposed. For example, Parker (1963) suggested that the feedback inhibition of photosynthesis by foliar sugars (including sucrose) may be one factor limiting winter carbon gain in evergreen conifers. Reductions in photochemical efficiency (F_v/F_m) could follow the inhibition of enzymatic dark reactions (Mohammed et al., 1995). In addition, because sucrose is the major substrate for the respiratory network, the negative relationship between P_{max} and sucrose concentration could also reflect an enhancement of respiration due to higher substrate concentrations (Farrar, 1985). The weak correlations reported here for a period with temperatures continuously below 0°C provide little support for a functional relationship among parameters. However, such relationships could become more apparent during thaws when respiration rates increase, foliar sucrose concentrations decrease, and net photosynthetic rates rise relative to the midwinter baseline levels reported here.

CONCLUSION

Data presented here highlight the marginal cold tolerance of red spruce relative to ambient minimum temperatures typically encountered. Because our data do not include the confounding influences of season or treatment, strong correlations between cold tolerance and sugar levels provide unique support for the possibility that high raffinose and sucrose concentrations play an important role in enhancing the foliar cold tolerance of red spruce. The weak correlation of sucrose concentration and P_{max} may result from the low tree to tree variability of these parameters in midwinter, when foliar sucrose levels are generally high and capacities for photosynthesis are uniformly low.

REFERENCES

Adams, G.T. 1996. Wintertime photostress in red spruce foliage. M.S. Thesis. Botany Dept., Univ. Of Vermont, Burlington, VT.

Alexander, J.A., J.R. Donnely and J.B. Shane. 1995. Photosynthetic and transpirational responses of red spruce understory trees to light and temperature. *Tree Physiol.* 15:393-398.

Anchorduguy, T.J., A.S. Rudolph, J.F. Carpenter and J.H. Crowe. 1987. Modes of interaction of cryoprotectants with membrane phospholipids during freezing. *Cryobiology* 24: 324-331.

Carpenter, J.F. and J.H. Crowe. 1988. The mechanism of cryoprotection of proteins by solutes. *Cryobiology* 25: 244-255.

DeHayes, D.H. 1992. Winter injury and developmental cold tolerance in red spruce.

In C. Eager and M.B. Adams (eds.), The Ecology and Decline of Red Spruce in the Eastern United States, 296-337. Springer-Verlag, New York.

Farrar, J.F. 1985. The respiratory source of CO_2. *Plant, Cell Environ.* 8:427-438.

Hendrix, D.L. 1993. Rapid extraction and analysis of nonstructural carbohydrates in plant tissue. *Crop Sci.* 33:1306-1311.

Hinesley, L.E., D.M. Pharr, L.K. Snelling, and S.R. Funderburk. 1992. Foliar raffinose and sucrose in four conifer species: relationship to seasonal temperature. *J. Amer. Soc. Hort. Sci.* 117:852-855.

Hirsh, A. 1987. Vitrification in plants as a natural form of cryoprotection. *Cryobiology* 24:214-228.

Koster, K.L and D.V. Lynch. 1992. Solute accumulation and compartmentation during cold acclimation of Puma rye. *Plant Physiol.* 98:108-113.

Krapp, A. and M. Stitt. 1995. An evaluation of direct and indirect mechanisms for the "sink regulation" of photosynthesis on spinach: changes in gas exchange, carbohydrates, metabolites, enzyme activities and steady-state transcript levels after cold-girdling source leaves. *Planta* 195:313-323.

Levitt, J. 1980. Responses of Plants to Environmental Stresses, Volume I: Chilling, Freezing, and High Temperature Stresses. Academic Press, New York.

Mohammed, G.H., W.D. Binder and S.L. Gillies. 1995. Chlorophyll fluorescence: A review of its practical forestry applications and instrumentation. *Scan. J. For. Res.* 10:383-410.

Neish, A.C. 1958. Seasonal changes in metabolism of spruce leaves. *Can. J. Bot.* 36:649-662.

Ögren, E. 1997. Relationship between temperature, respiratory loss of sugar and premature dehardening in dormant Scots pine seedlings. *Tree Physiol.* 17:47-51.

Ögren, E., T. Nilsson and L.-G. Sunblad. 1997. Relationship between respiratory depletion of sugars and loss of cold hardiness in coniferous seedlings over-wintering at raised temperatures: indications of different sensitivities of spruce and pine. *Plant, Cell Environ.* 20:247-253.

Parker, J. 1959. Seasonal variations in sugars of conifers with some observations on cold resistance. *For. Sci.* 5:56-63.

Parker, J. 1963. Causes of the winter decline in transpiration and photosynthesis in some evergreens. *For. Sci.* 9:158-166.

Sakai, A. and W. Larcher. 1987. Frost Survival of Plants: Responses and Adaptions to Freezing Stress. Springer-Verlag, New York.

Schaberg, P.G., R.C. Wilkinson, J.B. Shane, J.R. Donnelly and P.F. Cali. 1995. Winter photosynthesis of red spruce from three Vermont seed sources. *Tree Physiol.* 15:345-350.

Schaberg, P.G., J.B. Shane, P.F. Cali, J.R. Donnelly and G.R. Strimbeck. 1998. Photosynthetic capacity of red spruce during winter. *Tree Physiol.* 18:271-276.

Strimbeck, G.R. 1997. Cold Tolerance and Winter Injury in Montane Red Spruce. Ph.D. Dissertation. School of Natural Resources, Univ. of Vermont, Burlington, VT.

Effect of Two Years of Nitrogen Deposition on Shoot Growth and Phenology of Engelmann Spruce (*Picea engelmannii* Parry ex Engelm.) Seedlings

A. W. Schoettle

INTRODUCTION

As a result of anthropogenic combustion processes, ecosystems in the eastern and western United States and Europe have experienced elevated atmospheric deposition of nitrogen for most of this century and have begun to show symptoms of decline. If there is a cause and effect relationship between nitrogen deposition and ecosystem decline, one would expect that the current symptoms are a result of the cumulative effect of years of deposition. Deposition of anthropogenically produced nitrogenous compounds has increased along the Colorado Front Range in the past decades as a result of increased urbanization (Sievering et al., 1992, 1996; Williams et al., 1996). Annual pollutant deposition rates increase with elevation (Gilliam et al., 1996) due to the greater amount of precipitation at high elevations. There-

---

A. W. Schoettle is affiliated with the Rocky Mountain Research Station, 240 West Prospect Road, Fort Collins, CO 80526 (E-mail: Schoettl@lamar.colostate.edu).

[Haworth co-indexing entry note]: "Effect of Two Years of Nitrogen Deposition on Shoot Growth and Phenology of Engelmann Spruce (*Picea engelmannii* Parry ex. Engelm.) Seedlings." Schoettle, A. W. Co-published simultaneously in *Journal of Sustainable Forestry* (Food Products Press, an imprint of The Haworth Press, Inc.) Vol. 10, No. 1/2, 2000, pp. 181-189; and: *Frontiers of Forest Biology: Proceedings of the 1998 Joint Meeting of the North American Forest Biology Workshop and the Western Forest Genetics Association* (ed: Alan K. Mitchell et al.) Food Products Press, an imprint of The Haworth Press, Inc., 2000, pp. 181-189. Single or multiple copies of this article are available for a fee from The Haworth Document Delivery Service [1-800-342-9678, 9:00 a.m. - 5:00 p.m. (EST). E-mail address: getinfo@haworthpressinc.com].

fore, the health of high elevation Engelmann spruce (*Picea engelmannii* Parry ex Engelm.) and subalpine fir (*Abies lasiocarpa* (Hook.) Nutt.) forests may be at risk from sustained nitrogen inputs.

In the Colorado Rocky Mountains, the nitrogen deposition rate currently ranges from 3 to 6 kg N ha^{-1} y^{-1} at sites above 2700 m elevation. These deposition rates are low compared to maximum rates in Europe (58.8 kg N ha^{-1} y^{-1}, see Wright and Rasmussen, 1998), the eastern U.S. (10-20 kg N ha^{-1} y^{-1}, Gilliam et al., 1996), and the western U.S. (35-45 kg N ha^{-1} y^{-1}, Fenn et al., 1996). However, even the low nitrogen deposition rates in Colorado may exceed the nitrogen demand by high elevation plants since nitrogen requirement is limited by a short growing season. It has been hypothesized that excess nitrogen inputs may change shoot phenology by prolonging growth in fall (Friedland et al., 1984). This study quantified shoot growth and phenology of Engelmann spruce seedlings over two years in response to four levels of wet nitrogen deposition.

METHODS

Plant Material and Culture

Engelmann spruce seedlings were grown at the Mt. Sopris Nursery from seeds from a stand at 2888 m elevation on the White River National Forest, Colorado, USA. The seeds were planted in the spring of 1984. In June of 1986, the two-year-old seedlings were transplanted to 15 cm × 40 cm fiber pots with sieved forest soil (Darling series) collected from a spruce forest on the Fraser Experimental Forest in Fraser, Colorado. The plants were placed under an acrylic reinforced fiberglass open-air rain shelter in the understory of a *Pinus contorta* Dougl. forest at 3000 m elevation in southern Wyoming. During the growing seasons of 1986 and 1987, the plants were protected from ambient rain by the shelter and received only treatment solutions. At the end of October 1986 and 1987, the treatments were stopped, the shelter was removed, the pots were bermed with wood chips in a random arrangement with respect to treatment, and the plants were left to be covered with snow in winter.

Treatments

Two hundred and forty seedlings were randomly assigned to four nitrogen deposition treatments. The treatments were delivered in solution form to the surface of the soil for fifteen weeks per growing season. The treatment solutions were composed of reagent grade chemicals in combination with de-ionized water to simulate the rain chemistry that was collected during June-September in 1984 and 1985 in Rocky Mountain National Park, Colorado (Schoettle and Hubbard, 1992). The four treatments were prepared to represent ambient rain (1 ×), ambient rain containing ten times ambient nitrogen concentration (10 ×), ambient rain containing thirty times ambient nitrogen concentration (30 ×), and ambient rain containing fifty times ambient nitrogen concentration (50 ×) (Table 1). One hundred and seventy mL of solution was applied once a week to each pot and was the only water the seedlings received through the growing seasons. The quantity of treatment solution delivered to each pot was equal to the natural deposition rate of rainfall in subalpine forests at 3000 m in the central Rocky Mountains (National Atmospheric Deposition Program). The plants showed no sign of water stress.

Measurements

Bud break in spring of the second year was monitored weekly on the forty seedlings per treatment from mid-May until mid-July when all the seedlings had initiated growth. In both years, after growth was initiated, terminal shoot length and the length of the needles on the terminal shoot were recorded weekly until growth ceased. In fall of both years, the number of new shoots and lateral buds per seedling were counted.

Twenty plants per treatment were harvested after the first growing

TABLE 1. Summary of the nitrogen characteristics of treatment solutions. Deposition rates are for fifteen weeks of treatment each year between July and October. For complete chemical composition of simulated rain solutions, see Schoettle and Hubbard (1992).

Treatment	Nitrate concentration (mg L^{-1})	pH	Nitrogen deposition (kg N ha^{-1} growing $season^{-1}$)
1 ×	1.68	4.9	0.55
10 ×	16.82	4.2	5.48
30 ×	50.40	3.5	16.44
50 ×	84	2.9	27.38

season. Current-year and preexisting foliage from each seedling were oven dried at 65°C for 48 h. The mass per needle of current-year foliage was measured. The current-year and preexisting foliage were ground separately in a Wiley Mill (40 mesh) and analyzed for nitrogen by the micro-Kjeldahl method (Parkinson and Allen, 1975).

Net photosynthesis of one current-year lateral shoot from the second upper-most whorl of five seedlings per treatment was repeatedly measured, intact, from August 30 through November 11, 1987 with an ADC LCA-2 infrared gas analysis system (ADC, England). Measurements were made in the ambient environment when full sun was available and air temperature conditions were 15°C ± 3°C (standard deviation). Net photosynthesis is expressed on a total leaf surface area basis (Thompson and Leyton, 1971).

The data met the assumptions of normality and equal variance and were analyzed by analysis of variance. Orthogonal contrasts were used to test for linear, quadratic, and cubic relationships among the treatments (SAS, Cary, NC, USA).

RESULTS

One season of nitrogen additions caused a significant linear increase in foliar nitrogen in both new and preexisting foliage (Figure 1). There was no effect ($p > 0.05$) of the treatments on the number of needles produced per new shoot or the number of new shoots per seedling during the first growing season. There was a slight, but not statistically significant ($p = 0.10$), increase in the mass per needle and length of new needles at the end of the first season for those plants that received elevated nitrogen inputs.

Nitrogen additions significantly increased the number of new shoots and new lateral buds per seedling after the second growing season (Figure 2). The terminal shoot length of seedlings increased linearly with increased nitrogen addition (Figure 3). Average needle length of the terminal shoot was slightly, but not significantly, greater in response to nitrogen additions during the second growing season ($p = 0.06$).

Nitrogen additions altered some, but not all, of the measured phenological events during the second year of treatment. In spring after the first growing season, the initiation of growth of terminal buds was accelerated by nitrogen additions (Figure 4). The duration of terminal

FIGURE 1. Linear response ($p < 0.001$) of foliar nitrogen in new and preexisting foliage to nitrogen additions in the first growing season. Each point (± standard error) is an average of five samples each composed of foliage from four seedlings.

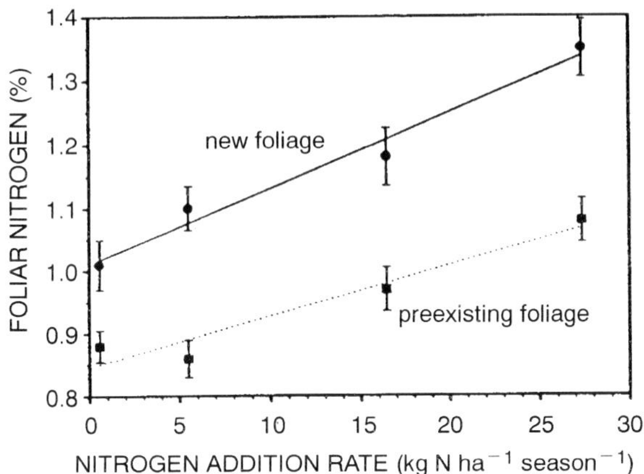

FIGURE 2. The effect of nitrogen additions on the number of new shoots (± standard error) and new lateral buds formed per seedling (± standard erro) in the second season of treatment. The relationships were linear ($p < 0.001$). The number of seedlings observed was 37, 37, 33, and 36 for the 1×, 10×, 30× and 50× treatments, respectively.

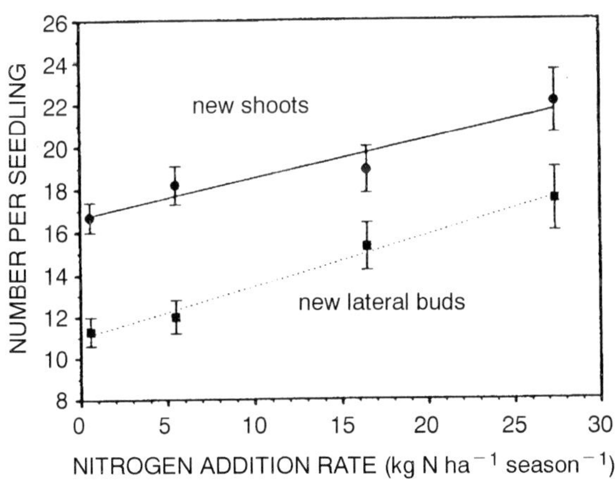

FIGURE 3. Length of the terminal shoot (± standard error) after the second season of nitrogen additions. The relationship was linear ($p < 0.001$). The number of seedlings observed was 37, 37, 33, and 36 for the 1×, 10×, 30× and 50× treatments, respectively.

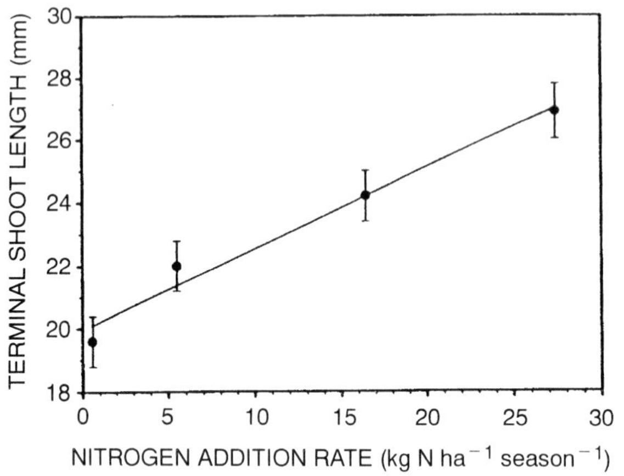

FIGURE 4. Percentage of seedlings whose terminal buds had burst by June 12, 1987 as a function of nitrogen addition treatment. The number of seedlings observed was 37, 37, 33, and 36 for the 1×, 10×, 30× and 50× treatments, respectively. The relationship was linear ($p < 0.05$).

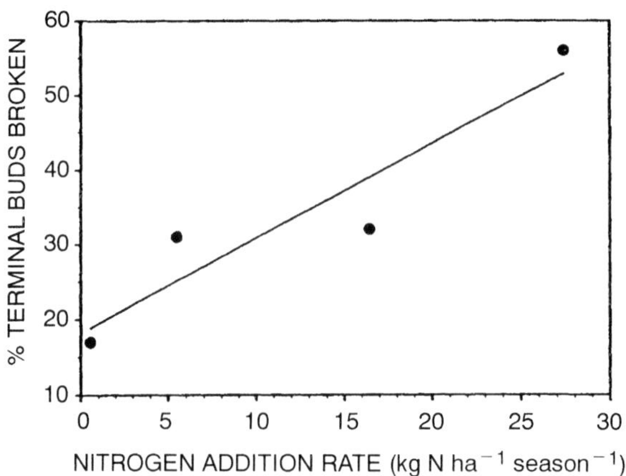

shoot elongation was not affected ($p > 0.05$) by the nitrogen treatments. The time course of the decline in net photosynthesis in the fall of the second year of treatment was not significantly affected by nitrogen additions (Figure 5).

DISCUSSION

During the first season of nitrogen additions, foliar nitrogen concentration in both the new and the preexisting foliage was significantly increased, demonstrating that the nitrogen delivered in the treatment solutions was available to the seedlings. As expected, there were no treatment effects on the number of needles or new shoots extended in the first growing season, because the number of needles and lateral buds per shoot are determined in the previous year for spruce.

An increase in the growth of seedlings that received elevated nitrogen became evident in the second year. More new shoots were initiated during the second growing season on seedlings that received nitrogen deposition treatments. The number of new lateral buds per seedling was also increased by nitrogen addition, as well, suggesting

FIGURE 5. The time course of the decline in net photosynthesis in the later portion of the second growing season. Measurements were repeatedly made on one shoot of each of five seedlings per treatment; the average standard error was 0.09 μmol m^{-2} s^{-1}. No readings were taken on plants treated with the 5.48 or 16.44 kg N ha^{-1} growing $season^{-1}$ treatments on Julian day 308.

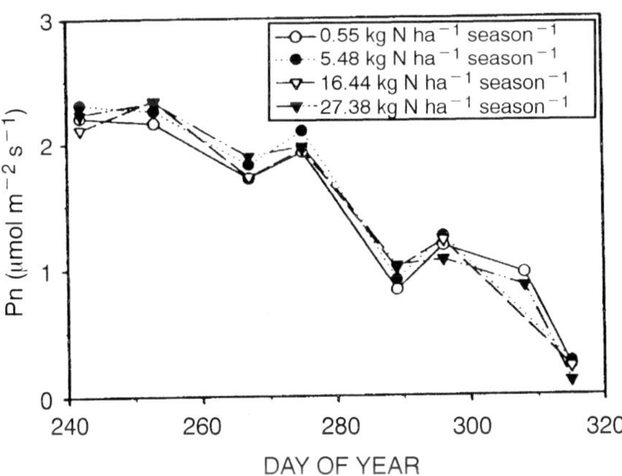

that the production of new shoots per year may continue to increase with sustained nitrogen treatment.

Nitrogen additions affected the phenology of Engelmann spruce seedlings during the second year. The growing season was lengthened by nitrogen additions. Contrary to the hypothesis of Friedland et al. (1984), the nitrogen additions in the present study extended the growing season for Engelmann spruce in the spring rather than fall. With increased nitrogen availability, growth was initiated earlier in the spring for Engelmann spruce, yet no effects were measured for the decline in physiological activity in the fall as estimated from net photosynthesis and shoot elongation. Although no effects were detected by Lumme and Smolander (1996) for Norway spruce (*Picea abies* Karst) seedlings until the nitrogen deposition rate approached the equivalent of 100 kg N ha^{-1} y^{-1}, they also observed that bud burst was accelerated by nitrogen addition. Acceleration of spring bud break by nitrogen additions indicates that nitrogen deposition at high elevations in the Rocky Mountains may expose seedlings to damage by late spring or early summer frost or cold periods.

CONCLUSIONS

These results suggest that the growth and phenology of Engelmann spruce seedlings are responsive to nitrogen inputs at deposition rates that could occur in Colorado. If nitrogen inputs continue to rise along the Front Range of Colorado, we may expect to see a change in the crown architecture of Engelmann spruce seedlings as well as an increase in damage to new foliage due to late frosts or abnormal cold periods in early summer. Further research is needed to assess Engelmann spruce seed source recommendations for regeneration plantings in areas with high nitrogen deposition and to evaluate the sensitivity of mature Engelmann spruce trees to nitrogen deposition.

REFERENCES

Fenn, M., M.A. Poth and D.W. Johnson. 1996. Evidence for nitrogen saturation in the San Bernardino Mountains in Southern California. Forest Ecology and Management 82:211-230.

Friedland, A.J., R.A. Gregory, L. Karenlampi and A.H. Johnson. 1984. Winter damage to foliage as a factor in red spruce decline. Can. J. For. Res. 14:963-965.

Gilliam, F.S., M.B. Adams and B.M. Yurish. 1996. Ecosystem nutrient responses to chronic nitrogen inputs at Fernow Experimental Forest, West Virginia. Can. J. For. Res. 26:196-205.

Lumme, I. and A. Smolander. 1996. Effect of nitrogen deposition level on nitrogen uptake and bud burst in Norway spruce (*Picea abies* Karst.) seedlings and N uptake by soil microflora. Forest Ecology and Management 89: 197-204.

National Atmospheric Deposition Program, "NADP/NTN Annual Data Summary 1984-1985," NADP Program Office, Colorado State University, Fort Collins, CO 80523.

Parkinson, J.A. and S.E. Allen. 1975. A wet oxidant procedure suitable for the determination of nitrogen and mineral nutrients in biological material. Communications in Soil Science and Plant Analysis 6: 1-15.

Schoettle, A.W. and R. Hubbard. 1992. A rain simulator for greenhouse use. Research Note RM-517, USDA Forest Service, Rocky Mountain Forest and Range Experiment Station, Fort Collins, CO. 4 p.

Sievering, H., D. Burton and N. Caine. 1992. Atmospheric deposition of nitrogen to alpine tundra in the Colorado Front Range. Global Biogeochemical Cycles 6:339-346.

Sievering, H., D. Rusch and L. Marques. 1996. Nitric acid, particulate and ammonium in the continental free troposphere: nitrogen deposition to an alpine tundra ecosystem. Atmospheric Environment 30:2527-2538.

Thompson, F.B. and L. Leyton. 1971. Method for measuring the leaf surface area of complex shoots. Nature 229:572.

Williams, M.W., J. Baron, N. Cain, R. Sommerfeld and R. Sanford. 1996. Nitrogen saturation in the Rocky Mountains. Environmental Science and Technology 30(2):640-646.

Wright, R.F. and L. Rasmussen. 1998. Introduction to the NITREX and EXMAN projects. Forest Ecology and Management 101:1-7.

Paper Birch Genecology and Physiology: Spring Dormancy Release and Fall Cold Acclimation

David G. Simpson
Wolfgang D. Binder
Sylvia L'Hirondelle

INTRODUCTION

Forest managers in British Columbia increasingly manage some forests as mixed species stands that include paper birch (*Betula papyrifera* Marsh.), which regenerates naturally or can be planted. This genecology study, in association with ongoing tree improvement field studies, is aimed at developing seed transfer guidelines and identifying faster growing sources of paper birch for use in BC forests. Two of the objectives discussed here are to examine the effects of seed origin on (1) dormancy release (bud flushing), and (2) growth cessation and cold acclimation.

David G. Simpson is affiliated with the BC Ministry of Forests, Kalamalka Research Station, Vernon, BC V1B 2C7, Canada.

Wolfgang D. Binder and Sylvia L'Hirondelle are affiliated with the BC Ministry of Forests, Glyn Road Research Station, Victoria, BC V8W 9C4, Canada.

[Haworth co-indexing entry note]: "Paper Birch Genecology and Physiology: Spring Dormancy Release and Fall Cold Acclimation." Simpson, David G., Wolfgang D. Binder, and Sylvia L'Hirondelle. Co-published simultaneously in *Journal of Sustainable Forestry* (Food Products Press, an imprint of The Haworth Press, Inc.) Vol. 10, No. 1/2, 2000, pp. 191-198; and: *Frontiers of Forest Biology: Proceedings of the 1998 Joint Meeting of the North American Forest Biology Workshop and the Western Forest Genetics Association* (ed: Alan K. Mitchell et al.) Food Products Press, an imprint of The Haworth Press, Inc., 2000, pp. 191-198. Single or multiple copies of this article are available for a fee from The Haworth Document Delivery Service [1-800-342-9678, 9:00 a.m. - 5:00 p.m. (EST). E-mail address: getinfo@haworthpressinc.com].

METHODS

Dormancy Release (Bud Flushing)

Seed collected from about 10 paper birch trees at each of 18 locations throughout the species' natural range in BC was used to produce one-year-old seedlings that were overwinter stored at $-2°C$. Spring bud flushing was determined in six forcing environments in a factorial of photoperiod (8 and 16 h) and temperature (7, 15, 25°C). The growing degree hours > 0°C for 50% bud flushing were estimated for each seed source and averaged for all forcing environments.

Growth Cessation and Fall Cold Acclimation

Seedlings of the 18 seed sources were located out-of-doors where they were exposed to natural photoperiod and temperature conditions at Prince George (54°N), Vernon (50.3°N) and Victoria (48.6°N). From late September through the end of October, chlorophyll fluorescence was measured weekly on the sixth leaf from the apex using either the PK Morgan CF-1000 (Vernon) or the Opti-Sciences OS-500 (Victoria) fluorometer (Binder et al., 1997; Mohammed et al., 1995). The fluorometers both measure dark-adapted Fv/Fm, but had different maximum values (Binder et al., 1997). Stem samples were freeze tested by freezing to a specific temperature at a cooling rate of 4 to $6°C\ h^{-1}$. Damage from freezing was assessed with electrolyte leakage (L'Hirondelle et al., 1992; Simpson 1994).

RESULTS

Dormancy Release (Bud Flushing)

There was no effect of forcing environment photoperiod or temperature on the degree days above 0°C for bud flushing. The average heat sum required for 50% bud flush ranged from 487 to 1963 degree hours. The physiographic variables, seed source latitude and elevation, explained more than 63% of the variability among the seed sources (Figure 1). The seed sources from 49-51°N flushed in less than 1000 degree hours whereas the more northerly seed sources required more

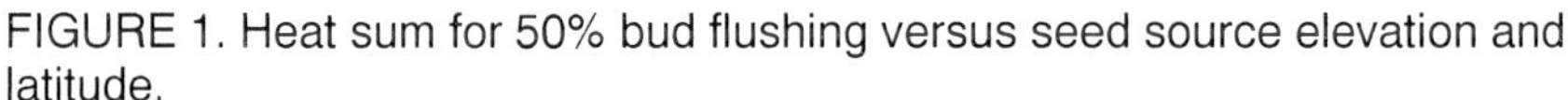

FIGURE 1. Heat sum for 50% bud flushing versus seed source elevation and latitude.

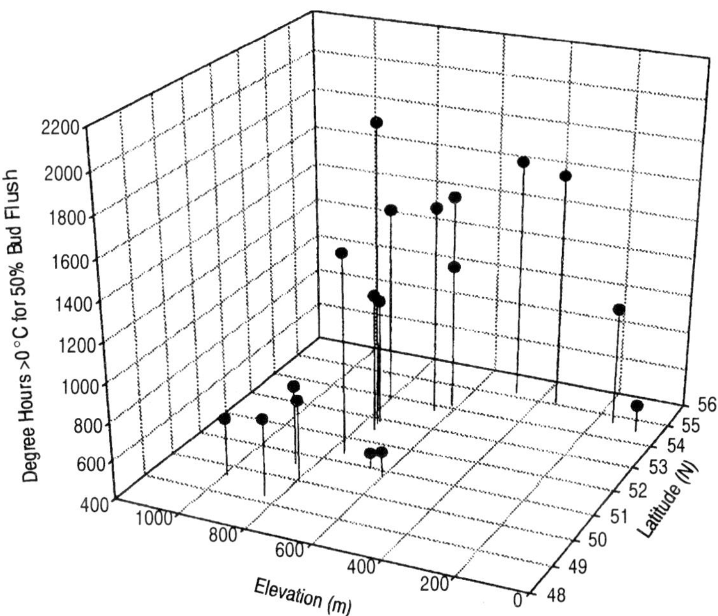

than 1000 degree hours. One low elevation (70 m) northern (54.5°N 128.57°W) collection was an exception to this trend, perhaps reflecting a maritime influence. An environmental variable, seed origin degree days above 0°C (from January to the end of March), was estimated from long-term weather station data nearest to the origin of each seed source. This single variable explained more than 52% of the bud flushing variation among the seed sources (Figure 2). Bud flushing evaluated at several field sites (not shown) in most cases reflected the results from controlled environment tests.

Growth Cessation and Fall Cold Acclimation

At Vernon from late September until mid-October (Julian day 270-285), the chlorophyll fluorescence variable Fv/Fm was between 0.60 and 0.75 indicating unimpaired photosynthetic potential (Figure 3a). Although differences existed among seed sources, these do not appear to be related to seed source latitude, longitude or elevation.

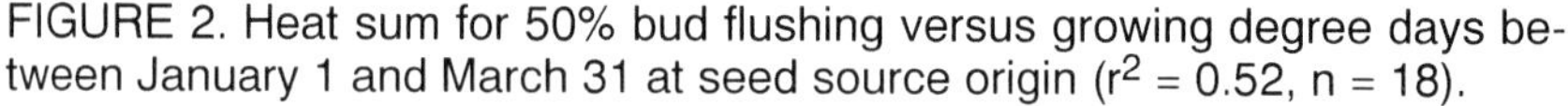
FIGURE 2. Heat sum for 50% bud flushing versus growing degree days between January 1 and March 31 at seed source origin ($r^2 = 0.52$, $n = 18$).

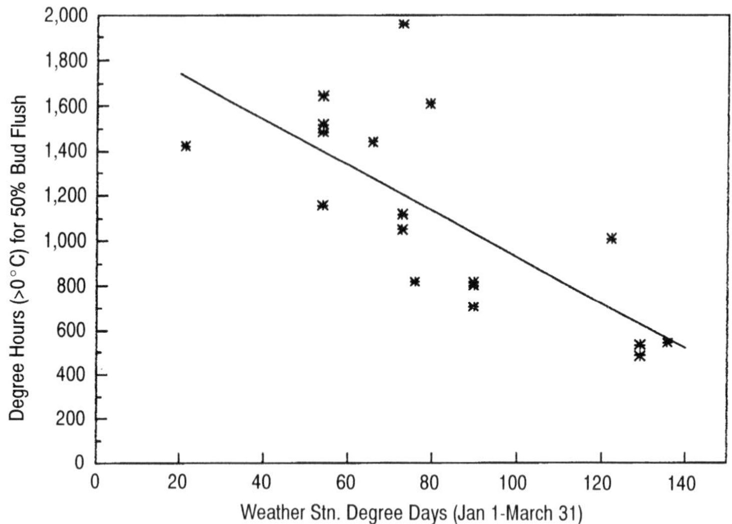

After Julian day 292, Fv/Fm of seedlings decreased indicating onset of foliage senescence, and variation among seed sources increased. Fv/Fm declined first in central seed sources followed by southeast then northwest. The difference between central and southeast sources was not significant.

At Victoria, quantum yield (QY of dark-adapted foliage = Fv/Fm) remained optimum (near 0.80) from the first measurement date, Julian day 255, until about day 285 (Figure 3b). The QY values of all seed sources began to decline between Julian days 286 and 300 and decreased quickly after the latter date. The QY decline after Julian day 280 was first for central sources followed by northwest then southeast.

Frost hardiness of stems from seedlings at Victoria and Prince George increased through the fall sampling period. For both locations, seed sources from higher latitudes and elevations had less damage after freezing and thus were more hardy. At Victoria, normalized seed source latitude plus elevation accounted for 79% of the variation in frost hardiness after freezing to $-15\,^{\circ}$C at the end of October (Figure 4). Furthermore, electrolyte leakage (normalized relative conductivity) results from Victoria correlated strongly with freezing damage averaged for five field sites (Figure 5). This suggests that cold hardiness

FIGURE 3. Mean Fv/Fm (Vernon) or quantum yield (Victoria) for paper birch seed sources. Critical threshold for Fv/Fm 0.5 and 60% for quantum yield are indicated.

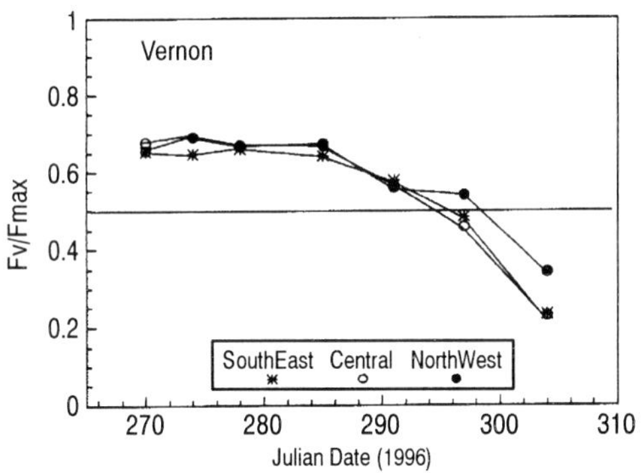

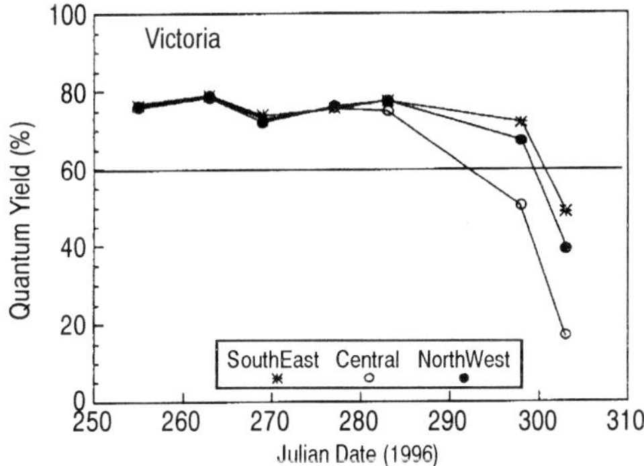

rankings based on laboratory measurements of seedlings from a common garden plot reflect actual field results.

Growing Season Length

Bud flushing and foliage senescence data can be used to estimate growing season length as follows. To a nominal average growing season length (L) of 100 frost-free days, a bud flushing value (b) and a leaf senescence value (s) are added (or subtracted).

$$L \text{ (days)} = 100 + b + s$$

FIGURE 4. Frost hardiness (October 28, 1996) of birch stem sections from Victoria versus seed source latitude and elevation (r^2 = 0.79, n = 16).

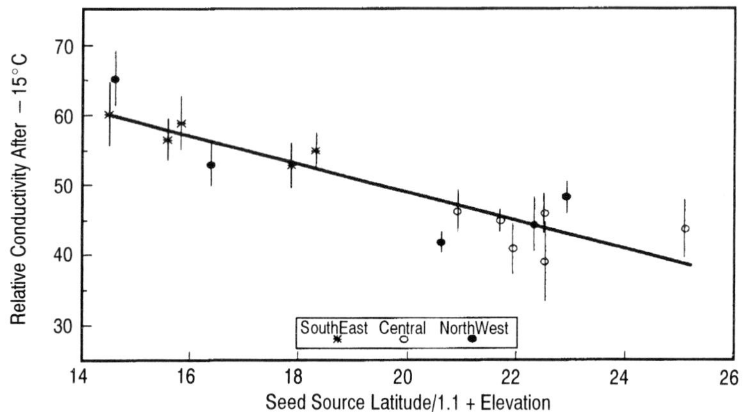

where:

b = (1121.67 − d)/h
s = (e − 295.56)
d = seed source degree hours > 0°C for 50% bud flush
e = day of year when Fv/Fm falls below 0.5 using the Morgan CF1000 fluorometer or a QY of 60% using the Opti Sciences OS 500
h = 100 degree hours > 0°C (nominal degree hours accumulated on an early spring day)
1121.67 = average d for all 18 seed sources
295.56 = average e for all 18 seed sources

Estimated growing season length for the 18 seed sources ranged from 89 (a central source) to 109 (a northwest source). This range of growing season lengths resulted from various combinations of days being added or subtracted at both ends of the growing season (Figure 6). When the estimated growing season lengths are compared with the seed source origin frost-free days, estimated from weather station data, a significant (r = 0.71) correlation was found.

DISCUSSION AND CONCLUSIONS

For the 18 seed sources tested, the heat sum above 0°C required for 50% bud flush ranged from 487 to 1963 growing degree hours. As a

FIGURE 5. Linear relationship between normalized percent damage (fall 1997) at all field plantings and normalized electrolyte leakage after −15°C at Victoria on October 28, 1996 (r = 0.71, n = 16).

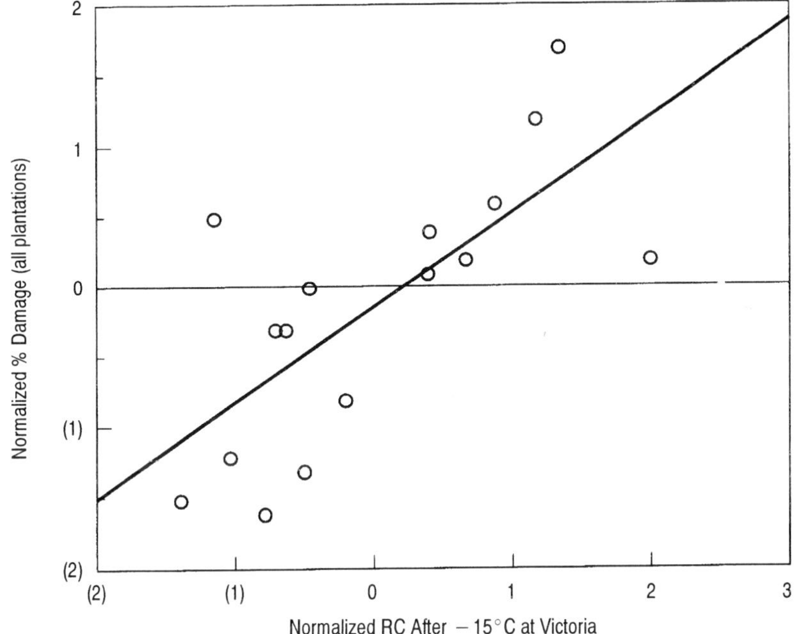

general rule, seed sources from 49-51°N latitude flushed in less than 1000 degree hours while more northerly sources took longer. We found that the seed source physiographic parameters latitude and elevation explained 63% of the bud flushing variation among the seed sources.

In the two geographically different (Vernon and Victoria, BC) common garden leaf senescence began about mid-October. Seed sources from the center of the province began leaf senescence before either southeast or northwest sources. We did not determine if the primary trigger for this event is day length or temperature (or a combination of both). Leaf senescence can easily be measured in the field with a fluorometer as loss of photosynthetic capacity.

Freezing tolerance was highest in the central (Prince George area) seed sources. Normalized seed source latitude plus elevation accounted for 79% of the variation among sources after freezing to −15°C at the

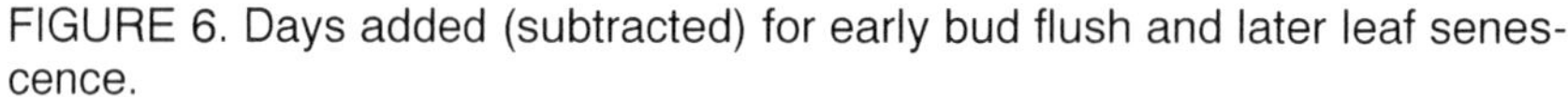
FIGURE 6. Days added (subtracted) for early bud flush and later leaf senescence.

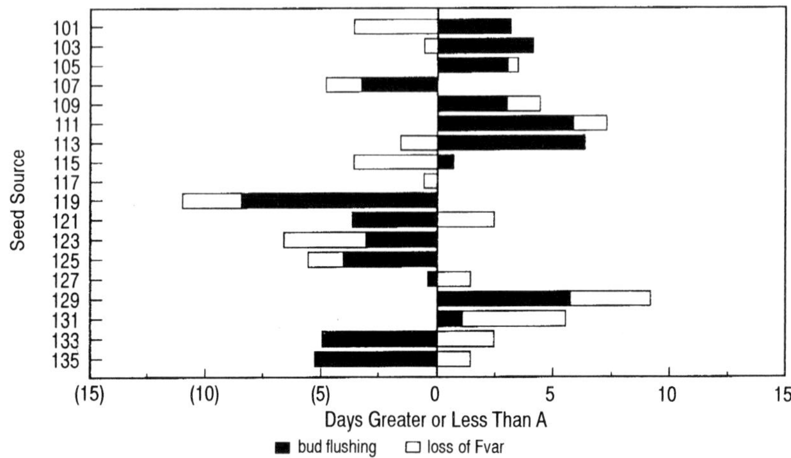

end of October. The timing and intensity of freezing will affect this value. Laboratory freeze injury tests from common garden plot samples correlated (r = 0.71) with actual freeze injury data collected from the field plantation sites.

From the phenological data collected, we estimated the growing season length of the 18 seed sources tested ranges from 89 to 109 days. This information, along with frost hardiness results, will be useful in developing seed transfer guidelines.

REFERENCES

Binder, W.D. and Fielder, P. 1996. Chlorophyll fluorescence as an indicator of frost hardiness in white spruce from different latitudes. New Forests 11: 233-253.

Binder, W.D., Fielder, P., Mohammed, G.H. and L'Hirondelle, S.J. 1997. Applications of chlorophyll fluorescence for stock quality assessments with different types of fluorometers. New Forests 13: 63-89.

L'Hirondelle, S.J., Jacobson, J.S. and Lassoie, J.P. 1992. Acid mist and nitrogen fertilization effects on growth, nitrate reductase activity, gas exchange, and frost hardiness of red spruce seedlings. New Phytologist 121: 611-622.

Mohammed, G.H., Binder, W.D. and Gilles, S. 1995. Chlorophyll fluorescence: A review of its practical forestry applications and instrumentation. Scand. J. For. Res. 10: 383-410.

Simpson, D.G. 1994. Seasonal and geographic origin effects on cold hardiness of white spruce buds, foliage and stems. Can. J. For. Res. 24: 1066-1070.

PART THREE: FRONTIERS OF SILVICULTURE

Plenary Address: Applications of Belowground Forest Biology

Alexander L. Friend

INTRODUCTION

Belowground forest biology has been underrepresented in the literature over the past 5 years. A literature search (Agricola, 1997) found 28,000 papers involving forests or trees. Of these, 5% addressed

Alexander L. Friend is affiliated with the Department of Forestry, Box 9681, Mississippi State University, MS 39762-9681 (E-mail: afriend@cfr.msstate.edu).

The author gratefully acknowledges the editorial inputs of M. D. Coleman, T. M. Hinckley, D. W. Johnson, E. J. Jokela, M. E. Kubiske, and B. E. Mahall.

Journal Article No. FO-095 of the Forest and Wildlife Research Center, Mississippi State University.

[Haworth co-indexing entry note]: "Plenary Address: Applications of Belowground Forest Biology." Friend, Alexander L. Co-published simultaneously in *Journal of Sustainable Forestry* (Food Products Press, an imprint of The Haworth Press, Inc.) Vol. 10, No. 3/4, 2000, pp. 199-212; and: *Frontiers of Forest Biology: Proceedings of the 1998 Joint Meeting of the North American Forest Biology Workshop and the Western Forest Genetics Association* (ed: Alan K. Mitchell et al.) Food Products Press, an imprint of The Haworth Press, Inc., 2000, pp. 199-212. Single or multiple copies of this article are available for a fee from The Haworth Document Delivery Service [1-800-342-9678, 9:00 a.m. - 5:00 p.m. (EST). E-mail address: getinfo@haworthpressinc.com].

leaves, while only 2% addressed roots. When the search was restricted to forest management, 3200 papers were found, with 1.8% addressed leaves and only 0.2% addressed roots. Thus, two to nine times more papers were found about leaves than about roots. An obvious reason for the bias toward leaves and aboveground biology over roots and belowground biology is the difficulty observing and studying this portion of the forest. Belowground forest biology spans a broad range of topics: primary producers (roots), symbionts (mycorrhizae and nitrogen-fixers), parasites and disease organisms, soil microflora and microfauna, and soil animals. However, this review focuses only on roots, because roots are the major conduits for carbon, water, and nutrients between above- and belowground realms. The purpose of this paper is to call attention to aspects of root biology that might offer applications to forest ecosystem management. Three outstanding phenomena are identified, each of which has applications to forest management: (1) the tremendous magnitude and diversity of carbon allocation to roots, (2) the physiological plasticity observed within root systems, and (3) the interactions that occur among root systems. In each case, the phenomenon is reviewed, and then applications are discussed.

BELOWGROUND CARBON ALLOCATION

The magnitude of belowground carbon inputs is tremendous. Belowground carbon allocation, including root respiration, root exudation, root growth and root turnover, is from 2 to 10 Mg C ha^{-1} yr^{-1} in world forests (Gower et al., 1996). This flux is put in perspective when one considers that an aboveground woody biomass production of 20 Mg ha^{-1} yr^{-1} (~ 10 Mg C ha^{-1} yr^{-1}) is high for world forests (Cannell, 1989). Similarly, 41 to 54% of total net primary production (NPP) occurs belowground in one-fourth of the world's forests (Vogt et al., 1996). Both belowground carbon allocation and belowground NPP are driven by fine root activity. These small diameter (0.1 to 2.0 mm) roots have short life spans; transfer carbon to microbes, symbionts, and herbivores; release CO_2 from respiratory processes; and often return more carbon to the soil than they accumulate in living biomass (Pregitzer and Friend, 1996). Mycorrhizae and symbiotic N-fixers extend similar fluxes to a level once removed from the tree but are at least as demanding of the tree for carbon. In fact, the mycorrhizal sink can be so strong it suppresses growth (e.g., Peng et al., 1993).

Quantification of the root absorbing surface provides an impressive metric for belowground carbon investment. Tree root systems commonly extend beyond 3 m in depth (Stone and Kalisz, 1991; Heilman et al., 1994a) and can extend horizontally more than eight times the radius of the canopy (Friend et al., 1991). Root length density (RLD, length of root per unit soil volume) is a sensitive indicator of nutrient absorbing capacity (Bowen, 1984) and can vary from < 0.1 cm cm^{-3} in developing stands to more than 6 cm cm^{-3} in species with relatively thin roots (Heilman et al., 1994a). Such a high RLD yields up to 60 km of roots per cubic meter of soil! RLD is greatest in the surface soil horizons, but temperate forests still deploy 18 to 30% of their root biomass below 50 cm depth (Jackson et al., 1996).

APPLICATIONS OF BELOWGROUND CARBON ALLOCATION

Given the large amount of carbon allocated to root systems, it is reasonable to ask whether this is wasteful of carbon that might be allocated to aboveground yield. There is some support for the concept of excessive root system size. For instance, root pruning has a less than expected effect on tree physiology (Teskey et al., 1985; Carlson et al., 1988). Also, there are arguments for how plant roots may have evolved as a sink for excess carbon (Barlow, 1994). However, genetics studies find that aboveground yield is not very sensitive to, or even consistently correlated with, the relative size of root systems (Heilman et al., 1994b). A more detailed approach to quantifying belowground structural-functional relationships is needed before the potential for selecting against root allocation can be productively applied. Two simpler applications of the large plant investment into root systems are enhancement of soil productivity (ecosystem restoration) and forest regeneration.

Ecosystem restoration and agroforestry each utilize belowground carbon fluxes to restore and maintain soil productivity. In addition to contributing soil organic matter, which improves soil water and nutrient holding capacity, tree roots may support vigorous nitrogen fixing symbionts and, due to deep rooting, may relocate displaced nutrients from deep to surface soil (Nair, 1993, pp. 271-272). For example, the presence of *Leucaena leucocephala*, a nitrogen-fixing tree, nearly doubled the amount of soil organic C and N, compared with conventional tillage, and provided 20 to 50% increases in soil C and N, compared with no-till agriculture (Nair, 1993, p. 128, after Lal, 1989). Similarly, *Byrsonima* cover resulted in more than twice the Ca, K, and base saturation in surface soil than open savanna or *Pinus* cover (Kellman, 1979). Restoration of temperate cropland to forest also results in marked increases in soil organic matter (Johnson, 1992). Although roots are not the sole contributors to soil enrichment under tree cover, a significant role is likely due to the tremendous activity of root turnover and the widespread distribution of roots in the soil profile. In addition to turnover, hydraulic lift is a remarkable mechanism by which roots may transfer water and nutrients from deep soil layers to surface soil layers via root systems. This phenomenon was discovered in desert shrubs (Caldwell and Richards, 1989) and recently was documented in trees (Dawson, 1998).

A strong link exists between high root investment and successful forest regeneration in stressful environments. This is evident in the deep or extensive root systems of plants adapted to dry or infertile environments (Fitter and Hay, 1987 p. 157). Genetic variation within a species has supported the concept that large root investment (Cannell et al., 1978) or large increases in root investment under stress (Li et al., 1991) are associated with better early stand growth under water- or nutrient-limited conditions. Even within an individual plant, cultural practices can enhance seedling performance by increasing root abundance or root regenerating ability. However, emphasis placed on producing seedlings with extensive root systems, or the ability to regenerate roots rapidly upon planting, must be balanced with emphasis on the handling of planting stock and seedling responses to environmental conditions at the time of planting (Simpson and Ritchie, 1997).

ROOT SYSTEM PLASTICITY

Root systems face a more heterogeneous environment than any other part of the plant, and are, therefore, very plastic in their physiology and distribution. They must adjust to large-scale temporal changes in temperature, water, nutrient, and oxygen availability with seasonal climatic changes, and to variations in site characteristics and soil parent material. Also, they must adjust to small-scale spatial variation in the distribution of soil resources with depth and in isolated patches throughout the soil (Stark, 1994). As a result, root system plasticity occurs at large and small scales and is driven by both environmental and genetic variation. The largest scale is in whole plant and ecosystem carbon allocation to roots and relative root mass. At this scale, the classic environmental controls are elevated atmospheric CO_2 (Curtis et al., 1994) and nutrient or water deficiency (Friend et al., 1994); each generally increases the investment into roots. Genetic influences at this scale result in variable changes in root investment in response to water stress (Dickmann et al., 1992) or nutrient stress (Li et al., 1991). At the small scale, roots are preferentially deployed into resource-rich patches of soil, particularly in response to N and P (George et al., 1997). Within days after being exposed to an enriched patch, roots begin to proliferate and increase ion uptake rates (Caldwell, 1994). The growth response is a combination of increased root initiation and increased elongation of existing roots, and is enhanced by nutrient stress in the tree (Friend et al., 1990; Hinckley et al., 1992). Enhanced proliferation under nutrient stress suggests that growth plasticity within root systems may partially compensate for nutrient stress by more intensive exploration of patches compared with non-stressed conditions. Ion uptake is similarly compensatory. Uptake rates (at the same nutrient concentration) are elevated for nutrient deficient trees compared with trees near nutrient sufficiency (Dighton et al., 1993). Hypoxic soil conditions elicit similar responses, with adventitious root formation away from the stress, and morphological changes (e.g., aerenchyma formation) in roots exposed to the stress, in flood-adapted trees (Kozlowski et al., 1991, p. 331).

APPLICATIONS OF ROOT SYSTEM PLASTICITY

Understanding genetic and environmental effects on tree root plasticity, at both scales, may lead to improvements in fertilization practices, diagnosis of nutrient deficiencies, and genetic improvement in planting stock. One of the more relevant findings is that roots seem to proliferate in response to N and P, but not to other macronutrients (Philipson and Coutts, 1977). Since roots proliferate in response to N and P, and these are the most limiting elements to forest growth, it follows that existing forest fertilization practices that rely on localized placement versus broadcast application are often successful. Examples include banding of P at planting (Pritchett and Fisher, 1987, p. 370), spot applications for newly planted trees (Ballard, 1984), and drip fertigation in intensive culture situations (Zsuffa et al., 1996). Potential advantages of localized applications over broadcast application are (1) concentration of fertilizer increases the availability of certain nutrients, e.g., P, (2) fertilizer response of competing herbaceous vegetation will be more localized and, therefore, easier to control, and (3) directed placement minimizes absorption by noncrop species. Other examples and advantages of localized placement should be explored for particular trees and soils, as both variables will affect compensatory root growth and uptake responses.

The metabolic and growth plasticity seen in tree roots also may have application to nutrient deficiency diagnosis. For example, root growth in N-enriched soil cores was less in fertilized stands than in control stands of *Pseudotsuga* (Hinckley et al., 1992). Similarly, ^{15}N uptake of excised roots decreased with increasing N addition to *Eucalyptus* stands, and was a better predictor of fertilizer response than foliar N (Jones and Dighton, 1993). The implication is that compensatory root growth or ion uptake could be used in addition to, or in place of, foliar analysis to predict stand response to fertilizer addition (see also Gleeson and Good, 1994).

If root production is "wasteful," as previously discussed, perhaps one way to reduce belowground carbon allocation and increase yield would be to select for more effective root allocation, i.e., root systems that evidence strong plasticity in response to soil nutrients. An ideal combination could be low overall root investment, increased root investment in response to nutrient stress, and strong preferential investment in the most fertile patches of soil. Such an approach may only be possible in highly intensive silvicultural systems. For example, a short-term screening study of *Populus* hybrids found both variation in overall root investment and variation in root sensitivity to patches of soil enrichment. Results showed that the two traits varied independently of one another (Friend et al., 1998). The effectiveness of such traits for improving root efficiency still needs detailed investigation. Nevertheless, the general concept of exploiting genetic variation in root traits, especially in intensive and clonal forestry, is very promising (see Dickmann and Keathley, 1996).

ROOT-ROOT INTERACTIONS

Root interactions are often a dominant part of plant interactions. Shoot growth responses associated with trenching around understory seedlings, reported since 1904, indicate that root competition can be a substantial retardant to growth even in overtopped conditions (Spurr and Barnes, 1980, p. 387). In *Pinus taeda* plantations, the earliest interference with regeneration is from herbaceous plants (Morris et al., 1993). During the establishment year, grass competition has a tremendous potential for depleting soil water and reducing pine root growth (Ludovici and Morris, 1997). Hardwood tree competition constitutes the next challenge to establishment, as shown by a root-exclusion study documented by Harty (1996). Two years after planting *Pinus taeda* seedlings in an area with a heavy hardwood competition, researchers imposed four treatments: (1) complete vegetation control within a 75-cm radius of the stem (CTL), (2) CTL + clipping of competing vegetation to a 75-cm height (CLIP), (3) CTL + trenching to 45-cm depth and installing a root exclusion barrier (BAR), and (4) CTL + removal of hardwood competition with herbicide (HERB). After two growing seasons (two years after planting), trees had grown from 1 to 3 m in height, CTL and CLIP did not differ in *Pinus* stem volume, but HERB and BAR had 66% and 93% greater volume than CTL, respectively. The large positive effect of controlling belowground competition and the small effect of controlling aboveground competition on *Pinus* growth illustrates the dominance of belowground competition in this system. Similar findings are reported in a diversity of forest ecosystems (Bi et al., 1992; Neary et al., 1990; Putz 1992; Riegel, 1995; Van Auken and Bush, 1997).

In addition to competition for soil resources, interference from competing vegetation may come from direct deterrence of root growth. Elegant work on desert shrubs (Mahall, 1998) documents that repressed root growth may result from inhibitory substances exuded into the rhizosphere, or from a root contact mechanism that is independent of diffusable inhibitory substances and includes self and non-self recognition. Forest trees have not been studied in enough detail to isolate such mechanisms, but root growth inhibition appears to be an important process. For example, *Pinus taeda* roots may inhibit the growth of nearby *Liquidambar styraciflua* roots (Friend et al., 1996) and result in a relative increase in deep rooting of *Liquidambar* (Jifon et al., 1995), when *Liquidambar* is grown in mixed stands, compared with *Liquidambar* monocultures.

Interactions among roots also may be positive (Callaway and Walker, 1997). Nitrogen fixation is the most obvious illustration (Binkley, 1986, pp. 133-160), but others also exist. For example, mixtures of *Pinus sylvestris* and *Picea sitchensis* appear to facilitate the performance of *Picea*, compared with pure *Picea* stands, by enhancing nutrient availability (McKay and Malcolm, 1988). Similarly, *Abies lasiocarpa* seedlings and saplings showed better growth under *Pinus albicaulis* than in open-grown settings, but only under stressful abiotic conditions. This led the authors to hypothesize that facilitation is more important under stressful conditions than non-stressful conditions (Callaway and Walker, 1997). Rarely are the particular causes known for facilitation. One potentially facilitative mechanism is hydraulic lift. Surface soil enrichments in water and nutrients were attributed to this process in *Acer saccharum* by Dawson (1998). An interesting contrast to hydraulic lift is the facilitation resulting from one species oxygenating the rhizosphere of another under flooded conditions, as was reported for *Typha latifolia* by Callaway and King (1996).

APPLICATIONS OF ROOT-ROOT INTERACTIONS

The degree to which root systems overlap in forests is large. For instance, the cubic meter of soil beneath a sample tree was found to be dominated (70% of total) by fine roots of surrounding trees (Friend et al., 1991). This suggests that the role of root interactions may be a large component of plant community dynamics. Since the balance between facilitation and competition depends on several biotic and abiotic factors (Callaway and Walker, 1997), some detailed investigations into the sensitivity of mixed-stand performance to variations in stand age, density, and environmental stress could be used to enhance the values of mixed species plantings. Current prospects include: soil enrichment from nitrogen-fixing species (Binkley, 1986), early cost recovery and high log value from planting complementary mixtures of fast and slow growing species (e.g., *Populus* and *Quercus*, Twedt and Portwood, 1997), increased productivity or stress resistance from the appropriate use of mixed-versus pure-culture (DeBell and Harrington, 1993; Miller and Reukema, 1993), successful regeneration from the use of nurse crops and a mechanistic understanding of competition

(Callaway and Walker, 1997; see also Kronzucker, 1997), and ecosystem stability (stress resistance, resilience, water quality) from using species mixtures that exploit varied but overlapping ecological niches (Burch et al., 1997; MacDonald, 1995). In all cases, particular species used, or cultural conditions, could be refined by using an understanding of belowground processes to maximize positive interactions.

Given the range of variation in positive and negative root-root interactions, it should be possible to select for root physiological traits within species that benefit intensive plantation forestry, e.g., through improved resource acquisition or suppression of competing species. Before such possibilities can be realistically considered, however, the sensitivity of soil resource acquisition by roots to variations in the biotic and abiotic environment needs to be better understood (e.g., Smethurst and Comerford, 1993). After the significance of variation in root properties is understood in the context of realistic soil conditions, efforts toward genetic selection or engineering may be more productively directed. Mathematical modeling will be an important component of such a process (e.g., Yanai et al., 1995).

CONCLUSIONS

- Tree root systems consume massive amounts of plant carbon and are extensively distributed in soil. Fluxes of carbon, nutrients, and water from fine roots to soil can be used to the benefit of ecosystem restoration and agroforestry. Rapid and extensive seedling root growth is an important component of successful forest regeneration. Further applications may result from exploring the relationship between belowground carbon allocation and aboveground yield.
- The spatial partitioning of root absorbing surface is very plastic and open to both environmental and genetic manipulation. The effective use of spot or banded fertilizer is one current application. Further applications may exist in diagnosis of nutrient deficiencies and in genetic manipulation of rooting traits to favor more effective nutrient acquisition.
- The importance of root versus shoot competition in forests is impressive, as is the diversity and complexity of root interactions (both positive and negative). Competition control is the most com-

mon application of this information. Further applications may exist in refining species mixes based on knowledge of below-ground interactions, and in genetic or cultural manipulation of root interactions to achieve ecosystem management objectives.

REFERENCES

Agricola. 1997. CD-ROM bibliographic database. SilverPlatter, Norwood, MA.

Ballard, R. 1984. Fertilization of plantations. In: pp. 327-360 G.D. Bowen and E.K.S. Nambiar (eds.). Nutrition of Plantation Forests. Academic Press, New York.

Barlow, P.W. 1994. The origin, diversity and biology of shoot-borne roots. In: pp. 1-23. T.D. Davis and B.E. Haissig (eds.). Biology of Adventitious Root Formation. Plenum Press, New York.

Bi, H.Q., N.D Turvey, and P. Heinrich. 1992. Rooting density and tree size of *Pinus radiata* (D-Don) in response to competition from *Eucalyptus obliqua* (Lherit). Forest Ecology and Management 49: 31-42.

Binkley, D. 1986. Forest Nutrition Management. John Wiley & Sons, New York.

Bowen, G.D. 1984. Tree roots and the use of soil nutrients. In: pp. 147-179. G.D. Bowen and E.K.S. Nambiar (eds.). Nutrition of Plantation Forests. Academic Press, New York.

Brisson, J., and J.F. Reynolds. 1997. Effects of compensatory growth on population processes: a simulation study. Ecology 78: 2378-2384.

Burch, W.H., R.H. Jones, P. Mou, and R.J. Mitchell. 1997. Root system development of single and mixed plant functional type communities following harvest in a pine-hardwood forest. Canadian Journal of Forest Research 27: 1753-1764.

Caldwell, M.M. 1994. Exploiting nutrients in fertile soil microsites. In: pp. 325-347. M.M. Caldwell and R.W. Pearcy (eds). Exploitation of Environmental Heterogeneity by Plants: Ecophysiological Processes Above- and Belowground. Academic Press, New York.

Caldwell, M.M. and J.H. Richards. 1989. Hydraulic lift: water efflux from upper roots improves effectiveness of water uptake by deep roots. Oecologia 79:1-5.

Callaway, R.M., and King, L. 1996. Oxygenation of the soil rhizosphere by *Typha latifolia* and its facilitative effects on other species. Ecology 77: 1189-1195.

Callaway, R.M., and L.R. Walker. 1997. Competition and facilitation: a synthetic approach to interactions in plant communities. Ecology 78: 1958-1965.

Cannell, M.G.R. 1989. Physiological basis of wood production: a review. Scandinavian Journal of Forest Research 4:459-490.

Cannell, M.G.R., F.E. Bridgewater, and M.S. Greenwood. 1978. Seedling growth rates, water stress responses and root-shoot relationships related to eight-year volumes among families of *Pinus taeda* L. Silvae Genetica 27: 237-248.

Carlson, W.C., C.A. Harrington, P. Farnum, and S.W. Hallgren 1988. Effects of root severing treatments on loblolly pine. Canadian Journal of Forest Research 18: 1376-1385.

Curtis, P.S., E.G. O'Neill, J.A. Teeri, D.R. Zak, and K.S. Pregitzer. 1994. Below-ground responses to rising atmospheric CO_2: implications for plants, soil biota and ecosystem processes. Plant and Soil 165: 1-6.

Dawson, T.E. 1998. Water loss from tree roots influences soil water and nutrient status, and plant performance. In: pp. 235-250. H.E. Flores, J.P. Lynch, and D. Eissenstat (eds.). Radical Biology: Advances and Perspectives on the Function of Plant Roots. American Society of Plant Physiologists, Rockville, MD.

DeBell, D.S., and C.A. Harrington. 1993. Deploying genotypes in short-rotation plantations: mixtures and pure cultures of clones and species. The Forestry Chronicle 69: 705-713.

Dickmann, D.I., and D.E Keathley. 1996. Linking physiology, molecular genetics, and the Populus ideotype. In: pp. 491-514. R.F. Stettler, H.D. Bradshaw, P.E. Heilman, and T.M. Hinckley (eds.). Biology of *Populus* and Its Implications for Management and Conservation. NRC Research Press, National Research Council of Canada, Ottawa.

Dickmann, D.I., Z. Liu, P.V. Nguyen, and K.S. Pregitzer. 1992. Photosynthesis, water relations, and growth of two hybrid *Populus* genotypes during a severe drought. Canadian Journal of Forest Research 22: 1094-1106.

Dighton, J., H.E. Jones, and J.M. Poskitt. 1993. The use of nutrient bioassays to assess the response of *Eucalyptus grandis* to fertilizer application. 1. Interactions between nitrogen, phosphorus, and potassium in seedling nutrition. Canadian Journal of Forest Research 23: 1-6.

Fitter, A.H., and R.K.M. Hay. 1987. Environmental physiology of plants, Second Edition. Academic Press, London.

Friend, A.L., P.C. Berrang, and J.R. Seiler. 1996. Contrasting N accumulation responses to CO_2 between *Pinus taeda* and *Liquidambar styraciflua*. Bulletin of the Ecological Society of America 77 (3):151.

Friend, A.L., M.D. Coleman, and J.G. Isebrands. 1994. Carbon allocation to root and shoot systems of woody plants. In: pp. 245-273. T.D. Davis and B.E. Haissig (eds.). Biology of Adventitious Root Formation. Plenum Press, New York.

Friend, A.L., M.R. Eide and T.M. Hinckley. 1990. Nitrogen stress alters root proliferation in Douglas-fir seedlings. Canadian Journal of Forest Research 20: 1524-1529.

Friend, A.L., J.A. Mobley, E.A. Ryan, and H.D. Bradshaw, Jr. 2000. Root growth plasticity of hybrid poplar in response to soil nutrient gradients. Journal of Sustainable Forestry 10(1/2): 133-140.

Friend, A.L., G., Scarascia-Mugnozza, J.G. Isebrands, and P.E. Heilman. 1991. Quantification of two-year-old hybrid poplar root systems: morphology, biomass, and ^{14}C distribution. Tree Physiology 8: 109-119.

George, E., B. Seith, C. Schaeffer, and H. Marschner. 1997. Responses of *Picea*, *Pinus* and *Pseudotsuga* roots to heterogeneous nutrient distribution in soil. Tree Physiology 17: 39-45.

Gleeson, S.K., and R.E. Good, 1994. Root foraging and limiting factors in the New Jersey Pinelands. Bulletin of the Ecological Society of America 75(2): 76-77.

Gower, S.T., S. Pongracic, and J.J. Landsberg. 1996. A global trend in belowground carbon allocation: can we use the relationship at smaller scales? Ecology 77: 1750-1755.

Harty, R.L. 1996. Belowground hardwood competition in loblolly pine (*Pinus taeda* L.) stands. M.S. Thesis. Mississippi State University, Mississippi State, MS.

Heilman, P.E., G. Ekuan, and D. Fogle. 1994a. Aboveground and belowground biomass and fine roots of 4-year-old hybrids of *Populus trichocarpa* × *Populus deltoides* and parental species in short-rotation culture. Canadian Journal of Forest Research 24: 1186-1192.

Heilman, P.E., G. Ekuan, and D.B. Fogle. 1994b. First-order root development from cuttings of *Populus trichocarpa* × *P. deltoides* hybrids. Tree Physiology 14: 911-920.

Hinckley, T.M., A.L. Friend and A.K. Mitchell. 1992. Response at foliar, tree, and stand levels to nitrogen fertilization: a physiological perspective. In: pp. 82-89. H.N. Chappell, G.F. Weetman, and R.E. Miller (eds.). Forest Fertilization: Sustaining and Improving Nutrition and Growth of Western Forests. Contribution 73, Institute of Forest Resources, University of Washington, Seattle.

Jackson, R.B., J. Canadell, J.R. Ehleringer, H.A. Mooney, O.E. Sala, and E.D. Schultze. 1996. A global analysis of root distribution for terrestrial biomes. Oecologia 108: 389-411.

Jifon, J.L., A.L. Friend, and P.C. Berrang. 1995. Species mixture and soil-resource availability affect the root growth response of tree seedlings to elevated atmospheric CO_2. Canadian Journal of Forest Research 25: 824-832.

Johnson, D.W. 1992. Effects of forest management on soil carbon storage. Water, Air, and Soil Pollution 64: 83-120.

Jones, H.E., and J. Dighton. 1993. The use of nutrient bioassays to assess the response of *Eucalyptus grandis* to fertilizer application. 2. A field experiment. Canadian Journal of Forest Research 23: 7-13.

Kellman, M. 1979. Soil enrichment by neotropical savanna trees. Journal of Ecology 67: 565-577.

Kozlowski, T.T., P.J. Kramer, and S.G. Pallardy. 1991. The Physiological Ecology of Woody Plants. Academic Press, Inc., San Diego.

Kronzucker, H.J., M.Y. Siddiqi, and A.D.M. Glass. 1997. Conifer root discrimination against soil nitrate and the ecology of forest succession. Nature 385: 59-61.

Lal, R. 1989. Agroforestry systems and soil surface management of a tropical alfisol. Parts I-VI. Agroforestry Systems 8:1-6; 7-29; 97-111; 113-132; 197-215; 217-238; 239-242.

Li, B., H.L. Allen, and S.E. McKeand. 1991. Nitrogen and family effects on biomass allocation of loblolly pine seedlings. Forest Science 37: 271-283.

Ludovici, K.H., and L.A. Morris. 1997. Competition-induced reductions in soil water availability reduced pine root extension rates. Soil Science Society of America Journal 61: 1196-1202.

MacDonald, G.B. 1995. The case for boreal mixwood management: an Ontario perspective. The Forestry Chronicle 71: 725-734.

Mahall, B.E. 1998. Inter-root communications and the structure of desert plant communities. In: pp. 265-280. H.E. Flores, J.P. Lynch, and D. Eissenstat (eds.) Radical Biology: Advances and Perspectives on the Function of Plant Roots. American Society of Plant Physiologists, Rockville, MD.

McKay, H.M., and D.C. Malcolm. 1988. A comparison of the fine root component of a pure and a mixed coniferous stand. Canadian Journal of Forest Research 18: 1416-1426.

Miller, R.E., and D.L. Reukema. 1993. Size of Douglas-fir trees in relation to distance from a mixed red alder–Douglas-fir stand. Canadian Journal of Forest Research 23:2413-2418.

Morris, L.A., S.A. Moss, and W.S. Garbett. 1993. Competitive interference between selected herbaceous and woody plants and *Pinus taeda* L. during two growing seasons following planting. Forest Science 39: 166-187.

Nair, P.K.R. 1993. An Introduction to Agroforestry. Kluwer Academic Publishers, Boston.

Neary, D.G., E.J. Jokela, N.B. Comerford, S.R. Colbert and T.E. Cooksey. 1990. Understanding competition for soil nutrients–the key to site productivity on southeastern Coastal Plain spodosols. In: pp. 432-450. S.P. Gessel, D.S. Lacate, G.F. Weetman, and R.F. Powers (eds.). Sustained Productivity of Forest Soils. Proceedings of the 7th North American Forest Soils Conference, Univ. British Columbia. Faculty of Forestry Publication, Vancouver.

Peng, S.B., D.M. Eissenstat, J.H. Graham, K. Williams, and N.C. Hodge. 1993. Growth depression in mycorrhizal Citrus at high-phosphorus supply–analysis of carbon costs. Plant Physiology 101: 1063-1071.

Philipson, J.J., and M.P. Coutts. 1977. The influence of mineral nutrition on the root development of trees. II. The effect of specific nutrient elements on the growth of individual roots of Sitka spruce. Journal of Experimental Botany 28: 864-871.

Pregitzer, K.S., and A.L. Friend. 1996. The structure and function of Populus root systems. In: pp. 331-354. R.F. Stettler, H.D. Bradshaw, P.E. Heilman, and T.M. Hinckley (eds.). Biology of *Populus* and Its Implications for Management and Conservation. NRC Research Press, National Research Council of Canada, Ottawa.

Pritchett, W.L., and R.F. Fisher. 1987. Properties and Management of Forest Soils, Second ed. John Wiley & Sons, New York.

Putz, F.E. 1992. Reduction of root competition increases growth of slash pine seedlings on a cutover site in Florida. Southern Journal of Applied Forestry 16: 193-197.

Riegel, G.M. 1995. The effects of aboveground and belowground competition on understory species composition in a *Pinus ponderosa* forest. Forest Science 41: 864-889.

Simpson, D.G., and G.A. Ritchie. 1997. Does RGP predict field performance? A debate. New Forests 13: 253-277.

Smethurst, P.J., and N.B. Comerford. 1993. Potassium and phosphorus uptake by competing pine and grass: observations and model verification. Soil Science Society of America Journal 57:1602-1610.

Spurr, S.H., and B.V. Barnes, 1980. Forest Ecology, Third Edition. John Wiley & Sons, New York.

Stark, J.M. 1994. Causes of soil nutrient heterogeneity at different scales. In: pp. 255-284. M.M. Caldwell and R.W. Pearcy (eds.). Exploitation of Environmental Heterogeneity by Plants: Ecophysiological Processes Above- and Belowground. Academic Press, New York.

Stone, E.L., and P.J. Kalisz. 1991. On the maximum extent of tree roots. Forest Ecology and Management 46: 59-102.

Teskey, R.O., C.C. Grier, and T.M. Hinckley. 1985. Relation between root system size and water inflow capacity of *Abies amabilis* growing in a subalpine forest. Canadian Journal of Forest Research 15: 669-672.

Twedt, D.J., and J. Portwood. 1997. Bottomland hardwood reforestation for neotropical migratory birds: are we missing the forest for the trees? Wildlife Society Bulletin. 25: 647-652.

Van Auken, O.W., and J.K. Bush. 1997. Growth of *Prosopis glandulosa* in response to changes in aboveground and belowground interference. Ecology 78: 1222-1229.

Vogt, K.A., D.J. Vogt, P.A. Palmiotto, P. Boon, J. O'Hara, and H. Asbjornsen. 1996. Review of root dynamics in forest ecosystems grouped by climate, climatic forest type and species. Plant and Soil. 187: 159-219.

Yanai, R.D., T.D. Fahey, and S.L. Miller. 1995. Efficiency of nutrient acquisition by fine roots and mycorrhizae. In: pp. 75-103. W.K. Smith and T.M. Hinckley (eds.). Resource Physiology of Conifers. Academic Press, San Diego.

Zsuffa, L., Giordano, E., Pryor, L.D., and Stettler, R.F. 1996. Trends in poplar culture: some global and regional perspectives. In: pp. 515-539. R.F. Stettler, H.D. Bradshaw, P.E. Heilman, and T.M. Hinckley (eds.). Biology of *Populus* and Its Implications for Management and Conservation. NRC Research Press, National Research Council of Canada, Ottawa.

Assessing the Controls on Soil Mineral-N Cycling Rates in Managed Coastal Western Hemlock Ecosystems of British Columbia

R. L. Bradley
B. D. Titus
K. Hogg
C. Preston
C. E. Prescott
J. P. Kimmins

INTRODUCTION

The practice of clearcutting in the coastal western hemlock (CWH) ecosystem of British Columbia and the effect of clearcutting on site

---

R. L. Bradley is affiliated with the Département de biologie, Université de Sherbrooke, Sherbrooke, Québec J1K 2R1, Canada.

B. D. Titus, K. Hogg, and C. Preston are affiliated with the Pacific Forestry Centre, 506 West Burnside Road, Victoria, B.C. V8Z 1M5, Canada.

C. E. Prescott and J. P. Kimmins are affiliated with the Department of Forest Sciences, University of British Columbia, 2357 West Mall Road, Vancouver, B.C. V6T 1Z4, Canada.

Address correspondence to R. L. Bradley at the above address (E-mail: robert.bradley@courrier. usherb.ca).

This project was supported by Forest Renewal of British Columbia research grant #PA97229-ORE.

[Haworth co-indexing entry note]: "Assessing the Controls on Soil Mineral-N Cycling Rates in Managed Coastal Western Hemlock Ecosystems of British Columbia." Bradley, R. L. et al. Co-published simultaneously in *Journal of Sustainable Forestry* (Food Products Press, an imprint of The Haworth Press, Inc.) Vol. 10, No. 3/4, 2000, pp. 213-219; and: *Frontiers of Forest Biology: Proceedings of the 1998 Joint Meeting of the North American Forest Biology Workshop and the Western Forest Genetics Association* (ed: Alan K. Mitchell et al.) Food Products Press, an imprint of The Haworth Press, Inc., 2000, pp. 213-219. Single or multiple copies of this article are available for a fee from The Haworth Document Delivery Service [1-800-342-9678, 9:00 a.m. - 5:00 p.m. (EST). E-mail address: getinfo@haworthpressinc.com].

productivity are being debated. Central to this issue is the possible post-clearcutting flush of soil mineral-N and concomitant increase in the soil NO_3^--to-NH_4^+ ratio, a phenomenon referred to as the "assart flush" (Rommell, 1935). The timing and magnitude of the assart flush appears to vary significantly between clearcut sites. It has been suggested that this has significant implications for ecological succession rates and pathways and for the sustainability of wood production (Kronzucker et al., 1997).

Soil mineral-N concentrations generally increase markedly in the post-disturbed ecosystem because of the great reduction in potential plant N uptake relative to the capacity of the soil to supply mineral-N. Many studies have also explored changes to soil moisture and temperature conditions following clearcutting, which may increase microbial activity and N turnover (e.g., Frazer et al., 1990). These factors can be categorized as exogenous in that they should change in a like manner across all clearcuts within a given region. There needs to be a closer examination of possible indigenous factors–that is, biochemical controls within the soil–that could explain why the assart flush varies distinctly between similar sites.

Perhaps the reason why indigenous factors have received less attention than exogenous factors is because most studies on post-clearcutting soil-N dynamics have measured net rates at which NH_4^+ and NO_3^- accumulate in soil. The net amounts of N mineralized during laboratory or *in situ* incubations (Prescott, 1997), accreted on ion exchange resins (Martin, 1985), collected in lysimeters (Feller and Olanski, 1995), or measured in stream water (Kimmins and Feller, 1976) are good indices of the bio-availability or potential leaching loss of mineral-N. However, these measurements do not identify the factors that control the internal cycling rates and pathways of soil mineral-N. There is a need to understand the controls on gross production and consumption rates of mineral-N.

An ongoing research program is striving to identify, quantify and describe the role of various biochemical factors in the soil that may control mineral-N supply before and after clearcutting. Different experiments have been designed to establish a continuum between laboratory and field conditions in order to compare soil-N dynamics based solely on inherent biochemical properties, and to evaluate some of the confounding effects of environmental factors impinging on field-level observations. Four main indigenous factors are considered the available carbon supply, the activity of free phenolic compounds, NH_4^+ and NO_3^- concentrations, and the ecophysiological status and functional diversity of the soil microbial community. Gross production and consumption rates of soil-NO_3^- and soil-NH_4^+ are being measured under various conditions, and in the presence of different amendments, by applying the principle of stable isotope pool dilution (Kirkham and Bartholomew, 1954). The objective of our research is to generate discussion around these indigenous factors in terms of the possible implications for natural succession, fertilizer applications and wood-waste disposal.

METHODS

Given the scope and magnitude of the project, we only present a brief summary of selected results from a study conducted at the Montane Alternative Silvicultural Systems (MASS) site, located near Campbell River, B.C. The site is located in a mid- to high-elevation (740 to 850 m) unevenaged (200- to 800-year-old) forest dominated by western hemlock (*Tsuga heterophylla* [Raf.] Sarg.) and amabilis fir (*Abies amabilis* Dougl.), with a smaller component of western red cedar (*Thuja plicata* Donn.) and yellow cedar (*Chamaecyparis nootkanensis*). We used three replicate plots of two silvicultural treatments–clearcut (CC) and shelterwood (SW)–and of an unharvested old-growth (OG) control treatment. The two silvicultural treatments had been harvested in 1993 and the shelterwood plots contained 30% of the original basal area.

Three bulk samples of forest floor (FH-layer) material were collected from each of the nine plots during the first week of June, 1997, and returned to the Pacific Forestry Centre (Canadian Forest Service, Victoria, B.C.) for laboratory analyses. Subsamples were dried (101°C) and analyzed for gravimetric moisture content. A second set of subsamples was immediately extracted (1 N KCl) and analyzed colorimetrically for NH_4^+-N and NO_3^--N concentrations (Keeney and Nelson, 1982). Net ammonification and nitrification rates were determined by re-analyzing mineral-N concentrations following a 20 week aerobic incubation (25°C), at constant moisture content. More subsamples were analyzed respirometrically for basal and substrate-induced respiration rates, from which various eco-physiological indices were derived (Anderson and Domsch, 1978; Anderson and

TABLE 1. Effect of clearcut and shelterwood harvesting on various measurements of soil-N cycling rates and microbial dynamics in a montane CWH ecosystem (MASS site); values within the same row followed by a different lower-case letter differ significantly (Duncan's Multiple Range Test, $p < 0.05$, n = 3).

Measurement	Old-growth	Shelterwood	Clearcut
Mineral-N ($\mu g\ g^{-1}$ soil)	0.13 b	2.01 b	42.8 a
Mineralized NH_4^+-N ($\mu g\ g^{-1}$ soil)	311 b	259 b	605 a
Mineralized NO_3^--N ($\mu g\ g^{-1}$ soil)	2.46 b	2.57 b	3.80 a
Gross NO_3^--N consumption rate ($\mu g\ g^{-1}$ soil d^{-1})	7.96 a	4.24 b	2.35 c
Basal respiration rate ($\mu g\ CO_2$-C g^{-1} soil h^{-1})	15.6 a	9.7 b	10.5 b
Microbial biomass (mg biomass-C g^{-1} soil)	5.82 a	4.73 b	4.51 b
qCO_2 ($\mu g\ CO_2$-C mg^{-1} biomass-C h^{-1})	2.75 a	2.08 b	2.01 b
Organic-nutrient demand index (unitless)	< 0.01 b	< 0.01 b	0.14 a

Domsch, 1993; Bradley and Fyles, 1995a). The gross rates of NO_3^- consumption were measured using ^{15}N-dilution methodology (Hart et al., 1994).

RESULTS AND DISCUSSION

Mineral-N concentrations of soil samples collected at the MASS site were very low in the old-growth and shelterwood plots, and an order of magnitude higher in the 3-year-old clearcuts (Table 1). Similarly, the amount of N mineralized during aerobic incubations was twice as high in forest floor from the clearcuts as in forest floor from the shelterwood and old-growth plots. We did not observe a "partial assart effect" in the forest floor from the shelterwood system. Contrary to what has commonly been reported for deciduous forest ecosystems in Northeastern U.S. (e.g., Vitousek et al., 1979), the assart flush on clearcut plots at the MASS site was not accompanied with an increase in the soil NO_3^--to-NH_4^+ ratio.

Much attention is given to the ability of a soil to supply NO_3^--N, because of the mobility of this nutrient and its potential leaching-loss from the ecosystem. It is important to remember that the net supply of this ion is the result of two opposing processes, namely the gross production and consumption of NO_3^--N. The same applies to the net production rate of NH_4^+-N. Gross rates of NO_3^--N consumption were significantly lower in the shelterwood than in the old-growth plots, and lower yet in the clearcuts (Table 1). It is, therefore, just as likely that higher mineral-N concentrations in the clearcuts were due to lower microbial immobilization rates than they were to higher nitrification rates. Since microbial immobilization of NO_3^--N is mainly dependent on the soil available-C supply, clearcuts should contain microbial communities that exhibit symptoms of C-deficiency. Basal respiration rates, which reflect the immediate supply of available-C at time of sampling, were significantly higher in the forest floor of the old-growth plots than in the forest floor of the other two treatments. Similarly, the average microbial biomass, considered an index of C-availability over a preceding time interval (Bradley and Fyles, 1995b), was significantly higher in the old-growth plots than in the other two treatments (Table 1). The metabolic quotient (qCO_2), which describes the amount of soil-C respired per unit microbial biomass, was also statistically higher in the old-growth plots than in the other treatments. The relative increase in forest floor respiration rates due to organic-nutrient amendments (i.e., Difco™ nutrient broth) was significantly higher in samples collected in clearcuts than in samples collected in the other treatments, suggesting a post-clearcutting shift in microbial functional diversity towards populations with a higher affinity to C-bonded nutrients. It is generally estimated that trees shunt approximately one-third of recently fixed-C as labile compounds into the soil via their roots (Norton et al., 1990; Bowden et al., 1993). Qualls and Haines (1992) found that dissolved organic-C (DOC) in throughfall was much more labile than DOC in soil solution. Therefore, the total removal of the forest canopy results in the near elimination of two important sources of labile-C to the soil, that is, rhizodeposition and throughfall. Our results indicate that soil microbial communities in this particular CWH ecosystem are more C-limited in clearcuts than in shelterwood and old-growth plots, and that this factor alone may represent a major control over the assart flush. Questions remain, however, on the impact of shelterwood harvesting which, on the one hand, displays the soil-C cycling characteristics of clearcuts, but on the other hand, many of the N-cycling characteristics of old-growth plots. It is hypothesized that microbial C-deficiency is the first symptom to appear following partial canopy removal whereas N-cycling is more resistant to change.

REFERENCES

Anderson, T.H. and K.H. Domsch. 1978. A physiological method for the quantitative measurement of microbial biomass in soils. Soil Boil. Biochem. 10: 215-221.

Anderson, T.H. and K.H. Domsch. 1993. The metabolic quotient for CO_2 (qCO_2) as a specific activity parameter to assess the effects of environmental conditions, such as pH, on the microbial biomass of forest soils. Soil Boil. Biochem. 25: 393-395.

Bowden, R.D., Nadelhoffer, K.J., Boone, R.D., Melillo, J.M. and J.B. Garrison. 1993. Contributions of aboveground litter, belowground litter, and root respiration to total soil respiration in a temperate mixed hardwood forest. Can. J. For. Res. 23: 1402-1407.

Bradley, R.L. and J.W. Fyles. 1995a. Growth of paper birch (*Betula papyrifera*) seedlings increases soil available C and microbial acquisition of soil-nutrients. Soil Biol. Biochem. 27: 1565-1571.

Bradley, R.L. and J.W. Fyles. 1995b. A kinetic parameter describing soil available carbon and its relationship to rate increase in C mineralization. Soil Biol. Biochem. 27: 167-172.

Feller, M.C. and P. Olanski. 1995. Influence of alternative timber harvesting regimes in montane coastal western hemlock zone forests on soil nutrient leaching: Initial results. B.C. Ministry of Forests, FRDA Report 238. Pp. 89-100.

Frazer, D.W., McColl, J.G. and R.F. Powers. 1990. Soil nitrogen mineralization in a clearcutting chronosequence in a northern California conifer. Soil Sci. Soc. Am. J. 54:1145-1152.

Hart, S.C., Stark, J.M., Davidson, E.A., and M.K. Firestone 1994. Nitrogen mineralization, immobilization, and nitrification. *In* R.W. Weaver et al. (eds.) Methods of soil analyses–Part 2. (3rd ed.). Soil Sci. Soc. Am., Madison. Pp. 985-1018.

Keeney, D.R. and Nelson, J.L. 1982. Nitrogen-inorganic forms. *In* A.L. Page, R.H. Miller and D.R. Keeney (eds.). Methods of Soil Analysis–Part 2. (2nd ed.). Agron. Monogr. 9. ASA and SSSA, WI. Pp. 643-698.

Kimmins, J.P. and M.C. Feller. 1976. Effect of clearcutting and broadcast slashburning on nutrient budgets, streamwater chemistry, and productivity in Western Canada. *In* XVI IUFRO World Congress Proc. Div. 1, Oslo, Norway, pp. 186-197.

Kirkham, D. and W.V. Bartholomew. 1954. Equations for following nutrient transformation in soil, utilizing tracer data. Soil Sci. Soc. Am. J. 18: 33-34.

Kronzucker, H.J., Siddiqi, M.Y. and A.D.M. Glass. 1997. Conifer root discrimination against soil nitrate and the ecology of forest succession. Nature 385: 59-61.

Martin, W.F. 1985. Post-clearcutting forest floor nitrogen dynamics and regeneration response in the coastal western hemlock wet subzone. Ph.D. thesis, U.B.C. Dept. For. Sc., Vançouver.

Norton, J.M., Smith and M.K. Firestone. 1990. Carbon flow in the rhizosphere of ponderosa pine seedlings. Soil Biol. Biochem. 22: 449-455.

Prescott. C.E. 1997. Effects of clearcutting and alternative silvicultural systems on rates of decomposition and nitrogen mineralization in a coastal montane coniferous forest. For. Ecol. Manag. 95: 253-260.

Qualls, R.G. and Haines, B.L. 1992. Biodegradability of dissolved organic matter in forest throughfall, soil solution, and stream water. Soil Sci. Soc. Am. J. 56: 578-586.

Rommell, L.G. 1935. Ecological problems of the humus layer in the forest. Cornell Agric. Exp. Sta. Memoir No. 170. p. 28.

Vitousek, P.M., Gosz, J.R., Grier, C.C., Melillo, J.M., Reiners, W.A. and R.L. Todd. 1979. Nitrate losses from disturbed ecosystems. Science 204: 468-474.

White/Red Pine Stand Response to Partial Cutting and Site Preparation

Darwin Burgess
Suzanne Wetzel
Jeff Baldock

INTRODUCTION

Eastern white pine (*Pinus strobus* L.) often grows in association with red pine (*Pinus resinosa* Ait.) in fire-adapted stands (Ahlgren, 1976). Both pines may be well suited to partial cutting because of their high value and white pine's mid-tolerance to shade and susceptibility to white pine weevil (*Pissodes strobi* Peck) damage when grown in

---

Darwin Burgess is Research Scientist, Natural Resources Canada, Canadian Forest Service, Pacific Forestry Centre, 506 West Burnside Road, Victoria, BC, Canada V8Z 1M5.

Suzanne Wetzel is Research Scientist, Natural Resources Canada, Canadian Forest Service, Ontario Forestry Centre, Sault Ste. Marie, Ontario, Canada P6A 5M7.

Jeff Baldock is Research Scientist, CSIRO Land and Water, Glen Osmond, SA 5064, Australia.

The authors acknowledge all others involved in this study, in particular Fred Pinto, Conifer Program Leader, Ontario Ministry of Natural Resources, North Bay, Ontario and senior technicians Craig Robinson, Chalk River, Ontario and Gord Brand, Sault Ste. Marie, Ontario both with the Canadian Forest Service.

The authors gratefully acknowledge financial support from the Northern Ontario Development Agreement (NODA) and Ecosystem Processes and Forest Practices Networks, Canadian Forest Service.

[Haworth co-indexing entry note]: "White/Red Pine Stand Response to Partial Cutting and Site Preparation." Burgess, Darwin, Suzanne Wetzel, and Jeff Baldock. Co-published simultaneously in *Journal of Sustainable Forestry* (Food Products Press, an imprint of The Haworth Press, Inc.) Vol. 10, No. 3/4, 2000, pp. 221-227; and: *Frontiers of Forest Biology: Proceedings of the 1998 Joint Meeting of the North American Forest Biology Workshop and the Western Forest Genetics Association* (ed: Alan K. Mitchell et al.) Food Products Press, an imprint of The Haworth Press, Inc., 2000, pp. 221-227. Single or multiple copies of this article are available for a fee from The Haworth Document Delivery Service [1-800-342-9678, 9:00 a.m. - 5:00 p.m. (EST). E-mail address: getinfo@haworthpressinc.com].

open areas. Although these pines are major commercial species in eastern North America, relatively little is known about their physiology (Wetzel and Burgess, 1994) or their ability to regenerate and develop in managed stands (Hibbs and Bentley, 1987). Natural regeneration of pine after harvesting is often unacceptably low (Wray, 1985) and debate continues about how these stands can best be regenerated (Carleton et al., 1996). Well-defined management strategies are needed to encourage pine regeneration to help ensure the sustainability of this valuable resource (Stiell et al., 1994).

The general objective of this study was to assess commercial thinning and site preparation treatments in natural pine stands as methods for obtaining pine regeneration. In this paper, the objectives were focused to provide a brief description of the field study and some early (the first two years after treatment) results on natural regeneration establishment and residual stand conditions.

MATERIALS AND METHODS

A fixed-plot forest cruise and soil survey were completed prior to treatment in three 110-year-old natural pine stands within the Petawawa Research Forest, Chalk River, Ontario, Canada (45° 57′ N, 77° 34′ W). All trees greater than 9.0 cm DBH (diameter at breast height, 1.3m) were tallied within 4-m wide transects spaced 40 m apart. Species composition was 87.5% white and red pine. Other tree species present were white spruce (*Picea glauca* [Moench] Voss) and some major pine competitors, namely balsam fir (*Abies balsamea* [L.] Mill.), red maple (*Acer rubrum* L.), trembling aspen (*Populus tremuloides* Michx.) and white birch (*Betula papyrifera* Marsh.).

The soil survey used the same transects as the forest cruise and characterized soil features including texture, drainage and depth (Anon, 1985). The soils were acidic (pH 3.8) podzols and brunisols with textures ranging from coarse to fine sand. Soil depth varied but was > 1.2 m across 62% of the study area. The study site is probably an ecosite type ES 11.2 within the new Ontario Forest Classification (FEC) System (Chambers et al., 1997). The vegetation and soils data were compiled and provided input for experimental design and plot layout.

A randomized complete block, split-plot design with four replicates was used for the experiment. Blocks were stratified on the basis of white pine basal area. Main treatment plots were one ha and surrounded by buffer area 25 m wide. Main plot treatments were (1) thinning to one-crown width between trees, (2) two-crown widths between trees, and (3) control (uncut) spacing. Tree marking and skid trail layout was completed prior to harvesting. Tree volumes were derived from metric volume equations (Honer et al., 1983).

A forest technician supervised winter logging activities during January to March 1994. Logging crews using chain saws and Timberjack 230 and 350 cable skidders moved tree-length logs to roadside. The Forest Engineering Research Institute of Canada (FERIC, Pointe Claire, Quebec, Canada) selected and evaluated two of three logging crews for productivity and recorded damage to residual trees.

The site preparation treatments were applied randomly to sub-plots (0.25 ha each) plus buffers. The four site preparation treatments were (a) blade scarification, (b) brush control using herbicide, (c) blade scarification and brush control, and (d) untreated (control). Scarification was completed in August 1994 by an experienced operator using a John Deere 350c bulldozer with a six-way blade attachment, which mixed forest floor material and exposed mineral soil. Vision® (n-phosphonomethyl) was applied in mid-September 1995 using backpack sprayers at a rate of 2% active ingredient.

Half of each sub-plot was planted with at least 250 white pine container-grown seedlings at a 2 m × 2 m spacing (plus an additional 50 seedlings planted at a 1 m × 1 m spacing for later destructive sampling studies), the remaining sub-plot areas were left to regenerate naturally. Seedlings from a local seed source were sown in early February 1994 in Hillsons-type containers (Spencer-Lemaire Industries Ltd., Edmonton, Alberta) in a breeding hall. In June 1994 they were moved to a shade hall where seedlings continued to grow until late fall and remained for the winter. Seedlings were outplanted in late spring 1995 when mean seedling height was 13.9 cm and mean root collar diameter was 3.7 mm.

Scarification treatments were characterized by assessing the length of transects covered by slash or exposed mineral soil. Wind-thrown trees were recorded. A regeneration survey was completed in the fall of 1995 and 1996. A total of 288 (2 m × 2 m) quadrats were examined, six randomly selected from within each thinning × site preparation × replicate combination.

RESULTS AND DISCUSSION

Partial cutting in mature pine stands is possible without causing serious damage to residual trees. The two logging crews had similar productivity and both crews were effective at partial cutting in this forest type. Experienced well-trained loggers and equipment operators were essential and financial incentives to encourage careful harvesting were included in the logging contract. Furthermore, harvesting operations were closely supervised. Only 2% of the residual trees were damaged during harvesting. Half of those damaged had minor wounds, smaller than 100 cm^2.

Although the tree markers gave preference to leaving large, healthy pine as residual trees, a large volume of pine was harvested (Table 1). About 62% of the total pine volume was removed in the one-crown and 81.6% in the two-crown thinning treatment. The intensity of thinning was greater here than in most operational thinning treatments because local forest managers had indicated that the current guideline of thinning to a half-crown spacing inhibited vigorous growth of natural pine regeneration.

Wind throw losses were small, overall, with the highest number occurring in the heaviest thinning treatment. Tree losses to wind throw were 1.75, 1.50 and 2.50 trees ha^{-1}, respectively, in the control, one-crown and two-crown thinned areas, two years after treatment. The two-crown thinning treatment could leave stands highly susceptible to wind throw. Uncut stands can also be damaged by unpredictable, high windstorms, which happened in one control plot area during the second year after thinning.

Scarification exposed 81.6% of the surface mineral soil in the con-

TABLE 1. Average stand attributes for pine after thinning. (Numbers in parentheses are S.D.)

Treatment	Average DBH (cm)	Basal Area (m^2 ha^{-1})	Total Volume (m^3 ha^{-1})	Merch. Volume[1] (m^3 ha^{-1})	Sawlog Volume[2] (m^3 ha^{-1})	No. Trees (ha^{-1})
Control	38.2 (2.5)	35.5 (1.8)	433.0 (22.6)	417.4 (21.8)	399.3 (22.0)	312.2 (25.2)
One-Crown	44.1 (1.2)	13.1 (0.7)	164.3 (8.6)	158.8 (8.3)	154.2 (7.9)	86.0 (6.1)
Two-Crown	48.1 (1.8)	6.2 (0.3)	79.8 (3.6)	77.2 (3.5)	75.5 (3.4)	34.5 (1.9)

[1] ≥ 9.0 cm DBH; ≥ 7.0 cm top Dia.
[2] ≥ 17.0 cm DBH; ≥ 15.0 cm top Dia.

trol, 77.5% in the one-crown and 75.0% in the two-crown thinned areas. Stiell (1985) recommended that at least 20%, and preferably 40%, of the mineral soil surface be exposed when scarifying a site to promote pine regeneration. The larger amount of slash in the thinned areas caused the reductions in area of exposed mineral soil. A careful and experienced equipment operator was essential to ensure that only the surface of the mineral soil was exposed and to minimize damage to residual trees.

Scarification increased pine regeneration, but also increased seedling numbers for some major competitors (Table 2). Scarification is known to promote pine regeneration (Burgess, 1996), but competition from red maple and balsam fir is known generally to inhibit the regeneration and development of white pine (Corbett, 1994). The amount of white pine regeneration was greatest in non-thinned areas where the number of white pine trees bearing seed was largest, a relationship noted in earlier work (Kittredge and Ashton, 1990). Scarification was completed just before seed fall in a good seed crop year for white pine. Unfortunately, this was not the case for red pine, a species known to have fewer years of good seed production.

Natural pine stocking (% of regeneration plots with at least one seedling) also was increased by scarification from 62 to 87%. Brush control had little effect on regeneration numbers (Table 2) and no effect on pine stocking. The brush control treatment may have had a

TABLE 2. Natural tree regeneration (thousands of seedlings ha^{-1}) after two growing seasons post-treatment.

		Control (uncut)		One-Crown Thinning		Two-Crown Thinning	
Species	Brush control	Non-scarified	Scarified	Non-scarified	Scarified	Non-scarified	Scarified
White pine[1]	No herbicide	15.6 b	248.3 a	5.7 b	64.7 ab	3.8 b	11.8 b
	Herbicide	20.8 b	168.6 ab	4.5 b	57.3 ab	1.9 b	16.0 b
Red pine	No herbicide	0.0 b	0.0 b	0.2 b	1.5 a	0.0 b	0.0 b
	Herbicide	0.0 b	0.0 b	0.0 b	0.2 b	0.0 b	0.0 b
Other	No herbicide	27.7 b	259.8 a	20.8 b	35.8 b	13.2 b	26.1 b
Species[2]	Herbicide	29.4 b	83.1 b	25.0 b	35.9 b	8.9 b	18.8 b
Total	No herbicide	43.3 b	508.1 a	26.8 b	102.0 b	17.0 b	37.9 b
	Herbicide	50.2 b	251.8 ab	29.4 b	93.4 b	10.8 b	34.8 b

[1] Number of seedlings within a species group followed by the same letter do not differ (α = 0.05) using the Tukey HSD significance test.
[2] Predominantly balsam fir, red maple, trembling aspen and white birch.

greater influence, especially on white pine regeneration success, if completed one year earlier during a good white pine seed year. Effects of brush control may be greater in future years. The planted white pine seedlings had a high survival rate in all treatments and averaged 94.1% overall.

It is too early to determine the eventual effect of these treatments on pine regeneration success. Scarification promoted the regeneration of pine and some major competing species. The number of natural pine seedlings necessary to sustain these pine forests remains unknown under these conditions. We hope to be in a better position, in future, to advise forest managers on thinning and site preparation techniques useful for regenerating and obtaining good early growth of pine. Three major questions being addressed are:

1. Can pine ecosystems be managed using a shelterwood silvicultural system and regenerated with sufficient pine seedlings to sustain them,
2. What environmental conditions are influenced by thinning and site preparation practices, and
3. How do these factors influence natural and planted pine regeneration development and site processes?

CONCLUSIONS

- Partial cutting in mature pine stands can be completed without causing significant damage to residual trees. As the intensity of thinning increased, losses to wind throw were greater.
- With scarification improving pine regeneration (both number and stocking) most in the non-harvested treatment, forest managers may be able to scarify within pine stands prior to thinning to encourage advance regeneration of white pine. Red pine has infrequent seed years and is more shade intolerant; and is, therefore, more difficult to regenerate naturally.
- Research is continuing to characterize the conditions necessary for establishment and good early growth of pine regeneration in understory conditions. Growth analysis of response for both natural and underplanted seedlings to environmental conditions (being monitored using environmental probes connected to dataloggers) is underway to describe changes resulting from thinning and site preparation treatments.

REFERENCES

Ahlgren, C.E. 1976. Regeneration of red and white pine following logging in northeastern Minnesota. J. For. 74: 135-140.

Anon. 1985. Field Manual for Describing Soils, 3rd Ed., Ontario Institute of Pedology and University of Guelph, Guelph, Ontario p. 42.

Burgess, D. 1996. Forests of the Menominee–a commitment to sustainable forestry. For. Chron. 72: 268-275.

Carleton, T.J., P.F. Maycock, R. Arnup and A.M. Gordon. 1996. *In situ* regeneration of *Pinus strobus* and *P. resinosa* in the Great Lakes forest communities of Canada. J. Vegetation Sci. 7: 431-444.

Chambers, B.A., B.J. Naylor, J. Nieppola, B. Merchant and P. Uhlig. 1997. Field guide to forest ecosystems of central Ontario. Ont. Min. Natural Resources, Ontario. SCSS Field Guide FG-01. p. 200.

Corbett, C.M. 1994. White pine management and conservation in Algonquin Park. For. Chron. 70: 435-436.

Hibbs, D.E. and W.R. Bentley. 1987. White pine management: volume and value growth. North. J. Appl. For. 4: 197-201.

Honer, T.G., M.F. Ker, and I.S. Alemdag. 1983. Metric timber sales for the commercial tree species of central and eastern Canada. Environment Canada, Maritimes For. Res. Centre, Fredericton, N.B. Inf. Rep. M-X-140. 139 p.

Kittredge, D.B. and P.M.S. Ashton. 1990. Natural regeneration patterns in even-aged mixed stands in southern New England. North. J. Appl. For. 7: 163-168.

Stiell, W.M. 1985. Silviculture of eastern white pine. In. Proc. Entomological Soc. Ont., Supplement to Vol. 116: 95-107.

Stiell, W.M., C.F. Robinson and D. Burgess. 1994. 20-year growth of white pine following a commercial improvement cut in pine mixedwoods. For. Chron. 70: 385-394.

Wetzel, S. and D. Burgess. 1994. Current understanding of white pine physiology. For. Chron. 70: 420-426.

Wray, D.O. 1985. Securing the future of white pine in Ontario. Proc. Ent. Soc. of Ont. No. 116 (Suppl.): 109-110.

Total, Hydrolyzable and Condensed Tannin Concentrations of Leaf Litters of Some Common Hardwoods of Eastern Canada at Two Sites of Contrasting Productivity

Benoît Côté

INTRODUCTION

A larger diversity and quantity of polyphenols is normally produced in leaves on mor sites than on mull sites (Coulson et al., 1960). These phenolic compounds can strongly affect the rate of litter decomposition (Berg, 1986). The capacity of polyphenols to precipitate proteins makes them a key factor in controlling soil N-availability (Howard and Howard, 1993). The objectives of this study were therefore to quantify and characterize leaf litter phenolics of some hardwoods on two sites of contrasting productivity.

MATERIALS AND METHODS

Site Description

Two sites were selected in southern Quebec. The first site was located at Saint-Hippolyte in the Lower Laurentians, approximately

Benoît Côté is Associate Professor, Department of Natural Resource Sciences, Macdonald Campus of McGill University, 21,111 Lakeshore, Ste-Anne-de-Bellevue, QC, Canada, H9X 3V9.

[Haworth co-indexing entry note]: "Total, Hydrolyzable and Condensed Tannin Concentrations of Leaf Litters of Some Common Hardwoods of Eastern Canada at Two Sites of Contrasting Productivity." Côté, Benoît. Co-published simultaneously in *Journal of Sustainable Forestry* (Food Products Press, an imprint of The Haworth Press, Inc.) Vol. 10, No. 3/4, 2000, pp. 229-234; and: *Frontiers of Forest Biology: Proceedings of the 1998 Joint Meeting of the North American Forest Biology Workshop and the Western Forest Genetics Association* (ed: Alan K. Mitchell et al.) Food Products Press, an imprint of The Haworth Press, Inc., 2000, pp. 229-234. Single or multiple copies of this article are available for a fee from The Haworth Document Delivery Service [1-800-342-9678, 9:00 a.m. - 5:00 p.m. (EST). E-mail address: getinfo@haworthpressinc.com].

100 km north of Montreal. The forest was classified as a sugar maple (*Acer saccharum* Marsh.) and yellow birch (*Betula alleghaniensis* Britt.) forest. Soils were Orthic Ferro-Humic Podzols with a moder humus form and a mean pH of 4.2 (0-15 cm depth). The second site was located in the Morgan Arboretum of McGill University on the Island of Montreal. Two stands were used: a sugar maple–hickory (*Carya* spp.) forest on a Sombric Brunisol with a mull humus form and a mean pH of 6.0 (0-15 cm depth), and a beech (*Fagus grandifolia* Ehrh.)–red maple (*Acer rubrum* L.) forest on a Humo-Ferric Podzol with a moder humus form and a mean pH of 4.0 (0-15 cm depth). Based on climate and rates of growth, the Morgan Arboretum site was considered more productive than the Saint-Hippolyte site. Leaf N concentrations in sugar maple were similar but below critical levels at both sites.

Sampling

Five species (beech, largetooth aspen [*Populus grandidentata* Michx.], sugar maple, red maple and basswood [*Tilia americana* L.]) and eight species (beech, largetooth aspen, sugar maple, red maple, basswood, bitternut hickory (*Carya cordiformis* [Wang.] K. Kock.), red oak (*Quercus rubra* L.), white ash (*Fraxinus americana* L.) and black walnut (*Juglans nigra* L.) were sampled at Saint-Hippolyte and the Morgan Arboretum, respectively. Four 16 m^2 plots were selected and sampled along a linear transect at each site. Leaves were handpicked from litter at different times during October 1997 to coincide with the peak of leaf fall for each species. Sampled leaf litter was not exposed to rain. Leaves were dried at room temperature for 48 hours before being ground in a cyclotec mill to pass through a 40-mesh screen.

Chemical Analysis

Leaf litter extracts were prepared by boiling 0.05 g of ground litter in 25 ml of hot distilled water (total phenolics and hydrolyzable tannins) or hot 50% aqueous methanol (condensed tannins and BSA precipitation) for 1 hour. The choice of the extractant was based on reproducibility of results. Total phenolics were determined using Folin-Denis reagent (Martin and Martin, 1982) and results were expressed in tannic acid equivalents. Hydrolyzable tannins were determined by measuring gallic (Bate-Smith, 1977) and ellagic (Bate-Smith, 1972) acid concentrations in leaf extracts. Condensed tannins were assessed using the proanthocyanidin method of Porter (1989). Results were expressed in catechin equivalents. The capacity of leaf extracts to precipitate protein was assessed using bovine serum albumin (BSA) and a ratio of sample to protein of 8 to 1 (Martin and Martin, 1982).

Statistics

A completely randomized design was used to measure tree species effects on phenolic concentrations and amounts of protein precipitated at each site. A two-way factorial design was used to measure the effect of tree species, site and their interaction. Comparisons of species means on individual sites were done using Duncan's multiple-range test whereas least square means and contrast methods were used when interactions between site and tree species were significant. All statistics were calculated for a probability level of 5%.

RESULTS

Large interspecific differences were observed in phenolic concentrations but site differences were generally small (Table 1). At the poor site, ellagic acid, gallic acid and total phenolics were maximum in red maple whereas proanthocyanidins were maximum in beech. At the rich site, ellagic acid, gallic acid and total phenolics were maximum in black walnut followed closely by red maple whereas proanthocyanidins were maximum in beech and black walnut. Basswood was consistently among the lowest in hydrolyzable and condensed tannins at both sites. Beech had the least hydrolyzable tannins at the poor site, and white ash had low levels of hydrolyzable and condensed tannins at the rich site. A site-species interaction was observed for proanthocyanidins due to the much higher concentrations in beech at the poor site.

Concentrations of ellagic acid were consistently higher than those of gallic acid. Total hydrolyzable tannins (ellagic + gallic acid concentrations) were generally more abundant than condensed tannins. Exceptions to this trend were beech, which had higher levels of proanthocyanidins than hydrolyzable tannins, and sugar maple and large-tooth aspen, which had similar amounts of both types of tannins.

TABLE 1. Extractable content in ellagic and gallic acids (hydrolizable tannins), proanthocyanidins (condensed tannins), and total phenolics, and the protein-precipitating capacity of leaf litter extracts of some hardwood tree species on two sites of contrasting productivity. (Mean ± SE; N = 4)

Tree species/site	Ellagic acid	Gallic acid	Proantho-cyanidins	Total phenolics	Precipitated protein
	($mg\ g^{-1}$)		(mg catechin equiv. g^{-1})	(mg tannic acid equiv. g^{-1})	(%)
Poor site					
Basswood	48 ± 8c	8 ± 1d	28 ± 2c	26 ± 1c	75 ± 4c
Beech	43 ± 2c	7 ± 0.5d	447 ± 40a	25 ± 1c	73 ± 3c
Largetooth aspen	88 ± 4b	19 ± 1c	67 ± 16bc	14 ± 1d	85 ± 2b
Red maple	127 ± 10a	60 ± 3a	92 ± 8b	117 ± 3a	99 ± 0.2a
Sugar maple	84 ± 9b	33 ± 3b	128 ± 15b	80 ± 3b	99 ± 0.4a
Rich site					
Basswood	46 ± 3e	13 ± 0.5e	18 ± 5d	38 ± 1de	80 ± 4b
Beech	74 ± 10c	22 ± 3d	182 ± 10a	38 ± 20de	90 ± 0.2a
Largetooth aspen	54 ± 5cde	16 ± 1de	72 ± 8bc	33 ± 2e	77 ± 0.8bc
Red maple	93 ± 7b	70 ± 1a	93 ± 20b	112 ± 3b	93 ± 3a
Sugar maple	68 ± 8cd	42 ± 4c	70 ± 3bc	62 ± 5c	95 ± 1a
White ash	49 ± 3de	13 ± 7e	24 ± 4d	31 ± 0.4e	70 ± 4c
Bitternut hickory	65 ± 4cd	21 ± 3de	54 ± 10c	57 ± 1c	73 ± 2bc
Black walnut	114 ± 3a	62 ± 3b	164 ± 8a	183 ± 2a	91 ± 3a
Red oak	63 ± 9cde	20 ± 3de	46 ± 6d	42 ± 3d	70 ± 3c

Means within sites followed by the same letter are not significantly different ($p < 0.05$)

Protein precipitation was maximum in red and sugar maple at the poor site, and in red and sugar maple, black walnut and beech at the rich site. Low levels of protein precipitation were observed in basswood and beech at the poor site, and in white ash, bitternut hickory and red oak at the rich site. A site-species interaction was observed for protein precipitation due to the much higher levels observed in beech at the rich site. Significant correlations ($p < 0.05$) were observed between protein precipitation and gallic acid ($r = 0.77$), ellagic acid ($r = 0.77$) and total phenolics ($r = 0.62$) but not with proanthocyanidins.

DISCUSSION

Based on concentrations of the different phenolics and the protein precipitation capacity of the leaf litters, litter quality would rank the lowest in red and sugar maple at the poor site, and in red maple, sugar maple, beech and black walnut at the rich site. Using the same variables, litter quality would rank the highest in basswood at the poor site and in basswood, largetooth aspen, white ash, bitternut hickory and red oak at the rich site. This ranking is generally consistent with the known contribution of the different species to soil fertility (Burns and Honkala, 1990), with the exception of red and sugar maple leaf litters that are not generally considered of poor quality. However, a recent study on earthworm feeding activities revealed that some maple leaf litters were of low palatability to large earthworms (Côté and Fyles, 1994). Red oak leaf litter may be relatively low in tannin concentration and astringency but the toughness of its litter could be a deterrent to soil detritivores (Côté and Fyles, 1994). The high levels of condensed tannins observed in the leaf litter of beech suggest that it may be closer to conifers in phenolic composition than other hardwoods which would be consistent with its soil acidification potential.

In terms of site effect, only beech showed a large site response with proanthocyanidins being higher and protein precipitation being lower at the poor site. The weak response of leaf litter phenolics to site productivity is not consistent with the literature and could be associated with the experimental design that prioritized species effects over site effects. Although the lower protein precipitation and higher proanthocyanidin levels of beech leaf litter at the poor site may seem contradictory, it could be explained by the higher hydrolyzable tannins of beech at the rich site. Protein precipitation correlated better with hydrolyzable tannins than proanthocyanidins in our study.

REFERENCES

Bate-Smith E.C. 1972. Detection and determination of ellagitannins. Phytochem. 11:1153-1156.

Bate-Smith E.C. 1977. Astringent tannins of *Acer* species. Phytochem. 16:1421-1426.

Berg B. 1986. Nutrient release from litter and humus in coniferous forest soils–a mini review Scand. J. For. Res. 1:359-369.

Burns R.M. and B.H. Honkala 1990. Silvics of North America: 2. Hardwoods. Agric. Handbook 654. U.S. Dept. Agric., For. Serv., Washington DC.

Côté B. and J.W. Fyles 1994. Leaf litter disappearance of hardwood species of southern Québec: Interaction between litter quality and stand type. Ecoscience 1:322-328.

Coulson C.B., R.I. Davies and D.A. Lewis 1960. Polyphenols in plant, humus and soil. I. Polyphenols of leaves. Litter and superficial humus from mull and mor sites. J. Soil Sci. 11:20-29.

Howard P.J.A. and D.M. Howard 1993. Ammonification of complexes prepared from gelatin and aqueous extracts of leaves and freshly fallen litter of trees on different soil types. Soil Biol. Biochem. 25:1249-1256.

Martin J.S. and M.M. Martin 1982. Tannin assays in ecological studies; precipitation of ribulose-1,5 bisphosphate carboxylase/oxygenase by tannic acid, quebracho, and oak foliage extracts. J. Chem. Ecol. 9:285-294.

Porter L.J. 1989. Tannins. In: p. 407. P. Dey and L. Harborne (eds). Methods in plant biochemistry. Plant phenolics. Academic Press, New York, N.Y.

Comparative Growth of Balsam Fir and Black Spruce Advance Regeneration After Logging

René Doucet
Louis Blais

INTRODUCTION

Black spruce (*Picea mariana* Mill. [B.S. P.]) is the most abundant conifer tree species of the Canadian boreal forest. Advance regeneration is plentiful under mature stands (Doucet, 1988), and is often relied upon to regenerate harvested stands. It has long been assumed that balsam fir (*Abies balsamea* [L] Mill.) would replace spruce if advance growth was favored (Roy, 1940; Blais, 1983), especially on upland sites. Maintenance of spruce is considered essential to the conservation of biodiversity (Ministère des Ressources naturelles, 1996) and spruce is also preferred by industry. This study compares the development of both species in the same stands after logging, to determine whether replacement of spruce by fir does indeed take place.

René Doucet and Louis Blais are affiliated with Direction de la recherche forestière, ministère des Ressources naturelles, 2700, rue Einstein, Sainte-Foy (Québec) G1P 3W8.

[Haworth co-indexing entry note]: "Comparative Growth of Balsam Fir and Black Spruce Advance Regeneration After Logging." Doucet, René, and Louis Blais. Co-published simultaneously in *Journal of Sustainable Forestry* (Food Products Press, an imprint of The Haworth Press, Inc.) Vol. 10, No. 3/4, 2000, pp. 235-239; and: *Frontiers of Forest Biology: Proceedings of the 1998 Joint Meeting of the North American Forest Biology Workshop and the Western Forest Genetics Association* (ed: Alan K. Mitchell et al.) Food Products Press, an imprint of The Haworth Press, Inc., 2000, pp. 235-239. Single or multiple copies of this article are available for a fee from The Haworth Document Delivery Service [1-800-342-9678, 9:00 a.m. - 5:00 p.m. (EST). E-mail address: getinfo@haworthpressinc.com].

METHODS

Three areas of the balsam fir-white birch (*Betula papyrifera* Marsh) ecological domain (Thibault and Hotte, 1985) were sampled. The Lac-des-Battures (B) area, was located 200 km north of Quebec City. The Lac-Tourangeau (T) area was 30 km due north of the first. And the Lac-au-Loup-Marin (L) area was near Baie-Comeau, about 300 km to the northeast. In each area, sites within a few km of each other were chosen at random. They had supported spruce or spruce-fir stands harvested 26 to 30 years before. Only upland sites (14 in all) were retained, because the objective was to study productive sites where the problem of replacement of spruce by fir would likely be more prevalent. Two sites had supported pure black spruce stands and the other 12, spruce-fir. Feathermosses dominated the ground cover. Drainage varied from rapid to somewhat poorly drained. Depth of the mor humus was about 10 cm in every case.

Each stand was sampled with 15, 10 m^2 plots located 7 m apart. Trees were tallied by species and 2 cm dbh (diameter at 1.30 m) classes, and a class between 50 cm height and 1 cm dbh. The tallest fir and spruce > 1.30 m in each plot were felled to measure height increments, identified by bud scars on the bole, and ring counts when necessary. Ages were determined from disks at ground level and dbh increments at 1.30 m. Growth increments by species within stands were compared with repeated measures analyses based on the mixed model with special parametric structure on the covariance matrices (Littell et al., 1996). Species were also compared by paired t-tests for 6-10, and 21-25 years average height increments.

RESULTS

All stands were well stocked, with 6,500 to 14,000 stems/ha > 1 cm dbh. Basal areas were between 13 and 26 m^2/ha. From 4,000 to 34,000 stems taller than 50 cm height but less than 1 cm dbh were also present. Spruce accounted for 17 to 65 percent of total basal area and was the main species in only five of the fourteen stands. Most stems felled were advance regeneration less than 1 m high at the time of harvest, but some had established after logging.

Fir usually reached maximum height increments 6 to 10 years after harvest (Table 1). Response by spruce was slower but, in most cases, increments 21-25 years after logging were significantly larger than those of fir (Table 1). Figure 1 shows stand L1 as an example of height growth for the whole period after logging. Significantly larger fir increments occurred in 1976 and 1978 and for spruce from 1984 onward. The same trend was present in most other stands. Dbh increments (not shown) behaved the same as heights.

DISCUSSION

Spruce growth is typical of what has been observed elsewhere (Boily and Doucet, 1993), but such an early decline of fir has not been reported before. Growth rates in some of the B and T stands declined, especially for fir, around 1975, when a spruce budworm outbreak occurred in the area (Morin and Laprise, 1990). But both species resumed normal growth rates well before the second decline of fir. The L area did not show any evidence of spruce budworm, but spruce

TABLE 1. Characteristics of the stems used for stem analysis.

Stand	Years after logging	Average dbh (mm)		Average height (cm)		Periodic annual increment (mm)					
						6-10 years			21-25 years		
		Spruce	Fir	Spruce	Fir	Spruce	Fir	p-value*	Spruce	Fir	p-value*
B1	27	79	84	583	559	177	212	0.5391	264	202	0.0003
B2	27	82	78	621	603	192	222	0.2821	252	172	0.0014
B3	26	65	72	511	538	167	256	0.0016	255	153	0.0001
B4	27	87	80	634	585	171	143	0.2525	250	186	0.0001
B5	27	87	93	622	656	192	195	0.9965	286	227	0.0107
T1	29	67	77	492	545	114	182	0.2033	213	166	0.1018
T2	27	59	48	508	377	174	252	0.0561	208	93	0.0001
T3	27	75	73	566	576	220	294	0.0238	222	162	0.0010
T4	28	54	76	400	584	102	183	0.1539	179	206	0.8894
T5	30	91	84	605	602	134	213	0.3165	267	215	0.5030
T6	30	86	88	517	560	86	126	0.5051	202	181	0.6478
L1	30	89	79	617	517	174	324	0.0022	297	150	0.0001
L2	26	86	56	607	424	187	147	0.1959	280	142	0.0001
L3	26	112	84	764	535	246	156	0.0011	318	133	0.0001

* Paired t-tests for differences between spruce and fir increments.

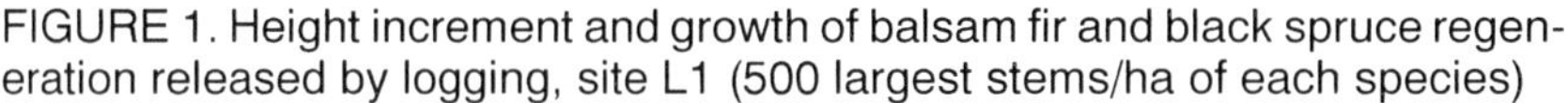
FIGURE 1. Height increment and growth of balsam fir and black spruce regeneration released by logging, site L1 (500 largest stems/ha of each species)

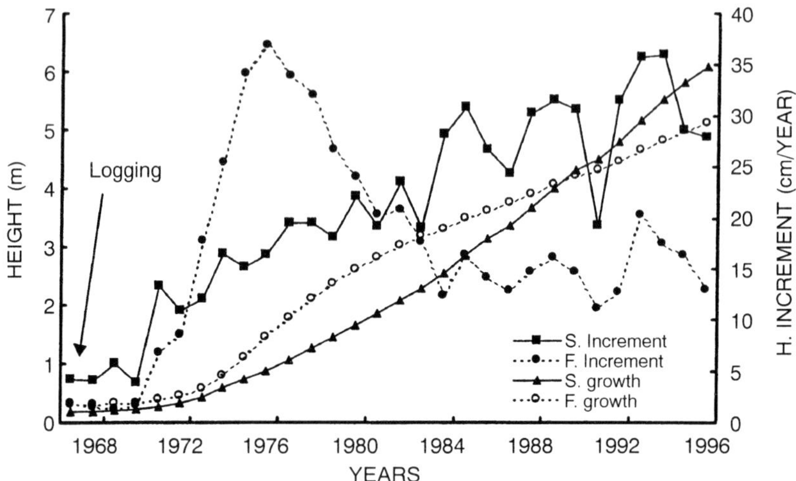

nevertheless grew faster than fir in later years. Doucet and Boily (1995) found that growth of fir declined in a younger stand that had become established after the spruce budworm outbreak, so that this does not explain the decline. Brumelis and Carleton (1988), using C/N ratio as a general indicator of site quality, found smaller height growth of fir than spruce on sites with high C/N ratios, but these were lowlands with a deep organic layer. As far as we know, this is the first study comparing fir and spruce growth trends in the same stands, on upland sites though explanations for the early decline of fir cannot be given from the present exploratory study, this trend may not be exceptional. It was found in three different areas, in the southern part of the boreal forest, on productive upland sites, so that it might be more likely in the heart of the boreal forest.

Black spruce and balsam fir are often present together as advance regeneration, both before (Doucet, 1988) and after (Pothier et al., 1995) harvesting. Thus, growth rates will determine which species will ultimately dominate. It seems that black spruce can reach the upper canopy when it is present as advance regeneration at the time of harvesting, even though its growth rate may be less than that of fir in the first few years. Moreover, management activities such as spacing and thinning could be used to increase the proportion of spruce in the stands. In the long term, this would create more structurally diverse stands, possibly increasing overall biodiversity.

REFERENCES

Blais, J. R. 1983. Trends in the frequency, extent, and severity of spruce budworm outbreaks in Eastern Canada. Can. J. For. Res. 13: 539-547.

Boily, J. and R. Doucet. 1993. Croissance juvénile de marcottes d'épinette noire en régénération après récolte du couvert dominant. Can. J. For. Res. 23: 1396-1401.

Brumelis, G. and T. J. Carleton. 1988. The vegetation of postlogged black spruce lowlands in central Canada. I. Trees and tall shrubs. Can. J. For. Res. 18: 1470-1478.

Doucet, R. 1988. La régénération préétablie dans les peuplements forestiers naturels au Québec. For. Chron. 64: 116-120.

Doucet, R. and J. Boily, 1995. Croissance en hauteur de la régénération d'épinette noire et de sapin baumier après la coupe. Ministère des Ressources naturelles, Direction de la recherche forestière, Note de recherche forestière n° 68.

Littell, R. C., G. A. Milliken, W. W. Stroup and R. D. Wolfinger. 1996. SAS® system for mixed models. SAS Institute Inc., Carey, NC.

Ministère des Ressources naturelles. 1996. Biodiversité du milieu forestier. Bilan et engagements du ministère des ressources naturelles. Ministère des Ressources naturelles, Quebec City.

Morin, H. and D. Laprise. 1990. Histoire récente des épidémies de la tordeuse des bourgeons de l'épinette au nord du lac Saint-Jean (Québec): une analyse dendrochronologique. Can. J. For. Res. 20: 1-8.

Pothier, D. 1996. Évolution de la régénération après la coupe de peuplements récoltés selon différents procédés de récolte. For. Chron. 72: 519-527.

Roy, H. 1940. Spruce regeneration: Quebec. For. Chron. 16: 10-20.

Thibault, M. and D. Hotte. 1985. Les régions écologiques du Québec méridional. Ministère de l'Énergie et des Ressources, Quebec City.

Influence of Forest Harvesting Intensity on Nutrient Leaching Through Soil in Southwestern British Columbia

M. C. Feller
R. Lehmann
P. Olanski

INTRODUCTION

Forest productivity in British Columbia depends partly on the availability of nutrients, particularly nitrogen, in the soil. Timber harvesting has been shown to affect ecosystem nutrient status and generally increase losses of nutrients through the soil via leaching (e.g., Perry et al., 1989). Theoretically, partial cutting, which leaves some trees standing and able to utilize soil nutrients following harvesting of adjacent trees, should result in decreased leaching losses of nutrients compared to clearcutting. However, neither Stark (1979) nor McLurkin et al. (1987) found consistent relationships between soil solution

---

M. C. Feller, R. Lehmann and P. Olanski are Faculty of Forestry, University of British Columbia, Vancouver, B.C., Canada, V6T 1Z4.

P. Olanski is also affiliated with the B.C. Ministry of Forests, Forest Sciences Branch, Bag 5000, 3726 Alfred Avenue, Smithers, B.C., Canada, V0J 2N0.

Many forestry students assisted with field and laboratory work.

The project has been funded by the Canadian Forestry Service through the Forest Regional Development Agreement program, Forest Renewal B.C., MacMillan Bloedel Ltd., and the National Science and Engineering Research Council of Canada.

[Haworth co-indexing entry note]: "Influence of Forest Harvesting Intensity on Nutrient Leaching Through Soil in Southwestern British Columbia." Feller, M. C., R. Lehmann, and P. Olanski. Co-published simultaneously in *Journal of Sustainable Forestry* (Food Products Press, an imprint of The Haworth Press, Inc.) Vol. 10, No. 3/4, 2000, pp. 241-247; and: *Frontiers of Forest Biology: Proceedings of the 1998 Joint Meeting of the North American Forest Biology Workshop and the Western Forest Genetics Association* (ed: Alan K. Mitchell et al.) Food Products Press, an imprint of The Haworth Press, Inc., 2000, pp. 241-247. Single or multiple copies of this article are available for a fee from The Haworth Document Delivery Service [1-800-342-9678, 9:00 a.m. - 5:00 p.m. (EST). E-mail address: getinfo@haworthpressinc.com].

nutrient levels and cutting intensity, although Stark reported that clearcutting generally had a greater impact on nutrient levels than did partial (shelterwood) cutting.

As a result of low nutrient availability, nutrient losses which accompany timber harvesting operations, and the unknown influence of timber harvest intensity on nutrient loss from the soil, the nutrient sustainability of different intensities of timber harvesting in the montane forests of Vancouver Island in southwestern British Columbia has been uncertain. The objective of this study was to assess the influence of forest harvesting intensity on soil nutrient leaching.

The Study Area

The study occurred in the Montane Alternative Silvicultural Systems (MASS) study site, located in British Columbia's Montane Moist Maritime Coastal Western Hemlock biogeoclimatic variant (Klinka et al., 1991) around 800 m elevation in central eastern Vancouver Island near the town of Campbell River. The old growth forests present were dominated by *Abies amabilis* and *Tsuga heterophylla* with some *Thuja plicata* and *Chamaecyparis nootkatensis*. The study was restricted to one ecosystem type–the mesic *Tsuga heterophylla, Abies amabilis-Rhytidiopsis robusta* site association (Banner et al., 1990) the dominant ecosystem within the study area. This ecosystem had 3-57 cm deep forest floors (average depth = 14 cm) over coarse textured mineral soils. Understory vegetation was dominated by regenerating tree seedlings and the shrub *Vaccinium* spp. over a dense moss layer.

The MASS study involved 4 different forest harvesting treatments as well as undisturbed old-growth forests. The harvesting treatments were–clearcut (a 69 ha cut), patch cut (1.5-2.0 ha cuts), green tree retention (clearcutting with reserves, leaving approximately 25 trees/ha), and shelterwood (leaving approximately 30% of the basal area after cutting). Further details about the site and harvesting treatments are given in Arnott and Beese (1997).

METHODS

To assess the influence of harvesting intensity on soil nutrient leaching, it was not necessary to utilize all the MASS treatments. The present study was established in August 1993, in undisturbed forests

(24 sampling sites utilizing the old growth forest and patchcut uncut forest), 50% basal area removal areas (12 sampling sites utilizing 2 clearcut and 2 patchcut forest-cutover edges), 71% basal area removal areas (14 sampling sites utilizing the shelterwood treatment), and 100% basal area removal areas (24 sampling sites utilizing the clearcut and patchcut cutover areas), with a total of 74 sampling sites. Sampling sites were established at 20-40 m intervals along randomly located straight line transects. The number of transects per treatment (sampling sites per transect) varied from 2(7) for the shelterwood to 4(3) for the forest cutover edges, to 12(2) for each of the old growth and cutover areas. The transects along the forest-clearcut edges could not be randomly located, but the locations of sampling sites along these transects were.

Each sampling site contained a forest floor leachate and a mineral soil leachate collector. The forest floor leachate collector was a tensionless lysimeter consisting of a plastic Buchner funnel (13 cm diameter) draining into a 4 L polyethylene bottle from which forest floor leachate was obtained by means of a tygon tube running from the bottom of the bottle to above the forest floor surface. Approximately 1-2 m distant from a forest floor leachate collector, on a line parallel to the slope, was a mineral soil leachate collector. The collector consisted of a Soil Moisture Inc. soil moisture tube inserted into the mineral soil such that the suction cup was located 30 cm below the forest floor–mineral soil interface, which is considered to be close to the effective rooting depth. In the field, each tube was evacuated to a pressure of 0.5 bar.

All leachate collectors were emptied once every 4 weeks or whenever possible during dry or snowpack periods. On these occasions, samples were taken from each collector and transported to the laboratory where they were analyzed for K, Mg, and Ca (by atomic absorption spectrophotometry using a Varian Spectra AA 10 instrument), and NH_4^+, NO_3^-, SO_4^{-2}, and PO_4^{-3} (by colorimetric methods on a Technicon TRAACS 800 continuous flow analyzer). Organic N and P were also measured on the TRAACS instrument after persulphate digestion.

Nutrient concentrations in the leachates were multiplied by water fluxes to estimate nutrient fluxes. This was done for each sampling period, then the results were summed to obtain annual water year (1 October-30 September) nutrient fluxes. Soil water fluxes were estimated from on-site measurements of throughfall in the old growth forests and precipitation, air temperature and solar radiation in a nearby clearcut. These measurements were used to estimate weekly evapotranspiration using an energy balance approach described by McNaughton and Black (1973). Evapotranspiration was then subtracted from precipitation during each sampling interval to give the soil water flux. Evapotranspiration from the old growth forests was estimated as 21% of precipitation over the 3 year sampling period. Soil water fluxes for the cutting treatments were estimated from calculations of evapotranspiration assuming throughfall increased in direct proportion to the basal area removal (e.g., Butcher, 1977). It was further assumed that the forest floor water flux was identical to that in the mineral soil. Although this is not strictly correct, it does not influence comparisons of forest floor or mineral soil fluxes among treatments.

Collectors were installed in the forest, 50% cut, and 100% cut treatments close to the end of harvesting operations, in July and August 1993. Collectors were installed in the 71% cut treatment during spring 1994. Data collection for the study ceased in October 1996 and thus data was available for 3 water years (1 October 1993-30 September 1996). Soil solution chemistry in the 71% cut treatment during the October 1993 to April 1994 period was assumed to be represented by that of May 1994. Data from subsequent years suggested this assumption was valid.

The influence of harvesting treatment on nutrient fluxes was determined using one way Analyses of Variance comparing each chemical flux for each treatment separately for the sum of the three post-treatment water years. All analyses were conducted using SYSTAT software on a computer.

RESULTS AND DISCUSSION

Forest floor nutrient fluxes were greater than mineral soil fluxes for all chemicals except NO_3^- (Table 1). The general patterns of fluxes are similar to those observed in other coniferous forests in British Columbia (Feller, 1977) and the U. S. Pacific Northwest (Edmonds et al., 1991; Sollins et al., 1980). In the undisturbed forests, the 3 year fluxes decreased in the order: K > Ca > Mg > P > N for forest floor leachate and Ca > K > Mg > N > P for mineral soil leachate (Table 1). These trends are identical to those found by Sollins et al. (1980) for

TABLE 1. Cumulative three year (1993/94-1995/96) average forest floor and mineral soil water (mm) and nutrient leachate (kg/ha/3 yr) fluxes in each of the forest cover situations studied in the MASS study area

Forest Cover	Water	K	Mg	Ca	NO_3^--N	NH_4^+-N	Organic-N	PO_4^{-3}-P	Organic-P	SO_4^{-2}-S
Forest floor leachate										
Forest	4105	129.1 (13.7)[b]	45.7 (6.6)	107.0 (15.8)	5.3 (2.8)	10.0 (4.0)	7.8 (0.7)	13.6 (1.9)	6.5 (0.6)	67.1 (3.0)[b]
50% cut	4412	151.6 (26.2)[ab]	42.5 (8.8)	80.6 (19.4)	6.7 (5.2)	17.4 (6.0)	9.4 (1.8)	18.0 (3.7)	7.5 (0.7)	76.6 (4.6)[ab]
71% cut	4732	167.4 (20.7)[ab]	42.0 (7.2)	109.5 (23.0)	6.7 (1.0)	14.2 (4.8)	8.5 (1.1)	21.2 (4.7)	11.3 (4.3)	78.9 (4.6)[ab]
100% cute	5072	216.6 (21.0)[a]	50.9 (6.9)	103.5 (15.0)	8.8 (5.0)	20.5 (5.3)	7.9 (0.9)	19.0 (2.5)	9.8 (1.0)	84.9 (5.0)[a]
Mineral soil leachate										
Forest	4105	25.5 (6.0)[b]	19.4 (2.3)	27.6 (8.2)	1.1 (0.4)[b]	0.2 (0.1)	2.8 (0.5)	0.5 (0.1)	0.9 (0.1)	15.1 (2.7)
50% cut	4412	55.1 (17.1)[ab]	22.1 (3.6)	28.9 (9.8)	3.4 (1.4)[ab]	1.9 (1.4)	3.9 (1.2)	0.4 (0.2)	1.0 (0.1)	16.7 (4.3)
71% cut	4732	57.0 (7.6)[a]	35.6 (8.8)	48.8 (17.4)	4.3 (2.5)[ab]	0.8 (0.5)	1.8 (0.6)	0.5 (0.2)	1.3 (0.2)	22.7 (3.5)
100% cut	5072	40.9 (6.7)[ab]	34.6 (13.4)	49.7 (23.5)	7.9 (2.7)[a]	0.4 (0.1)	4.1 (0.8)	0.9 (0.3)	1.8 (0.7)	24.2 (6.0)

Standard errors are in parentheses
Numbers of observations (n) are–forest floor, n = 24, 12, 14, and 24 for the forest, 50% cut, 71% cut, and 100% cut areas, respectively.
Mineral soil, n = 18, 11, 13, and 19 for the forest, 50% cut, 71% cut, and 100% cut areas, respectively.
n is less for mineral soil than forest floor due to soil moisture tube malfunctions.
For a given type of leachate and a given chemical, different superscripts indicate significantly different ($P < 0.05$) fluxes.
An absence of superscripts indicates no sitgnificant differences among fluxes.

old growth *Pseudotsuga menziesii* forests in western Oregon, except for P and N being reversed in the Oregon forest floor leachate. Other studies in the region have reported concentrations only.

Although there was a general tendency for nutrient fluxes to increase with the intensity of harvesting, particularly in the case of K, Mg, NO_3^-, and SO_4^{-2}, this tendency was not evident for Ca and organic-N in forest floor leachate or NH_4^+ in mineral soil leachate (Table 1). The few statistically significant differences between treatments were restricted to K and SO_4^{-2} in forest floor leachate and K and NO_3^- in mineral soil leachate. Differences in fluxes between the forest and harvested areas were all < 25 kg/ha/3 yrs for mineral soil leachate and < 30 kg/ha/3 yrs for forest floor leachate, except for forest floor leachate K, for which fluxes differed by up to 88 kg/ha/3 yrs.

Relatively large increases in K and NO_3^- leaching, compared to those of other ions, have been found following clearcutting of other coniferous forests (e.g., Dahlgren and Driscoll, 1994). These have generally been attributed to the high leachability of K from decomposing plant material and increased nitrification following disturbance. Increased NO_3^- leaching in the present study, with the greatest increases on the clearcut areas and no significant differences between the forest and partially cut areas, is consistent with the results of Prescott's (1997) study of N mineralization in the same study area.

Although sampling only occurred for the first 3 post-treatment years, additional harvesting-induced leaching losses in subsequent years are likely to have been relatively small. Other studies have suggested that forest harvesting increases soil solution nutrient concentrations for only 4-5 years (e.g., Dahlgren and Driscoll, 1994) and streamwater nutrient fluxes in coastal British Columbia for only 2-3 years (Feller and Kimmins, 1984). Even if harvesting-induced leaching losses were doubled, losses of the critical nutrient–nitrogen–would still be < 15 kg/ha for all treatments. Thus, leaching losses of nutrients from the mineral soil were relatively low, regardless of harvesting treatment, a result that is consistent with other studies (e.g., Sollins and McCorison, 1981; Stark, 1979).

CONCLUSIONS

The partial cutting treatments at the study site can reduce nutrient leaching losses compared to clearcutting. However, none of the forest harvesting methods used, clearcutting included, is likely to cause excessive nutrient leaching. Harvesting-induced nutrient leaching is unlikely to influence the sustainability of forest management in the study area.

REFERENCES

Arnott, J. T. and W. J. Beese. 1997. Alternatives to clearcutting in B.C. coastal montane forests. Forestry Chron. 73: 670-678.

Banner, A., R. N. Green, A. Inselberg, K. Klinka, D. S. McLennan, D. V. Meidinger, F. C. Nuszdorfer, and J. Pojar. 1990. Site classification for coastal British Columbia. B. C. Ministry of Forests, Victoria, B. C.

Butcher, T.B. 1977. Impact of moisture relationships on the management of *Pinus pinaster* Ait. plantations in Western Australia. Forest Ecol. Manage. 1: 97-107.

Dahlgren, R. A., and C. T. Driscoll. 1994. The effects of whole-tree clear-cutting on soil processes at the Hubbard Brook Experimental Forest, New Hampshire, USA. Plant Soil 158: 239-262.

Edmonds, R. L., T. B. Thomas, and J. J. Rhodes. 1991. Canopy and soil modification of precipitation chemistry in a temperate rain forest. Soil Sci. Soc. Amer. J. 55: 1685-1693.

Feller, M. C. 1977. Nutrient movement through western hemlock-western red cedar ecosystems in southwestern British Columbia. Ecology 58: 1269-1293.

Feller, M. C., and J. P. Kimmins. 1984. Effects of clearcutting and slash burning on streamwater chemistry and watershed nutrient budgets in southwestern British Columbia. Water Resour. Res. 20: 29-40.

Klinka, K., J. Pojar, and D. V. Meidinger. 1991. Revision of biogeoclimatic units of coastal British Columbia. Northwest Sci. 65: 32-47.

McLurkin, D. C., P. D. Duffy, and N. S. Nelson. 1987. Changes in forest floor water and quality following thinning and clearcutting of 20-year old pine. J. Environ. Qual. 16: 237-241.

McNaughton, K. G. and T. A. Black. 1973. A study of evapotranspiration from a Douglas-fir forest using the energy balance approach. Water Resour. Res. 9: 1579-1590.

Perry, D. A., R. Meurisse, B. Thomas, R. Miller, J. Boyle, J. Means, C. R. Perry, and R. F. Powers. 1989. Maintaining the long-term productivity of Pacific Northwest forest ecosystems. Timber Press, Portland, Oregon.

Prescott, C. E. 1997. Effects of clearcutting and alternative silvicultural systems on rates of decomposition and nitrogen mineralization in a coastal montane coniferous forest. Forest Ecol. Manage. 95: 253-260.

Sollins, P., and F. M. McCorison. 1981. Nitrogen and carbon solution chemistry of an old growth coniferous forest watershed before and after cutting. Water Resour. Res. 17: 1409-1418.

Sollins, P., C. C. Grier, F. M. McCorison, K. Cromack, Jr., and R. Fogel. 1980. The internal element cycles of an old-growth Douglas-fir ecosystem in western Oregon. Ecol. Monogr. 50: 261-285.

Stark, N. M. 1979. Nutrient losses from timber harvesting in a larch/Douglas-fir forest. USDA Forest Service Res. Pap. INT-231. Intermountain Forest and Range Experiment Station, Odgen, Utah.

Intensive Management of Ponderosa Pine Plantations: Sustainable Productivity for the 21st Century

Robert F. Powers
Phillip E. Reynolds

INTRODUCTION

Natural forest reserves and type conversions are shifting the burden of growing wood at superior levels to plantations. This causes two questions to arise: can plantations produce more, and can higher productivity be sustained? Managers often see productivity as a compromise between a site potential conditioned by climate and soil, and a current yield constrained by genetic potential, tree stocking and age, weed competition, forest pests, and economics. But site potential also is malleable. It can be lowered or raised by practices that alter the soil resource (Nambiar, 1996; Powers et al., 1997). The "Garden of Eden" experiment was established in 1986 to see how both current and potential productivity of planted ponderosa pine (*Pinus ponderosa* var.

Robert F. Powers is Principal Research Silviculturist and Science Team Leader, USDA Forest Service Pacific Southwest Research Station, Redding, CA 96001 USA.

Phillip E. Reynolds is Research Scientist, Natural Resources, Canada, Canadian Forest Service, Sault Ste. Marie, Ontario P6A 5M7, Canada.

[Haworth co-indexing entry note]: "Intensive Management of Ponderosa Pine Plantations: Sustainable Productivity for the 21st Century." Powers, Robert F., and Phillip E. Reynolds. Co-published simultaneously in *Journal of Sustainable Forestry* (Food Products Press, an imprint of The Haworth Press, Inc.) Vol. 10, No. 3/4, 2000, pp. 249-255; and: *Frontiers of Forest Biology: Proceedings of the 1998 Joint Meeting of the North American Forest Biology Workshop and the Western Forest Genetics Association* (ed: Alan K. Mitchell et al.) Food Products Press, an imprint of The Haworth Press, Inc., 2000, pp. 249-255. Single or multiple copies of this article are available for a fee from The Haworth Document Delivery Service [1-800-342-9678, 9:00 a.m. - 5:00 p.m. (EST). E-mail address: getinfo@haworthpressinc.com].

ponderosa) can be altered silviculturally in a Mediterranean climate. First-decade findings are summarized here.

METHODS

Insecticides, herbicides, and fertilizers were applied repeatedly to eight plantations across a broad gradient of site qualities in California (Table 1). Each plantation contained 3 replications of 8 treatments assigned randomly to 24-0.04-ha plots. Treatments were full factorial combinations of acephate or dimethoate and glyphosate or hexazinone applied annually at manufacturers' recommended rates, and a 9-nutrient mix (applied biennially at an exponential rate) delivering from 36 kg boron ha^{-1} to 1,074 kg nitrogen (N) ha^{-1} over 6 years. Trees of superior families were planted at 2.4 m spacing. Site and treatment details are those described by Powers and Ferrell (1997). Dimensions and foliar chemistry of the innermost 20 trees per plot were measured in staggered years, and soil chemistry was measured between years 8 and 10. Soil fauna (Moldenke, 1992), plant water relations and net assimilation (Reynolds and Powers, 2000) were assessed less regularly. Data were examined by analysis of variance ($\alpha = 0.05$) with mean separations by Tukey's test or Fisher's LSD. Yield projections were made to age 50 using the growth simulator SYSTUM-1 (Ritchie and Powers, 1993).

RESULTS

Through the First 10 Years

Neither repeated fertilization nor insecticide treatment had an overall effect on volume growth. Despite the broad range of physiological

TABLE 1. Physical characteristics of eight Garden of Eden plantations in California.

Location	Site index	Yr. planted	Elevation	Annual ppt.	Geology	Soil texture
Chester	20	1987	1465	890	Volcanic ash	Cindery
Elkhorn	17	1988	1490	1015	Schist	Loamy skeletal
Erie Point	24	1987	1370	1700	Schist	Loamy skeletal
Feather	30	1988	1220	1780	Basalt	Loam
Jaws	23	1988	1005	1035	Schist	Loam
Pondosa	20	1988	1175	760	Volcanic ash	Sandy loam
Tickey	28	1987	1280	1525	Basalt	Clay loam
Whitmore	23	1986	730	1140	Basalt	Clayey

stress conditions created by site quality and treatment, insect activity was negligible and will not, therefore, insecticide effects be discussed further. Herbicides, with or without fertilization, had a major impact, however, usually tripling volume growth across all sites. Leaf water potentials were raised significantly by weed control–particularly on drier sites such as Whitmore, where summer leaf water potential (ψ) averaged 0.1 MPa greater in weed-free plots (Reynolds and Powers, 2000). However, effects dissipated as trees approached crown closure. Weed control also improved tree nutrition as early as the first year (Powers and Ferrell, 1996), with effects lasting through year 9 on average and poorer sites (Table 2). The effect was particularly strong at Whitmore, the lowest and droughtiest plantation. Without weed control on drier sites, foliar concentrations of N (sometimes phosphorus [P]) declined steadily to deficiency levels by year 5 (Powers and Ferrell, 1996).

Fertilization improved nutrient uptake by pines the first year but weed growth also increased, generally masking early nutritional benefits to pines by year 5 (Powers and Ferrell, 1996). But with the massive fertilizer increment in year 6, both uptake (particularly of N, Table 2) and growth (Figure 1) surged again in trees on average and better sites. Combined with herbicides, fertilization improved tree

TABLE 2. Nutrient concentration in current-year foliage of 9-year-old ponderosa pine relative to site index and silvicultural treatment. Annual precipitation at Elkhorn, Whitmore, and Feather averages 1015, 1140, and 1780 mm, respectively.

Location Site Index	Treatment	Foliar concentration of–									
		N	P	K	Ca	Mg	S	Cu	Fe	Zn	B
		g kg^{-1}					mg kg^{-1}				
Elkhorn (17 m)	Control	8.8a*	1.11a	8.9a	1.15a	0.99a	552a	3.28a	38.5a	26.2	28.3a
	Fertilizer	12.4b	1.24a	9.5ab	1.25ab	1.02a	628ab	3.12a	36.3a	26.3	59.3b
	Herbicide	9.8a	1.37b	11.4c	1.30ab	1.17a	611ab	3.38a	38.0a	28.3	35.3a
	Herb. + Fert.	12.6b	1.44b	10.4bc	1.47b	1.11a	709b	3.55a	36.7a	29.8	72.3b
Whitmore (23 m)	Control	8.5a	0.78a	4.6a	1.02a	0.80a	524a	2.47a	36.3a	176.3ab	20.3a
	Fertilizer	10.7b	0.89ab	4.9a	1.04ab	0.75a	605b	3.17b	34.0a	197.0b	19.7a
	Herbicide	11.8c	0.83a	6.8b	1.06b	0.97b	710c	3.45b	35.2a	156.2a	25.3a
	Herb. + Fert.	12.5d	0.98b	6.8b	1.32c	0.94b	752c	3.38b	57.2a	204.0b	23.5a
Feather (30 m)	Control	11.0a	1.11a	7.7a	1.48a	1.25a	720a	3.95a	42.3a	175.7	23.5a
	Fertilizer	12.8b	1.15a	8.9a	1.81b	1.15b	781b	3.83a	32.8a	137.8	24.2a
	Herbicide	10.7a	1.14a	7.8a	1.65ab	1.34a	756ab	3.88a	35.0a	185.5	23.2a
	Herb. + Fert.	12.5b	1.09a	8.8a	1.69ab	1.15b	791b	4.08a	34.3a	131.2	22.2a

*Plantation means followed by differing letters within a column differ significantly at $\alpha = 0.05$.

FIGURE 1. Cumulative height growth for control (C), herbicide, (H), and herbicide + fertilizer (HF) treatments through the first 10 years at three Garden of Eden plantations of site indices 17, 23, and 30 m at 50 years (adapted from Powers and Ferrell, 1996).

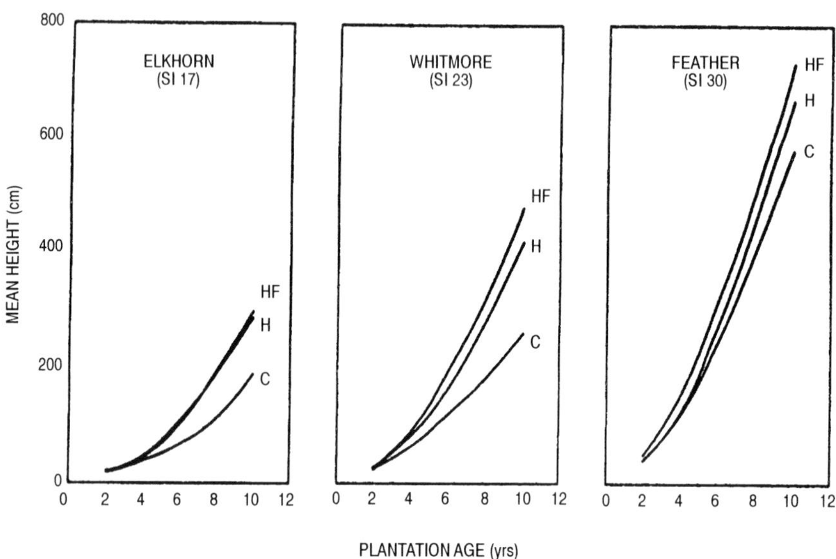

growth and uptake of most nutrients on all sites. Although trees receiving combined treatments grew faster, their leaf ψ was not affected until late summer, when ψ was 0.2 MPa lower at Whitmore (which is droughty) and 0.2 Mpa higher at Feather (which is more mesic).

Soil organic carbon and N were not affected by repeated vegetation control, contrasting with Busse et al.'s (1996) longer-term findings in eastern Oregon. Soil arthropods in 5- and 6-year-old plantations were fewer and less diverse than in adjacent natural stands (Moldenke, 1992). Other than a lack of litter-dwelling arthropods (a continuous litter layer had not yet formed by year 6), functional guilds (arthropods grouped by their roles in ecosystem processes) were not eliminated by treatment.

Projections to Age 50

Stand growth projected to 50 years suggests a sizable gain from weed control that persists on poorer sites and diminishes on better sites. On the best site (Feather), effects dissipated quickly, accounting for only a 12% yield gain by age 50. In contrast, fertilization leads to notable gains on average and better sites, appearing progressively sooner as site quality improves (Figure 2). On poorer sites, early weed control projects to a doubling of volume growth by age 50 but fertilization offers no further advantage.

DISCUSSION

Weed control with herbicides was the most effective early treatment across all site qualities, but response was variable. Weed competition for water and nutrients was greatest on the poorer sites–a key fact where drought dominates site quality. There, weed control spells the difference between plantation success and failure. Responses to herbi-

FIGURE 2. SYSTUM-1 projections of cumulative yields for several Garden of Eden plantations. Treatment codes follow those in Figure 1, as well as fertilization (F). Site indices (SI) are in meters at 50 years.

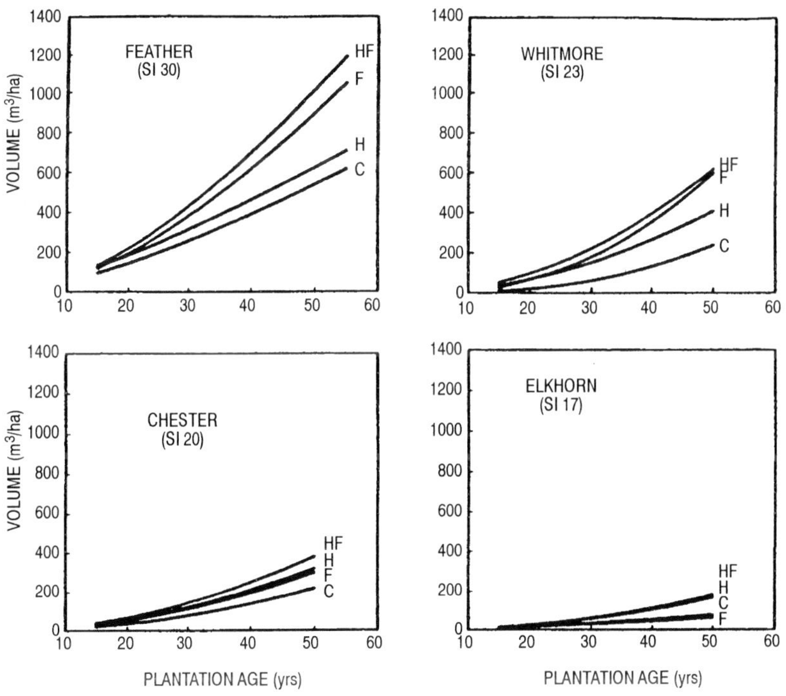

cides were less striking on better sites where deeper soils and lower vapor pressure deficits (Reynolds and Powers, 2000) buffer them against midsummer drought. Weed control improved soil moisture and nutrient availability on poor sites, but stresses returned as stands approached crown closure and moisture again became the primary limiting factor.

Nutrient uptake–particularly N–was increased substantially by fertilization on all sites through year 9 (Table 2), but growth was increased only where site quality was average or better (Figure 1). Growth response is limited where drought dominates production. Stands on sites with more mesic moisture regimes usually develop incipient nutrient deficiencies, especially N and P, as they approach crown closure, making them prime candidates for prolonged fertilization response if rates are balanced in quality and quantity.

CONCLUSIONS

Massive incremental loadings of nine nutrients coupled with low rates of nutrient leaching common to summer-dry climates (Frazer et al., 1990; McColl and Powers, 1984; Powers, 1992), point to nutrient retention and to fundamental increases in site potential on average and better sites. Because functional guilds of soil arthropods were resilient to treatment and because litter processors should return once a forest floor develops, biotic regulation of detrital processing, nutrient cycling, and soil physical maintenance should not be impaired. Therefore, the long-term projections of treatment responses shown in Figure 2 seem realistic if severe water stress is avoided by timely thinnings. Ponderosa pine plantations are capable of productivity levels much greater than currently realized. Such levels should be sustainable if soil organic matter and structure are maintained.

REFERENCES

Busse, M.D., P.H. Cochran and J.W. Barrett. 1996. Changes in ponderosa pine site productivity following removal of understory vegetation. Soil Sci. Soc. Am. J. 60:1614-1621.

Frazer, D.W., J.G. McColl and R.F. Powers. 1990. Soil nitrogen mineralization in a clearcutting chronosequence in a northern California conifer forest. Soil Sci. Soc. Am. J. 54:1145-1152.

McColl, J.G. and R.F. Powers. 1984. Consequences of forest management on soil-tree relationships. In: pp. 379-412. G.D. Bowen and E.K.S. Nambiar (eds.). Nutrition of plantation forests. Acad. Press, New York.

Moldenke, A.R. 1992. Non-target impacts of management practices on the soil arthropod community of ponderosa pine plantations. In: pp. 78-103. Habitat, fiber, society: balance or bias? Proc., 13th Ann. For. Veg. Mgt. Conf., Redding, CA.

Nambiar, E.K.S. 1996. Sustained productivity of forests is a continuing challenge to soil science. Soil Sci. Soc. Am. J. 60:1629-1642.

Powers, R.F. 1992. Fertilization response of subalpine Abies forests in California. In: pp. 114-126. H.N. Chappell, G.F. Weetman and R.E. Miller (eds.). Forest fertilization: sustaining and improving nutrition and growth of western forests. Inst. For. Resour. Contrib. 74, Univ. Washington, Seattle.

Powers, R.F. and G.T. Ferrell. 1996. Moisture, nutrient, and insect constraints on plantation growth: the "Garden of Eden" experiment. NZ J. For. Sci. 26:126-144.

Powers, R.F., A.E. Tiarks, J.A. Burger, and M.C. Carter. 1997. Sustaining the productivity of planted forests. In: pp. 97-134. M.C. Carter (ed.). Growing trees in a greener world: industrial forestry in the 21st century. 35th Forest. Symp., Sch. Forest., Wildlife & Fisheries, Louisiana State Univ., Baton Rouge.

Reynolds, P.E. and R.F. Powers. 2000. Gas exchange for managed ponderosa pine stands positioned along a climatic gradient. J. Sustainable For. 10(3/4):257-265.

Ritchie, M.F. and R.F. Powers. 1993. User's guide for SYSTUM-1 (Version 2.0): A simulator of growth trends in young stands under management in California and Oregon. Gen. Tech. Rep. PSW-GTR-147. PSW Res. Stn., USDA For. Serv., Albany, CA. 45 p.

Gas Exchange for Managed Ponderosa Pine Stands Positioned Along a Climatic Gradient

Phillip E. Reynolds
Robert F. Powers

INTRODUCTION

The "Garden of Eden" experiment was set up to determine how intensive silviculture affects sustainable productivity of ponderosa pine (*Pinus ponderosa* var. *ponderosa*) plantations (Powers and Ferrell, 1996). Plantation productivity is constrained (Powers and Reynolds, 2000) by both site potential (i.e., soils, climate) and by existing site conditions (e.g., inter- and intraspecific competition, age, density, genetics, herbivory, and disease). By managing these conditions to the maximum extent possible, it is possible to improve site potential.

In California, where ponderosa pine is the most widely planted tree, moisture availability and secondarily, nutrient availability, are the most common factors limiting seedling growth (Powers and Ferrell,

Phillip E. Reynolds is Research Scientist, Natural Resources Canada, Canadian Forest Service, 1219 Queen Street East, Sault Ste. Marie, Ontario, Canada P6A 5M7 (E-mail: preynold@nrcan.gc.ca).

Robert F. Powers is Principal Research Silviculturist and Science Team Leader, USDA Forest Service Pacific Southwest Research Station, Redding, CA 96001 USA.

[Haworth co-indexing entry note]: "Gas Exchange for Managed Ponderosa Pine Stands Positioned Along a Climatic Gradient." Reynolds, Philip E., and Robert F. Powers. Co-published simultaneously in *Journal of Sustainable Forestry* (Food Products Press, an imprint of The Haworth Press, Inc.) Vol. 10, No. 3/4, 2000, pp. 257-265; and: *Frontiers of Forest Biology: Proceedings of the 1998 Joint Meeting of the North American Forest Biology Workshop and the Western Forest Genetics Association* (ed: Alan K. Mitchell et al.) Food Products Press, an imprint of The Haworth Press, Inc., 2000, pp. 257-265. Single or multiple copies of this article are available for a fee from The Haworth Document Delivery Service [1-800-342-9678, 9:00 a.m. - 5:00 p.m. (EST). E-mail address: getinfo@haworthpressinc.com].

1996; Shainsky and Radosevich, 1986). In California's Mediterranean summer-dry climate, the growing season usually begins in early April as soils warm, and extends into July until soil drought dominates the site (Oliver et al., 1983). Woody shrubs, especially manzanita (*Arctostaphylos* sp. L.) and ceanothus (*Ceanothus* sp. L.) are generally regarded as pine's strongest competitors for soil moisture and nutrients. By managing these competitors, pine water potential can be increased, nutrient uptake improved, and growth increased on a variety of sites. McDonald and Fiddler (1990) observed that pines grown without competing shrubs had predawn water potentials 0.7 MPa higher than those surrounded by shrubs. Minimum potentials for pines growing with competing shrubs were reached by mid-morning during summer, but delayed until mid-afternoon for those grown without shrubs. Since pines close their stomates to conserve water when potentials drop below –1.2 MPa (Lopushinsky, 1969), usually just after sunrise during hot summer months, extending the period of higher water potential when stomates remain open is crucial to maximizing active periods of photosynthesis. In addition, photosynthetic rates are generally increased with greater N availability and uptake (Mitchell and Hinckley, 1993).

Whereas site conditions, and to some extent soils, can be managed, climate cannot. For droughty sites, periods of active photosynthesis during hot summer months may be greatly reduced compared with other sites with greater annual precipitation and more moderate temperatures. If however, annual precipitation is received mostly in the form of snow, the total number of days of active growth when photosynthesis is feasible may also be restricted. The present study was initiated to quantify climatic variability for pine sites positioned along an environmental gradient and to determine the affect of climatic variability on pine physiological response for managed and unmanaged ponderosa pine stands.

METHODS

This study is a component of the Garden of Eden Project, located throughout northern California, and has been reviewed in considerable detail elsewhere (Powers and Ferrell, 1996). The project was initiated in 1985 to study the effects of vegetation and insect control, fertilization, and combinations thereof on plantation development. For the project, a complete factorial design is being used at eight locations scattered throughout the Sierra, Cascade, and Coastal ranges of northern California. Each location consists of 24 contiguous, rectangular plots measuring 19.5 m × 21.9 m each. Three replications of the eight factorial treatment combinations per plantation were assigned in a completely randomized design. Treatments consisted of controls (C), fertilized (F) only, vegetation control (H) only, vegetation control + fertilized (HF), insecticides (I) only, fertilized + insecticides (FI), vegetation control + insecticides (HI), and vegetation control + fertilized + insecticides (HFI). Treatments were applied repeatedly (i.e., as needed for insecticides, annually for herbicides, and biennially using a ramp schedule for fertilizers) until crown closure (i.e., approximately 9 years). For purposes of this experiment, only the first four treatments (i.e., no insecticides) were used.

The eight sites were established between 1985 and 1988. Three of these sites, having widely varying characteristics, were chosen for this study. They included: (1) Whitmore (732 m, hot and dry, annual precipitation ~ 114 cm. yr^{-1}, snow rare, planted 1986), (2) Feather (1219 m, wet and warm, precipitation ~ 178 cm. yr^{-1}, snow occasional, planted 1988), and (3) Chester (1463 m, cool and dry, precipitation ~ 89 cm. yr^{-1}, snow normal for 5 or 6 months, planted 1987). All plantations occur on soils derived from volcanic material, and the Chester site is characterized by a high coarse cinder content. Soils at Whitmore are clays and those at Feather are loams. The Whitmore and Chester sites were brushfields that were cleared prior to planting. Feather was a natural stand of pine that was logged prior to planting. The highest estimated site index (base age 50 years) for all eight sites was 30 m at Feather. Whitmore ranked intermediate at 23 m and Chester was near the bottom at 20 m (i.e., lowest was 17 m).

Five pines spanning the range of heights were randomly selected and tagged at the centre of each of the twelve treatment plots at each location. Beginning in February 1995, stomatal conductance (G_S), transpiration (E), and net assimilation (NA) for tagged pines (upper crown, new foliage) and associated environmental parameters [photosynthetically-active radiation (PAR), air temperatures, and vapour pressure deficits (VPD)] were measured using a Li-Cor LI-6200 portable photosynthesis unit (Li-Cor Inc., Lincoln, NE). Subsequent measurements occurred in May, June, August, and December. Projected needle areas were determined with a Delta-T leaf area meter and corrected to actual surface areas using a ratio of 2.8 as recommended by Cannell (1982). Water use efficiencies (WUE) were calculated using the formulas (actual WUE = NA/E, Larcher 1980 and intrinsic WUE = NA/G_s, Farquhar and Richards, 1984). Mesophyll conductance was calculated using the formula: G_m = NA/CINT (Leverenz, 1981). Plant water potentials for shoots were measured with a model 3005 plant water status console (Soilmoisture Equipment Corp., Santa Barbara, CA).

To examine overall site differences, data for the four treatments were pooled prior to analysis. Possible differences in PAR for the three sites were evaluated using standard analysis of variance (ANOVA) procedures (Snedecor and Cochran, 1967). Site-related differences in air temperature were examined using analysis of covariance (ANCOVA) with PAR as a covariate. Site differences in plant water potential and VPD were studied using multiple analysis of variance (MANOVA) with air temperature and VPD as covariates for plant water potential and with air temperature and plant water potential as covariates for VPD. Gas exchange data, including G_s, E, NA, actual WUE, intrinsic WUE, and G_m were analyzed using MANOVA with PAR, air temperatures, plant water potential, and VPD as covariates. Treatment means were compared by Tukey's Test at 5% level of significance. Statistical analysis and graphics were performed using CSS Statistica (StatSoft, Tulsa, OK) versions 3.1 (DOS) and 5.1 (Windows).

RESULTS

PAR was highest at Chester in June (Table 1). Air temperatures tended to be highest at Whitmore, intermediate at Chester, and lowest at Feather throughout the measurement period (Table 1). At all sites, the lowest plant water potentials occurred in June and August, when Whitmore was significantly lower than the other two sites (Table 1). In May and December, the three sites differed, with the highest water potentials observed at Whitmore, and the lowest at Chester. Vapour pressure deficits were generally highest at Whitmore and Chester throughout the measurement period; VPD was lowest at Feather (Table 1).

Throughout most of the measurement period, G_s rates were higher or tended to be highest at Feather (Table 2). E and NA rates were higher at Feather in August when these parameters were lower at the

TABLE 1. Seasonal site differences for photosynthetically-active radiation (PAR), air temperatures, plant water potential, and vapour pressure deficits (VPD).

Variable	Site	Feb.	May	Jun.	Aug.	Dec.
PAR ($\mu mol \cdot s^{-1} \cdot m^{-2}$)	Whitmore	788 a	675 b	1220 b	1288 a	429 b
	Feather	415 b	755 b	1250 b	1345 a	772 a
	Chester	–	899 a	1571 a	1325 a	718 a
Air temperature (°C)	Whitmore	11.97 b	20.54 a	28.21 a	29.48 a	17.59 a
	Feather	14.74 a	18.54 b	18.88 c	28.30 b	15.81 b
	Chester	–	18.15 b	24.16 b	29.20 b	13.59 c
Midday water potential (MPa)	Whitmore	–	– 10.07 a	– 15.95 b	– 17.40 b	– 8.70 a
	Feather	–	– 10.70 b	– 14.55 a	– 15.89 a	– 11.75 b
	Chester	–	– 11.44 c	– 14.81 a	– 15.74 a	– 15.15 c
Vapour pressure deficit (kPa)	Whitmore	9.87 a	14.49 b	26.95 a	32.95 a	8.55 b
	Feather	9.49 b	12.13 c	16.96 c	26.10 b	8.34 b
	Chester	–	15.32 a	23.27 b	33.93 a	10.22 a

Mean values are for all treatments. Monthly values followed by the same letter are not significantly different at the 5% level according to Tukey's Test.

Whitmore and Chester (i.e., droughtier) sites (Table 2). Seasonally, maximum NA rates were observed at Whitmore and Chester in June when the highest E rates for these sites were observed and when E rates at these sites were higher than at Feather. Overall, NA rates at Whitmore were highest during winter, spring and early summer months. Actual WUE was highest at Feather during late spring through August and higher at Feather than at Chester in May, June, and August, and higher at Feather than at Whitmore in June (Table 2). In February, actual WUE and NA were higher at Whitmore than at Feather. By contrast, intrinsic WUE tended to be highest at Whitmore and at Chester for most measurement dates and was higher at Whitmore than at Feather in February and higher at Chester than at Feather in May (Table 2). G_m was generally lowest for Whitmore, except in May, and tended to be highest and similar for the Feather and Chester sites (Table 2).

DISCUSSION

Environmental parameters and related physiological responses varied significantly between sites and seasonally. PAR tended to be some-

TABLE 2. Seasonal site differences for stomatal conductance (G_s), transpiration (E), net assimilation (NA), actual water use efficiency (WUE = NA/E), intrinsic water use efficiency (WUE = NA/G_s), and mesophyll conductance (G_m = NA/CINT).

Variable	Site	Feb.	May	Jun.	Aug.	Dec.
G_s ($mol \cdot m^{-2} \cdot s^{-1}$)	Whitmore	0.072 a	0.095 a	0.051 a	0.035 b	0.114 a
	Feather	0.069 a	0.105 a	0.058 a	0.059 a	0.114 a
	Chester	–	0.068 b	0.052 a	0.036 b	0.057 b
E ($mmol \cdot m^{-2} \cdot s^{-1}$)	Whitmore	0.49 a	0.90 a	1.46 a	1.15 c	0.87 a
	Feather	0.50 a	0.85 a	1.03 b	1.70 a	0.94 a
	Chester	–	0.86 a	1.40 a	1.38 b	0.66 b
NA ($\mu mol \cdot m^{-2} \cdot s^{-1}$)	Whitmore	4.10 a	3.88 a	4.42 a	2.83 b	3.51 a
	Feather	2.60 b	3.76 a	4.44 a	4.45 a	3.54 a
	Chester	–	3.04 b	4.55 a	3.06 b	2.61 b
WUE = NA/E ($\mu mol\ CO_2 \cdot mmol^{-1}\ H_2O$)	Whitmore	9.43 a	4.36 ab	3.11 b	2.53 ab	4.04 a
	Feather	5.61 b	4.53 a	4.34 a	2.78 a	3.98 a
	Chester	–	3.67 b	3.34 b	2.22 b	4.11 a
WUE = NA/G_s ($\mu mol\ CO_2 \cdot mmol^{-1}\ H_2O$)	Whitmore	66.22 a	42.30 ab	90.17 a	88.46 a	33.50 b
	Feather	40.70 b	37.90 b	81.10 a	77.74 a	38.09 ab
	Chester	–	47.03 a	88.80 a	87.45 a	47.56 a
G_m = NA/CINT ($\mu mol\ CO_2 \cdot m^{-2} \cdot s^{-1}$)	Whitmore	11.36 a	10.90 b	14.52 b	10.41 b	6.96 b
	Feather	7.68 b	12.93 a	18.32 a	14.04 a	9.45 a
	Chester	–	10.42 b	18.21 a	12.70 a	8.30 ab

Mean values are for all treatments. Monthly values followed by the same letter are not significantly different at the 5% level according to Tukey's Test.

what higher at Chester in June, probably because of high elevation and due to increasing midday solar angle early in the summer. Air temperatures varied with elevation, tending to be highest throughout the year at the lowest elevation Whitmore site. Although slightly higher in elevation, the Chester site tended to be warmer in June than Feather, probably because of less precipitation and lower humidity (unpublished data). Hot droughty conditions at Whitmore during late summer produced the lowest water potentials and highest vapour pressure deficits for the three sites. Early winter and spring rains, accompanied by cooler temperatures, resulted in higher water potentials, lower VPD, and generally higher NA at the Whitmore site. This seasonal shift clearly suggests that most gains in net productivity at low elevation sites probably occur during winter, spring and early summer months.

Vapour pressure deficits tend to increase as temperatures climb seasonally or diurnally and as water stress increases. High VPD and low plant water potential combine to reduce gas exchange rates. In this study, VPD was highest in August at the two droughty sites, Whitmore and Chester. Harrington et al. (1995) also observed that VPD was highest at drier sites. Very low precipitation at Chester, coupled with greater soil drainage (i.e., high cinder content), likely produced droughty conditions equal to those at Whitmore.

G_s, E, NA, actual WUE, and G_m were generally highest at Feather where precipitation was greatest. Concurrently, pine growth and foliar nutrient concentrations were also greatest at this site (Powers and Reynolds, 2000; Powers and Ferrell, 1996). These findings are consistent with those reported by others that gas exchange tends to increase as plant water stress is elevated or as availability of water increases (Jiang et al., 1995; Lieffers et al., 1993). In addition, Harrington et al. (1995) also observed that actual WUE was lower at drier, low-elevation sites, and decreased as VPD increased with decreasing elevation. By contrast, intrinsic WUE was highest for the two droughty sites, Whitmore and Chester. Lajtha and Getz (1993), studying NA and intrinsic WUE in pinyon pine-juniper communities along an elevation gradient, also observed that both species had higher WUE at the lowest, and presumably driest, sites. Combined, these two methods for calculating WUE show that ponderosa pine lowers G_s on droughty sites and increases G_s and E on sites where water is plentiful. Plantations on sites such as Feather, therefore, offer the best option for maximizing sustained productivity.

In this study, G_m tended to be similar for the two higher altitude sites, but higher at Feather. At Feather, foliar nutrient concentrations were highest, presumably because of greater uptake due to greater water availability. Water potential data for Chester in June and August suggests that water potentials were similar to Feather, and higher than those at Whitmore. Greater foliar N concentrations at Feather resulting from greater water availability, vegetation control, and fertilization could have contributed to this response (Powers and Reynolds, 2000; Powers and Ferrell, 1996). In studies involving fertilization or release from competing vegetation, it has generally been observed that G_s, E, NA, and G_m may all increase in response to higher foliar N (Lieffers et al., 1993; Mitchell and Hinckley, 1993). However, nutrient influences appear to be primarily on G_m rather than G_s.

CONCLUSIONS

The following apply for measurements made in 1995 (i.e., plantation ages, 8 yrs at Feather, 9 yrs at Chester, 10 yrs at Whitmore). Average crown closure occurred during yr 9 (Powers and Ferrell 1996). In 1995, crown closure had occurred at Whitmore and Feather, but not at Chester.

- The three sites differed significantly climatically. The Whitmore (lowest elevation) and Chester (highest elevation) sites were droughtiest while the Feather (high elevation) site was wettest. Temperatures were highest at Whitmore and lowest at Feather. In early summer, radiation intensity was greatest at Chester
- Because of these climatic differences the three sites differed in their seasonal phenology. Most net productivity occurred during winter, spring, and early summer at Whitmore when temperatures were cooler and rainfall frequent. At Chester, net productivity was restricted to May through December because of snow cover the balance of the year, with maximum net assimilation in June. At Feather, climatic conditions were generally favourable for net productivity year-round, with maximum net assimilation in August.
- Because of favourable year-round climate at Feather, G_S, E, NA, and actual WUE were highest at this site. Concurrently, growth and foliar nutrient concentrations were highest at this site.
- Because of summer droughty conditions, intrinsic WUE was highest at the Whitmore and Chester sites.
- The two methods of calculating WUE show that ponderosa pine lowers G_S on droughty sites and increases G_S and E on sites where water is plentiful.
- We conclude that plantations on sites such as Feather, therefore, offer the best option for maximizing sustained productivity.

REFERENCES

Cannell, M.C.R. 1982. World forest biomass and primary production data. Academic Press, London.

Farquhar, G.D. and R.A. Richards. 1984. Isotopic composition of plant carbon correlates with water-use efficiency of wheat genotypes. Aust. J. Plant Physiol. 11: 539-552.

Harrington, R.A., J.A. Fownes, F.C. Meinzer and P.G. Scowcroft. 1995. Forest growth along a rainfall gradient in Hawaii: *Acacia koa* stand structure, productivity, foliar nutrients, and water-and nutrient-use efficiencies. Oecologia 102: 277-284.

Jiang, Y., S.E. MacDonald and J.J. Swiazek. 1995. Effects of cold storage and water stress on water relations and gas exchange of white spruce (*Picea glauca*) seedlings. Tree Physiol. 15:267-273.

Lajtha, K. and J. Getz. 1993. Photosynthesis and water-use in pinyon-juniper communities along an elevation gradient in northern New Mexico. Oecologia 94: 95-101.

Leverenz, J.W. 1981. Photosynthesis and transpiration in large forest-grown Douglas fir diurnal variation. Can. J. Bot. 59: 349-356.

Lieffers, V.J., A.G. Mugasha and S.E. MacDonald. 1993. Ecophysiology of shade needles of *Picea glauca* saplings in relation to removal of competing hardwoods and degree of prior shading. Tree Physiol. 12: 271-280.

Lopushinsky, W. 1969. Stomatal control in conifer seedlings in response to leaf moisture stress. Bot. Gaz. 134: 258-263.

McDonald, P.M. and G.O. Fiddler. 1990. Ponderosa pine seedlings and competing vegetation: Ecology, growth, and cost. USDA Forest Service Research Paper, PSW-199.

Mitchell, A.K. and T.M. Hinckley. 1993. Effects of foliar nitrogen concentration on photosynthesis and water use efficiency in Douglas fir. Tree Physiol. 12: 403-410.

Oliver, W.W., R.F. Powers and J.N. Fiske. 1983. Pacific ponderosa pine. Pp. 48-52. In: Silvicultural systems for the major forest types of the United States. R.M. Burns (ed.). Agric. Handb. 445, USDA Forest Service, Washington, DC.

Powers, R.F. and P.E. Reynolds. 2000. Intensive management of ponderosa pine plantations: ensuring sustainable productivity in the 21st century. J. Sustainable For. 10(3/4):249-255.

Powers, R.F. and G.T. Ferrell. 1996. Moisture, nutrient, and insect constraints on plantation growth: the "Garden of Eden" experiment. New Zealand J. For. Sci. 26: 126-144.

Shainsky, L.J. and S.R. Radosevich. 1986. Growth and water relations of *Pinus ponderosa* seedlings in competitive regimes with *Arctostaphylos patula* seedlings. J. Appl. Ecol. 23: 957-966.

Snedecor, G.W. and W.G. Cochran. 1967. Statistical methods. The Iowa State University Press, Ames, Iowa.

Microclimate Changes Following Alternative Conifer Release Treatments Continued Through Three Post-Treatment Growing Seasons

Phillip E. Reynolds
F. Wayne Bell
R. A. Lautenschlager
James A. Simpson
Andrew M. Gordon
Donald A. Gresch
Donald A. Buckley
John A. Winters

INTRODUCTION

Below- and near-ground microclimate is affected by the amount of vegetation that is present on a forest site. Changes in light, tempera-

---

Phillip E. Reynolds is Research Scientist, and Donald A. Buckley is Retired Technician, Natural Resources Canada, Canadian Forest Service, 1219 Queen Street East, Sault Ste. Marie, Ontario, Canada P6A 5M7 (E-mail: preynold@nrcan.gc.ca).

F. Wayne Bell and R. A. Lautenschlager are Research Scientists, and John A. Winters is Forestry Technician, Ontario Ministry of Natural Resources, Ontario Forest Research Institute, 1235 Queen Street East, Sault Ste. Marie, Ontario, Canada P6A 5N5.

James A. Simpson is Graduate Student, Andrew M. Gordon is Professor, and Donald A. Gresch is Former Technician, Department of Environmental Biology, University of Guelph, Guelph, Ontario, Canada N1G 2W1.

[Haworth co-indexing entry note]: "Microclimate Changes Following Alternative Conifer Release Treatments Continued Through Three Post-Treatment Growing Seasons." Reynolds, Phillip E. et al. Co-published simultaneously in *Journal of Sustainable Forestry* (Food Products Press, an imprint of The Haworth Press, Inc.) Vol. 10, No. 3/4, 2000, pp. 267-275; and: *Frontiers of Forest Biology: Proceedings of the 1998 Joint Meeting of the North American Forest Biology Workshop and the Western Forest Genetics Association* (ed: Alan K. Mitchell et al.) Food Products Press, an imprint of The Haworth Press, Inc., 2000, pp. 267-275. Single or multiple copies of this article are available for a fee from The Haworth Document Delivery Service [1-800-342-9678, 9:00 a.m. - 5:00 p.m. (EST). E-mail address: getinfo@haworthpressinc.com].

ture, relative humidity and moisture produced by different conifer release practices affect plant succession, habitat quality, and microsite suitability for crop production (Freedman et al., 1993; Horsley, 1994; Jobidon, 1992; Reynolds et al., 1997a and b). Removal of vegetation increases solar radiation reaching the forest floor, results in soil warming (Brand and Janas, 1988; Karakatsoulis et al., 1989; Wood and von Althen, 1993), and raises soil moisture levels (Bosch and Hewlett, 1982). Such conditions are generally favourable for seedling growth (Karakatsoulis et al., 1989; Lieffers et al., 1993; Newton et al., 1992; Radosevich and Osteryoung, 1987).

This study documents below- and near-ground microclimatic changes associated with several alternative conifer release treatments in the boreal mixedwood forest type. Objectives were to quantify treatment-related microclimatic changes and to determine the duration of the changes.

METHODS

This study is a component of the Fallingsnow Ecosystem Project, located near Thunder Bay, Ontario, and has been reviewed elsewhere (Lautenschlager et al., 1997, 1998). The study was initiated in 1993 and uses a randomized complete block design, with three 28 to 52 ha spruce blocks that were cut and operationally planted with white spruce [*Picea glauca* (Moench) Voss] bareroot seedlings four to seven years before the study began. Each block contains five post-harvest treatment plots including two herbicides (i.e., glyphosate, Trade name = Vision, manufactured by Monsanto and triclopyr, Trade name = Release, manufactured by Dow-Elanco), two brush cutting alternatives (i.e., motor-manual release with brushsaws and mechanical release with a tractor mounted Silvana Selective cutting head), and a control (no treatment). Unharvested controls adjacent to each block constituted a sixth treatment. The various release treatments were designed to control overstory trembling aspen (*Populus tremuloides* Michx.) and other non-conifer species (Bell et al., 1997; Thompson et al., 1997).

Plant covers (i.e., coniferous tree, deciduous tree, shrub, herb, grass, and fern) were measured prior to treatments and in each post-treatment season (August) within 360 circular sub-plots (4 m^2, r = 1.13 m, 24 plots/treatment/block). These circular sub-plots were located in close proximity to 75 sub-plot (5/treatment/block) crop tree locations (Reynolds et al., 2000).

All meteorological stations or soil temperature/moisture sensors were located in association with the 75 sub-plot crop tree locations. Generally, these locations also corresponded with previously existing lysimeter locations (Simpson et al., 1997). Beginning in May 1994, a maximum of twelve LiCor weather stations (Li-Cor Inc., Lincoln, NE) were established in the control, brushsaw, and glyphosate treatment plots on each of the three white spruce blocks and in the unharvested forest plots adjacent to these blocks. Weather stations were equipped to monitor photosynthetically-active radiation (PAR), air temperature, and relative humidity (RH) at 0.25 and 2 m above the forest floor and soil temperature at 5 and 15 cm depth (Reynolds et al., 1997a). Li-Cor LI-1000 dataloggers were programmed to monitor sensors continuously and to record integrated (PAR), mean, maximum, and minimum values on a 3-hour basis starting at midnight. Li-Cor quantum and temperature sensors and Vaisala humidity sensors were used. Fiberglass thermistor/resistance soil cells (ELE International, Lake Bluff, IL) were installed at each of the 75 sub-plot crop tree locations at two depths (15 and 30 cm). Additional soil cells (5 at each depth) were installed at each of the three unharvested forest plots. Soil moisture and temperature were read bimonthly until snowfall each year. Cell resistance data were converted to soil moisture data (%) using calibration curves developed for the Fallingsnow site.

Annual vegetation cover, meteorological and soil temperature/moisture data were examined using analysis of variance (ANOVA) procedures where each block was treated as a replicate (Steel and Torrie, 1980). Mean PAR, air and soil temperature, and RH values (Li-Cor stations) were calculated from daily maximum values for the period July 22 through August 21. Treatment means were compared by Tukey's HSD Test at 5% level of significance. Statistical analysis was performed using CSS Statistica (StatSoft, Tulsa, OK).

RESULTS

Both herbicide treatments significantly (P = 0.002) reduced woody deciduous cover (i.e. trees + shrubs) in 1994 (Table 1). Maximum woody deciduous covers occurred in 1996 for all release treatments, but cover was lowest for the Vision treatment. Woody deciduous cover

TABLE 1. Plant covers of competing vegetation by treatment prior to treatments (1993) and during the first (1994), second (1995), and third (1996) growing seasons after alternative conifer release treatments in 1993. Mean values are based upon a replicated block design.

Cover (%)	Year	Control	Brushsaw	Silvana	Vision	Release
N =	All	3	3	3	3	3
Woody deciduous	1993	49.4 a	47.8 a	55.1 a	59.6 a	51.2 a
	1994	70.3 a	52.4 ab	59.5 ab	28.1 b	38.0 b
	1995	56.6 a	49.9 a	55.9 a	35.3 a	41.9 a
	1996	70.4 a	59.5 a	66.0 a	42.1 a	55.0 a
Non-woody	1993	48.9 a	57.6 a	48.1 a	48.6 a	45.4 a
	1994	68.8 a	75.4 a	73.4 a	72.0 a	67.4 a
	1995	57.8 a	63.0 a	67.7 a	72.4 a	61.4 a
	1996	60.6 a	64.9 a	64.3 a	75.9 a	66.4 a

Values are means. Numbers within rows followed by the same letter are not significantly different at the 5% level according to Tukey's Test

declined in 1995 for the control, brushsaw, and Silvana treatments (Table 1), probably because of drought conditions. Cumulative precipitation (~186.5 mm) was approximately 25% less in 1995 than in 1994 for the period June 1 through September 5. In 1995, no rainfall occurred between June 7 and June 24 and again between July 24 and August 8. The 1996 growing season was wetter than 1994, and nearly 150% wetter than 1995. Non-woody deciduous cover (i.e., herbaceous + grass + fern) increased in 1994 for all treatments (Table 1). Non-woody deciduous cover also declined in 1995, except for the Vision treatment, where it increased from 49% in 1993 to a maximum of 76% in 1996.

Treatment-related microclimatic differences were more pronounced in 1996 than in 1995 (Table 2), specifically for PAR (2 m), air temperature (2 and 0.25 m), and RH (2 m). PAR was higher for the brushsaw and glyphosate treatments than for the control. Nearly all other differences were restricted to these two treatments and the unharvested forest.

Similarly, for soil cells, the only treatment differences observed in 1995 or 1996, with one exception, involved all release treatments vs. the unharvested forest treatment (Figure 1). In both years, soil temperatures for the release treatments were higher and soil moisture levels were lower than for the unharvested forest. Temperature differences occurred on June 8 ($P = 0.0003$), June 24 ($P = 0.007$), July 7 ($P =$

TABLE 2. Treatment-related differences in microclimate during the second (1995) and third (1996) growing seasons after alternative conifer release treatments were applied in 1993. Data are mean values based upon daily maximum observations (Li-Cor stations) for the period July 22 through August 21. Values are based upon a replicated block design.

Variable	Location	Year	Control	Brushsaw	Vision	Forest
N =		1995	2	2	2	
		1996	3	3	3	3
PAR (μmol $s^{-1}.m^{-2}$)	0.25 m	1995	214 a	916 a	663 a	747 a
		1996	196 a	508 a	573 a	326 a
	2 m	1995	302 b	2036 a	1950 a	1141 ab
		1996	1026 b	1759 a	1823 a	688 b
Air temp. (°C)	0.25 m	1995	22.99 a	27.58 a	28.47 a	24.06 a
		1996	21.52 b	22.91 ab	24.95 a	19.71 b
	2 m	1995	25.22 a	24.76 a	25.18 a	23.27 a
		1996	21.24 ab	22.14 a	22.0 a	19.91 b
Relative humidity (%)	0.25 m	1995	73.79 a	73.28 a	68.18 a	79.08 a
		1996	89.23 a	85.51 a	81.25 a	92.31 a
	2 m	1995	69.85 a	62.64 a	64.54 a	71.10 a
		1996	79.08 ab	73.23 b	73.94 b	82.74 a
Duff temp. (°C)	~5 cm	1995	18.30 a	19.14 a	19.20 a	15.83 a
		1996	16.36 a	17.77 a	17.64 a	15.88 a
Soil temp. (°C)	15 cm	1995	14.68 a	15.64 a	16.30 a	14.18 a
		1996	14.31 a	15.13 a	15.43 a	13.88 a

Values are means. Numbers within rows followed by the same letter are not significantly different at the 5% level according to Tukey's Test

FIGURE 1. Seasonal variation in (A) soil temperatures (°C) and (B) moisture (%) at 15 cm depth in 1995 after alternative conifer release treatments applied in 1993. Legend: Control = solid circle, brushsaw = open circle, Silvana = diamond, Vision = square, Release = triangle, Unharvested forest = asterisk.

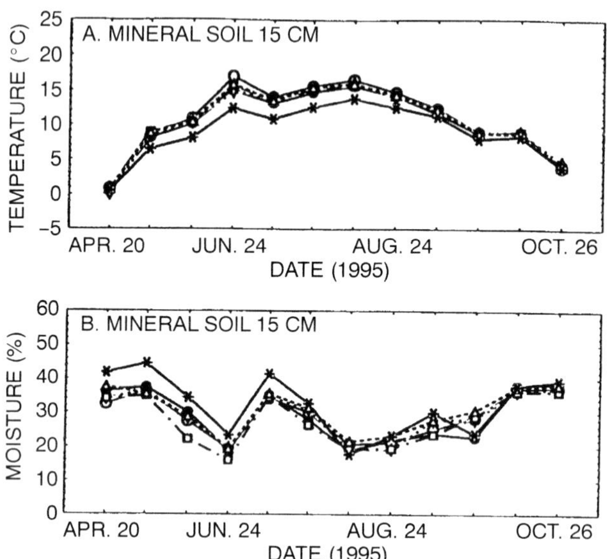

0.042), July 25 (P = 0.048), and August 10 (P = 0.034). Moisture differences occurred on July 7 (P = 0.009) and September 22 (P = 0.017). On September 22, soil moisture for the triclopyr treatment was higher than for the control.

DISCUSSION

Three years after alternative conifer release treatments were applied, some treatment-related microclimate differences continue to persist, but are greatly diminished compared with the first post-treatment growing season (Reynolds et al., 1997a). The first treatment-related differences to disappear were those associated with RH and soil temperature. Treatment differences related to light (PAR) remained the strongest. Differences among release treatments for soil moisture disappeared in 1995. In 1994, soil moisture levels for all release treatments were greater than for controls (Reynolds et al., 1997a). By

1995, soil moisture levels for release treatments were generally less than for controls. The reduction in differences in microclimatic variables among alternative release treatments corresponds with a rapid and progressive revegetation of treated plots. Leaf area (LAI) of new and resprouting vegetation was lowest in 1994, and highest in 1996 (Reynolds et al., 2000).

Despite a decrease in LAI in 1995, induced by drought conditions, enhanced seedling growth was presumably sufficient to reduce soil moisture levels below those observed on controls. Fewer microclimatic differences were observed in 1995, probably because of the severe reduction in LAI for all treatments (Reynolds et al., 2000). This reduction was sufficient to eliminate any significant difference in leaf area among treatments. Without leaf area differences among treatments, most microclimatic differences also disappeared. With more favourable precipitation conditions in 1996, treatment-related LAI differences reappeared, as did more microclimatic differences. Of the four release treatments, glyphosate (Vision) caused the greatest reduction in vegetation cover and resulted in successional vegetation and microclimate that is the most dissimilar to the control. Since woody vegetation recovery in the glyphosate treatments is expected to occur very slowly (Bell et al., 1997), treatment differences will likely persist until spruce crown closure.

The various microclimatic changes produced microsite conditions favourable seedling growth, and released seedlings exhibited higher photosynthetic rates and growth than controls (Reynolds et al., 2000).

CONCLUSIONS

After the first three post-treatment growing seasons:

- Except for PAR, microclimatic changes induced by conifer release were short-lived, generally fading or disappearing after the first post-treatment growing season, and as progressive revegetation of treated plots occurred.
- Microclimatic differences between release treatments and controls were greatest for the glyphosate treatment where succession to non-woody vegetation occurred.
- By 1996, most treatment-related differences in microclimate were associated with the glyphosate and the unharvested forest treatments.

REFERENCES

Bell, F.W., R.A. Lautenschlager, R.G. Wagner, D.G. Pitt, J.W. Hawkins and K.R. Ride. 1997. Motor-manual, mechanical, and herbicide release affect early successional vegetation in northwestern Ontario. For. Chron. 73: 61-68.

Bosch, J.M. and J.D. Hewlett. 1982. A review of catchment experiments to determine the effects of vegetation changes on water yield and evapotranspiration. J. Hydrol. 55: 3-23.

Brand, D.G. and P.S. Janas. 1988. Growth and acclimation of planted white pine and white spruce seedlings in response to environmental conditions. Can. J. For. Res. 18: 320-329.

Freedman, B., R. Morash and D. MacKinnon. 1993. Short-term changes in vegetation after the silvicultural spraying of glyphosate herbicide onto regenerating clearcuts in Nova Scotia, Canada. Can. J. For. Res. 23: 2300-2311.

Horsley, S.B. 1994. Regeneration success and plant species diversity of allegheny hardwood stands after roundup application and shelterwood cutting. North. J. Appl. For. 1194) 109-116.

Jobidon, R. 1992. Measurement of light transmission in young conifer plantations: A new technique for assessing herbicide efficacy. North. J. Appl. For. 9: 112-115.

Karakatsoulis, J., J.P. Kimmins and R.E. Bigley. 1989. Comparison of the effects of chemical (glyphosate) and manual conifer release on conifer seedlings physiology and growth on Vedder Mountain, British Columbia. Pp. 168-188. In: Proceedings of the Carnation Creek Herbicide Workshop. P.E. Reynolds (ed.). FRDA Report 063, B.C. Ministry of Forests, Victoria, B.C. 349 p.

Lautenschlager, R.A., F.W. Bell, R.G. Wagner and J.A. Winters. 1997. The Fallingsnow Ecosystem Project: Comparing conifer release alternatives in northwestern Ontario. For. Chron. 73:35-38.

Lautenschlager, R.A., F.W. Bell, R.G. Wagner, P.E. Reynolds and J.M. Woodcock. 1998. The Fallingsnow Ecosystem Project: Documenting the environmental consequences of conifer release. J. For. (In Press).

Lieffers, V.J., A.G. Mugasha and S.E. MacDonald. 1993. Ecophysiology of shade needles of *Picea glauca* saplings in relation to removal of competing hardwoods and degree of prior shading. Tree Physiol. 12: 271-280.

Newton, M., E.C. Cole, M.L. McCormack, Jr. and D.E. White. 1992. Young spruce-fir forests released by herbicides II. Conifer response to residual hardwoods and overstocking. North. J. Appl. For. 9: 130-135.

Radosevich, S.R. and K. Osteryoung. 1987. Principles governing plant-environment interactions. Pp. 105-156. In: Vegetation management for conifer production. J.D. Walstad and P.J. Kuch (eds.) John Wiley & Sons, N.Y.

Reynolds, P.E., J.A. Simpson, R.A. Lautenschlager, F.W. Bell, A.M. Gordon, D.A. Gresch and D.A. Buckley. 1997a. Alternative conifer release treatments affect below- and near-ground microclimate. For. Chron. 73: 75-82.

Reynolds, P.E., F.W. Bell, J.A. Simpson, A.M. Gordon, R.A. Lautenschlager, D.A. Gresch and D.A. Buckley. 1997b. Alternative conifer release treatments affect leaf area index of competing vegetation and available light for seedling growth. For. Chron. 73: 83-89.

Reynolds, P.E., F.W. Bell, J.A. Simpson, R.A. Lautenschlager, A.M. Gordon, D.A.

Gresch and D.A. Buckley. 2000. Alternative conifer release treatments affect competition levels, available light, net assimilation rates, and growth of white spruce seedlings. J. Sustainable For. 10(3/4):277-286.

Steel, R.G.D. and J.H. Torrie. 1980. Principles and procedures of statistics: A biometrical approach. McGraw Hill Book Co.

Simpson, J.A., A.M. Gordon, P.E. Reynolds, R.A. Lautenschlager, F.W. Bell, D.A. Gresch and D.A. Buckley. 1997. Influence of alternative conifer release treatments on soil nutrient movement. For. Chron. 73: 69-73.

Thompson, D.G., D.G. Pitt, B. Staznik, N.J. Payne, D. Jaipersaid, R.A. Lautenschlager and F.W. Bell. 1997. On-target deposit and vertical distribution of aerially released herbicides. For. Chron. 73: 47-59.

Wood, J.E. and F.W. von Althen. 1993. Establishment of white spruce and black spruce in boreal Ontario: effects of chemical site preparation and post-planting weed control. For. Chron. 69(5): 554-560.

Alternative Conifer Release Treatments Affect Competition Levels, Available Light, Net Assimilation Rates, and Growth of White Spruce Seedlings

Phillip E. Reynolds
F. Wayne Bell
James A. Simpson
R. A. Lautenschlager
Andrew M. Gordon
Donald A. Gresch
Donald A. Buckley

INTRODUCTION

Conifer growth often does not respond significantly to conifer release until two or three years after treatment (Michael, 1985). There-

Phillip E. Reynolds is Research Scientist, and Donald A. Buckley is Retired Technician, Natural Resources Canada, Canadian Forest Service, 1219 Queen Street East, Sault Ste. Marie, Ontario, Canada P6A 5M7 (E-mail: preynold@nrcan.gc.ca).

F. Wayne Bell and R. A. Lautenschlager are Research Scientists, Ontario Ministry of Natural Resources, Ontario Forest Research Institute, 1235 Queen Street East, Sault Ste. Marie, Ontario, Canada P6A 5N5.

James A. Simpson is Graduate Student, Andrew M. Gordon is Professor, and Donald A. Gresch is Former Technician, Department of Environmental Biology, University of Guelph, Guelph, Ontario, Canada N1G 2W1.

[Haworth co-indexing entry note]: "Alternative Conifer Release Treatments Affect Competition Levels, Available Light, Net Assimilation Rates, and Growth of White Spruce Seedlings." Reynolds, Phillip E. et al. Co-published simultaneously in *Journal of Sustainable Forestry* (Food Products Press, an imprint of The Haworth Press, Inc.) Vol. 10, No. 3/4, 2000, pp. 277-286; and: *Frontiers of Forest Biology: Proceedings of the 1998 Joint Meeting of the North American Forest Biology Workshop and the Western Forest Genetics Association* (ed: Alan K. Mitchell et al.) Food Products Press, an imprint of The Haworth Press, Inc., 2000, pp. 277-286. Single or multiple copies of this article are available for a fee from The Haworth Document Delivery Service [1-800-342-9678, 9:00 a.m. - 5:00 p.m. (EST). E-mail address: getinfo@haworthpressinc.com].

fore, a measure of conifer response to release treatments that provides an earlier indication of potential growth response is desired. Post-treatment stomatal conductance (G_S), transpiration (E), and net assimilation (NA) rates for released seedlings might provide such an indicator if these parameters could be linked with conifer growth during the first three years after release. The photosynthetic process in plants, whether they be seedlings or competing vegetation, is driven by the environment. Following conifer release, changes in light, temperature, moisture, relative humidity, evaporative demand, and nutrients affect NA, G_S, and E rates (Bosch and Hewlett, 1982; Brand and Janas, 1988; Eastman and Camm, 1995; Karakatsoulis et al., 1989; Lieffers et al., 1993; Newton et al., 1992; Reynolds et al., 1997a and b, 2000; Wood and von Althen, 1993). Perhaps the most important of these environmental parameters is light (Jobidon, 1992; Reynolds et al., 1997b), assuming that moisture and nutrients are not limiting.

All reforestation sites receive light that is captured by either the seedlings or the competing vegetation. If the competing vegetation is large enough to overtop and shade the seedlings, most of the available light energy will be diverted to NA and growth of the competitors rather than the seedlings. If the competing vegetation is controlled, and shading eliminated, then most of the available light will be available for seedling NA and growth. In short, the amount of competition present on the site alters the acquisition of site resources, limits the photosynthetic process of seedlings, and the subsequent growth of these seedlings. Alternative conifer release strategies may affect the amount of competition present, which is expected to influence shading and available light for seedling photosynthesis, and in the long-term, seedling growth. The present study was initiated to measure G_S, E, and NA rates and growth response of released white spruce [*Picea glauca* (Moench) Voss] seedlings and to determine if *in situ* rates of G_S, E, and NA are reasonable predictors of future growth.

METHODS

This study is a component of the Fallingsnow Ecosystem Project (48° 8-13′N, 89° 49-53′W) located about 60 km southwest of Thunder Bay, Ontario, which has been reviewed elsewhere (Lautenschlager et al., 1997, 1998). The project was initiated in 1993 and uses a randomized complete block design, with three 28 to 52 ha blocks that were cut and planted with white spruce four to seven years before the study began. Each block contains five post-harvest treatment plots including two herbicides (i.e., glyphosate, Trade name = Vision, manufactured by Monsanto and triclopyr, Trade name = Release, manufactured by Dow-Elanco), two cutting alternatives (i.e., motor-manual release with brushsaws and mechanical release with a tractor mounted Silvana Selective cutting head), and a control (no treatment). The various release treatments were designed to control overstory trembling aspen (*Populus tremuloides* Michx.) and other non-conifer species (Bell et al., 1997; Thompson et al., 1997).

In the summer of 1994, five seedlings were selected and tagged at each of 75 sub-plot locations (5/treatment/block). Seedlings were selected if aspen stems surrounded them prior to release in 1993. In August 1994, 1995, and 1996, leaf area indices (LAI) of competing vegetation within a 1 m radius of each seedling were determined using a Li-Cor LAI-2000 Plant Canopy Analyzer (Li-Cor Inc., Lincoln, NB). LAI was measured around the base of each seedling (crown edge, optical sensor facing out from the seedling, 90° lens cap) while facing north, south, east, and west. Measurements occurred in the morning (before 0900 h) or evening hours (1800 to 2000 hrs) and on overcast days to avoid interference from direct sunlight.

In late July 1996, G_S, E, and NA in the upper crown (new foliage) of seedlings and associated environmental parameters [photosynthetically-active radiation (PAR), air temperatures, and vapour pressure deficits (VPD)] were measured using a Li-Cor LI-6200 portable photosynthesis unit (Li-Cor Inc., Lincoln, NB). Three locations per treatment (most accessible) were visited daily between 1000 and 1800 hrs EST. Each block was measured on a separate but similar day. Projected needle areas were determined with a Li-Cor LI-3100 plant area meter and corrected to actual surface areas using a ratio of 2.5 as recommended by Barker (1968). Water use efficiencies (WUE) were calculated using the formulas: actual WUE = NA/E (Larcher, 1980) and intrinsic WUE = NA/G_S (Farquhar and Richards, 1984). Height and groundline diameter (GLD) of measured seedlings, along with maximum height of competing vegetation within a 1 m radius of each seedling, were determined in September of each year. Stem volume was computed using the formula of a cone (Avery, 1975).

Yearly LAI data were examined using standard analysis of variance (ANOVA) procedures where each block was treated as a replicate

(Steel and Torrie, 1980). Competition height and environmental data were analyzed using analysis of variance (ANOVA) procedures; air temperature and VPD were also examined via analysis of covariance (ANCOVA) with PAR or air temperature as covariates, respectively (Snedecor and Cochran, 1967). G_S, E, NA, and WUE data were analyzed using multiple analysis of variance (MANOVA) with environmental parameters as covariates (Landsberg et al., 1975; Potvin et al., 1990). Photosynthesis data were further examined using non-linear regression techniques (Causton and Dale, 1990). Seedling growth differences were studied using ANCOVA with plantation age or pre-treatment height (1993) as a covariate or using MANOVA with both as covariates (Snedecor and Cochran, 1967). Treatment means were compared by Tukey's Test for Unequal N at 5% level of significance. Stem volume was correlated with pre-treatment height by means of linear regression. Statistical analysis and graphics were done with CSS Statistica (StatSoft, Tulsa, OK) versions 3.1 (DOS) and 5.1 (Windows).

RESULTS

Compared with controls, all release treatments showed reduced LAI throughout the first three (1994, 1995, 1996) post-treatment growing seasons (Table 1). LAI was lower for the Vision treatment in 1994 and 1996. LAI declined in 1995 for all treatments because of drought (25% less rainfall than in 1994 during the active growing season, Reynolds et al., 2000), but increased again in 1996 beyond 1994 levels.

TABLE 1. Leaf area indices (LAI) of competing vegetation by treatment during the first (1994), second (1995), and third (1996) growing seasons after alternative conifer release treatments in 1993.

Year	Control	Brushsaw	Silvana	Vision	Release
	N = 3	N = 3	N = 3	N = 3	N = 3
1994	3.83 ± 0.17 a	2.80 ± 0.37 ab	2.75 ± 0.29 ab	2.06 ± 0.28 b	2.86 ± 0.02 ab
1995	2.99 ± 0.43 a	1.64 ± 0.16 a	1.69 ± 0.34 a	1.62 ± 0.31 a	2.21 ± 0.46 a
1996	4.42 ± 0.36 a	3.77 ± 0.23 ab	3.50 ± 0.12 ab	2.85 ± 0.21 b	3.76 ± 0.22 ab

Values are means. Numbers within rows followed by the same letter are not significantly different at the 5% level according to Tukey's Test.

The maximum height of competing vegetation surrounding seedlings was lower for all release treatments than for the control treatment, and lowest for the glyphosate (Vision) treatment (Table 2). Mean PAR values for all release treatments differed from the control, but did not differ among release treatments (Table 2). Mean air temperature values at the time of measurement did not differ among treatments, but mean VPD was lowest for the Silvana treatment (Table 2).

All release treatments had greater G_S and NA rates than the control, but did not differ from each other for G_S and actual WUE (Table 2). For NA and intrinsic WUE, differences between glyphosate (Vision) and all other release treatments were observed (Table 2). Non-linear predictions of light-saturated photosynthetic rates were highest for release treatments (Figure 1). Pre-treatment (1993) and 3rd-year post-treatment seedling heights did not differ among treatments (Table 2). Stem diameters and stem volumes of released seedlings were greater than for controls, but there were no differences among treatments (Table 2, Figure 1).

DISCUSSION

The four release treatments reduced the leaf area and height of competitive vegetation, and changed species composition (Bell et al., 1997). LAI and competition height were reduced most by the glyphosate (Vision) treatment where succession to non-woody vegetation cover occurred (Reynolds et al., 2000). Reductions in competing vegetation were accompanied by increased PAR, higher soil moisture, higher air and soil temperatures, and lower relative humidity (Reynolds et al., 1997a and b, 2000). These favourable environmental changes coupled with increased rates of soil nitrogen (N) mineralization (Reynolds et al., 1998a), benefited seedlings by increasing *in situ* G_S, E, and NA rates, and ultimately, seedling growth. Improved moisture and nutrient status combined with increased light, generally results in increased photosynthesis (Eastman and Camm, 1995; Lieffers et al., 1993) and ultimately improved crop survival and growth (Newton et al., 1992).

Since photosynthetic gains must proceed growth gains, we hypothesized that *in situ* rates of G_S, E, and NA, in reference to the amount of shading from competing vegetation, might be an early indicator of later treatment-related differences in growth. Although treatment-re-

TABLE 2. Treatment-related differences in competition height, environmental parameters, and white spruce seedling gas exchange and growth during the third (1996) growing season after alternative conifer release treatments.

Treatment	Control	Brushsaw	Silvana	Vision	Release
	N = 34	N = 35	N = 31	N = 35	N = 33
Maximum competition height (cm)	577.79 ± 26.77 a	184.66 ± 6.84 b	221.87 ± 9.09 b	119.28 ± 7.97 c	197.06 ± 17.85 b
Photosynthetically-active radiation (PAR) ($\mu mol \cdot s^{-1} \cdot m^{-2}$)	147.9 ± 22.18 b	845.9 ± 91.45 a	557.7 ± 64.78 a	948.1 ± 87.06 a	689.8 ± 77.89 a
Air temperature (°C)	23.68 ± 0.52 a	25.46 ± 0.61 a	23.70 ± 0.50 a	25.12 ± 0.55 a	24.84 ± 0.63 a
Vapour pressure deficit (VPD) (kPa)	1.69 ± 1.13 a	1.72 ± 0.89 a	1.49 ± 0.79 b	1.77 ± 0.96 a	1.72 ± 1.12 a
Stomatal conductance (G_s) ($mol \cdot m^{-2} \cdot s^{-1}$)	0.0762 ± 0.0036 b	0.0939 ± 0.0038 a	0.0960 ± 0.0045 a	0.0939 ± 0.0034 a	0.0908 ± 0.0047 a
Transpiration (E) ($mmol \cdot m^{-2} \cdot s^{-1}$)	1.26 ± 0.08 b	1.62 ± 0.10 a	1.42 ± 0.09 b	1.67 ± 0.09 a	1.53 ± 0.11 a
Net assimilation (NA) ($\mu mol \cdot m^{-2} \cdot s^{-1}$)	2.26 ± 0.22 c	4.34 ± 0.25 b	4.44 ± 0.22 b	5.57 ± 0.25 a	4.49 ± 0.29 b
Water use efficiency (WUE = NA/E)	1.87 ± 0.19 b	2.82 ± 0.14 a	3.32 ± 0.18 a	3.46 ± 0.12 a	3.21 ± 0.25 a
Water use efficiency (WUE = NA/G_s)	32.27 ± 3.67 c	47.55 ± 2.53 b	49.20 ± 3.25 b	60.93 ± 2.90 a	55.85 ± 6.00 b
Seedling height (cm)	158.90 ± 10.50 a	164.60 ± 8.24 a	168.61 ± 6.92 a	161.56 ± 7.81 a	170.70 ± 8.76 a
Seedling groundline diameter (cm)	28.23 ± 1.68 b	40.13 ± 2.32 a	38.75 ± 1.85 a	41.17 ± 2.00 a	38.99 ± 1.95 a
Seedling stem volume (cm^3)	460.94 ± 91.71 b	916.44 ± 148.52 a	781.92 ± 101.59 a	899.89 ± 153.92 a	844.36 ± 125.30 a

FIGURE 1. Treatment-related relationships (A) for photosynthesis (NA) and photosynthetically-active radiation (PAR) and (B) third year post-treatment stem volume and pre-treatment height. Legend: Control = solid circle and solid black line, Brushsaw = open circle and solid grey line, Silvana = diamond and dotted line, Vision = square and dashed/dotted line, Release = triangle and dashed line. Predictive equations for A are: Control, y = 4.055160x/87.55253 + x; Brushsaw, y = 5.997033x/220.8893 + x; Silvana, y = 6.824082x/238.3603 + x; Vision, y = 6.832783x/159.1110 + x; Release, y = 6.687729x/87.55253 + x. Predictive equations for B are: Control y = −694.907 + 12.239x, r = 0.885; Brushsaw, y = −939.925 + 20.616, r = 0.748; Silvana, y = −933.588 + 18.762x, r = 0.894; Vision, y = −1632.393 + 29.943x, r = 0.860; Release, y = −880.930 + 19.694x, r = 0.892. All relationships are significant.

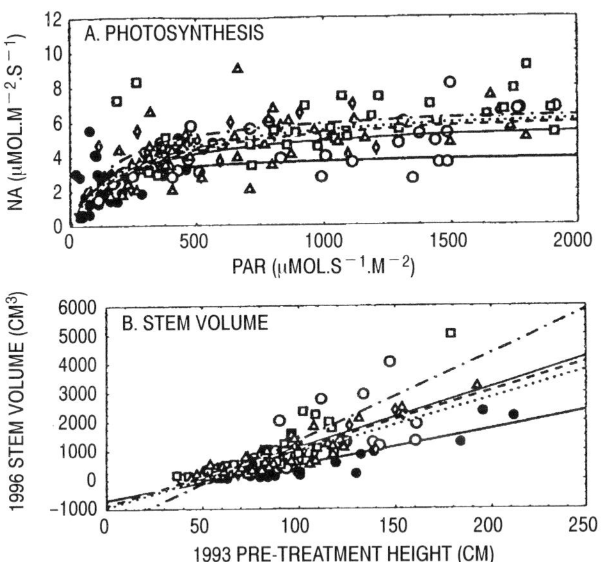

lated PAR, photosynthetic, and growth differences became apparent during the first post-treatment growing season (1994), PAR and photosynthetic differences were evident earlier in the summer than those for growth (Reynolds et al., 1998b). *In situ* treatment-related differences in PAR, photosynthesis, and growth have persisted through 1996. Because treatment-related PAR, G_S, E, and NA differences appear to be similar to those for growth, this trend suggests that these parameters may provide good indicators of both short-term and longer-term treatment-related growth responses. However, these *in situ* differences in G_S, E, and NA rates are not necessarily indicative of physiological adaptation of the seedlings to their new environments. To test whether such adaptation may have occurred, it will be necessary to evaluate gas exchange rates of foliar samples collected from all treatments under a common standardized light environment.

Long-term growth of seedlings from both cutting treatments could drop below that for the two herbicide treatments if regrowth of brush overtops the seedlings. Such overtopping will increasingly shade seedlings, and should reduce NA. The least regrowth is expected to occur for the glyphosate (Vision) treatment, where the original deciduous woody community has been replaced with a low-growing non-woody community (Bell et al., 1997; Reynolds et al., 2000).

CONCLUSIONS

After the first three post-treatment growing seasons:

- Both cutting and herbicide treatments, significantly increased available light (PAR) and seedling photosynthesis and growth (up to $2\times$) over that for untreated seedlings.
- Treatment-related differences in available light (PAR) and photosynthesis were evident during the first post-treatment growing season and have persisted through three seasons of study. These differences appear to be good early indicators of treatment-induced growth differences.
- Longer-term growth of seedlings released using brush cutting or triclopyr (Release) may decline because of anticipated regrowth of woody vegetation and overtopping, but sustained growth of seedlings released with glyphosate (Vision) is expected because of succession to a low-growing non-woody community (Bell et al., 1997; Reynolds et al., 2000).

REFERENCES

Avery, T.E. 1975. Natural resources measurements. McGraw-Hill, New York.

Barker, H. 1968. Methods of measuring leaf surface area of some conifers. Forestry Branch, Departmental Publication No. 1219. Ottawa.

Bell, F.W., R.A. Lautenschlager, R.G. Wagner, D.G. Pitt, J.W. Hawkins and K.R. Ride. 1997. Motor-manual, mechanical, and herbicide release affect early successional vegetation in northwestern Ontario. For. Chron. 73: 61-68.

Bosch, J.M. and J.D. Hewlett. 1982. A review of catchment experiments to determine the effects of vegetation changes on water yield and evapotranspiration. J. Hydrol. 55: 3-23.

Brand, D.G. and P.S. Janas. 1988. Growth and acclimation of planted white pine and white spruce seedlings in response to environmental conditions. Can. J. For. Res. 18: 320-329.

Causton, D.R. and M.P. Dale. 1990. The monomolecular and rectangular hyperbola as empirical models of the response of photosynthetic rate to photon flux density, with applications to three *Veronica* species. Ann. Bot. 65: 389-394.

Eastman, P.A.K. and E.L. Camm. 1995. Regulation of photosynthesis in interior spruce during water stress: changes in gas exchange and chlorophyll fluorescence. Tree Physiol. 15: 229-235.

Farquhar, G.D. and R.A. Richards. 1984. Isotopic composition of plant carbon correlates with water-use efficiency of wheat genotypes. Aust. J. Plant Physiol. 11: 539-552.

Jobidon, R. 1992. Measurement of light transmission in young conifer plantations: A new technique for assessing herbicide efficacy. North. J. Appl. For. 9: 112-115.

Karakatsoulis, J., J.P. Kimmins and R.E. Bigley. 1989. Comparison of the effects of chemical (glyphosate) and manual conifer release on conifer seedling physiology and growth on Vedder Mountain, British Columbia. Pp. 168-188. In: Proceedings of the Carnation Creek Herbicide Workshop. P.E. Reynolds (ed.). FRDA Report 063, B.C. Ministry of Forests, Victoria, B.C. 349 pp.

Landsberg, J.J., C.L. Beadle, P.V. Biscoe, D.R. Butler, B. Davidson, L.D. Incoll, G.B. James, P.G. Jarvis, P.J. Martin, R.E. Neilson, D.B.B. Powell, E.M. Slack, M.R. Thorpe, N.C. Turner, B. Warrit and W.R. Watts. 1975. Diurnal energy, water and CO_2 exchanges in an apple (*Malus pumila*) orchard. J. Appl. Ecol. 12: 659-684.

Larcher, W. 1980. Physiological plant ecology. Springer-Verlag, New York.

Lautenschlager, R.A., F.W. Bell, R.G. Wagner and J.A. Winters. 1997. The Fallingsnow Ecosystem Project: Comparing conifer release alternatives in northwestern Ontario. For. Chron. 73:35-38.

Lautenschlager, R.A., F.W. Bell, R.G. Wagner, P.E. Reynolds and J.M. Woodcock. 1998. The Fallingsnow Ecosystem Project: Documenting the environmental consequences of conifer release. J. For. 96:20-27.

Lieffers, V.J., A.G. Mugasha and S.E. MacDonald. 1993. Ecophysiology of shade needles of *Picea glauca* saplings in relation to removal of competing hardwoods and degree of prior shading. Tree Physiol. 12: 271-280.

Michael, J.L. 1985. Growth of loblolly pine treated with hexazinone, sulfometuron methyl, and metsulfuron methyl for herbaceous weed control. South. J. Appl. For. 9: 20-26.

Newton, M., E.C. Cole, M.L. McCormack, Jr. and D.E. White. 1992. Young spruce-fir forests released by herbicides II. Conifer response to residual hardwoods and overstocking. North. J. Appl. For. 9: 130-135.

Potvin, C., M.J. Lechowicz and S. Tardif. 1990. The statistical analysis of ecophysiological response curves obtained from experiments involving repeated measures. Ecol. 71: 1389-1400.

Reynolds, P.E., J.A. Simpson, R.A. Lautenschlager, F.W. Bell, A.M. Gordon, D.A.

Buckley and D.A. Gresch. 1997a. Alternative conifer release treatments affect below- and near-ground microclimate. For. Chron. 73: 75-82.

Reynolds, P.E., F.W. Bell, J.A. Simpson, A.M. Gordon, R.A. Lautenschlager, D.A. Gresch and D.A. Buckley. 1997b. Alternative conifer release treatments affect leaf area index of competing vegetation and available light for seedling growth. For. Chron. 73: 83-89.

Reynolds, P.E., F.W. Bell, R.A. Lautenschlager, J.A. Simpson, A.M. Gordon, D.A. Gresch, D.A. Buckley, and J.A. Winters. 2000. Microclimate changes following alternative conifer release treatments continue through three post-treatment growing seasons. J. Sustainable For. 10(3/4):267-275.

Reynolds, P.E., N.V. Thevathasan, J.A. Simpson, A.M. Gordon, R.A. Lautenschlager, F.W. Gell, D.A. Gresch, and D.A. Buckley. 1998a. Alternative conifer release treatments affect microclimate and soil nitrogen mineralization. For. Ecol. Mgmt. (Submitted).

Reynolds, P.E., J.A. Simpson, F.W. Bell, A.M. Gordon, R.A. Lautenschlager, D.A. Gresch and D.A. Buckley. 1998b. First-year physiological and growth response of white spruce to different chemical and mechanical release treatments. For. Chron. (Submitted).

Snedecor, G.W. and W.G. Cochran. 1967. Statistical methods. The Iowa State University Press, Ames, Iowa.

Steel, R.G.D. and J.H. Torrie. 1980. Principles and procedures of statistics: A biometrical approach. 2nd ed. McGraw-Hill Book Company, NY.

Thompson, D.G., D.G. Pitt, B. Staznik, N.J. Payne, D. Jaipersaid, R.A. Lautenschlager and F.W. Bell. 1997. On-target deposit and vertical distribution of aerially released herbicides. For. Chron. 73: 47-59.

Wood, J.E. and F.W. von Althen. 1993. Establishment of white spruce and black spruce in boreal Ontario: effects of chemical site preparation and post-planting weed control. For. Chron. 69(5): 554-560.

Natural Regeneration of Grey Alder (*Alnus incana* [L.] Moench.) Stands After Harvest

L. Rytter
L. Sennerby-Forsse
A. Alriksson

INTRODUCTION

The interest in hardwood forestry has increased considerably in Sweden in recent years. The access to timber and pulpwood from domestic hardwoods is limited, and has lead to an increase in import of deciduous wood. The present forest policy emphasizes the importance of hardwoods in the forest for biodiversity purposes. However,

---

L. Rytter is affiliated with The Forestry Research Institute of Sweden, Ekebo, S-268 90 Svalöv, Sweden.

L. Sennerby-Forsse is Professor, The Forestry Research Institute of Sweden, Glunten, S-751 83 Uppsala, Sweden.

A. Alriksson is affiliated with the Swedish University of Agricultural Science, Department of Forest Soils, P.O. Box 7001, S-750 07, Uppsala, Sweden.

The field work was performed at the Department of Short Rotation Forestry in the university, and the authors wish to thank Richard Childs, Eva-Marie Fryk, Christina Segerqvist, and others for doing the various samplings and measurements.

This study was jointly supported by the National Board for Industrial and Technical Development in Sweden, the Swedish University of Agricultural Sciences, and the Forestry Research Institute of Sweden.

[Haworth co-indexing entry note]: "Natural Regeneration of Grey Alder (*Alnus incana* [L.] Moench.) Stands After Harvest." Rytter, L., L. Sennerby-Forsse, and A. Alriksson. Co-published simultaneously in *Journal of Sustainable Forestry* (Food Products Press, an imprint of The Haworth Press, Inc.) Vol. 10, No. 3/4, 2000, pp. 287-294; and: *Frontiers of Forest Biology: Proceedings of the 1998 Joint Meeting of the North American Forest Biology Workshop and the Western Forest Genetics Association* (ed: Alan K. Mitchell et al.) Food Products Press, an imprint of The Haworth Press, Inc., 2000, pp. 287-294. Single or multiple copies of this article are available for a fee from The Haworth Document Delivery Service [1-800-342-9678, 9:00 a.m. - 5:00 p.m. (EST). E-mail address: getinfo@haworthpressinc.com].

one obstacle for promoting hardwood silviculture among forest owners is the widespread doubt about profitability, where the cost of regeneration is one important part. The knowledge and information about inexpensive and effective regeneration methods of hardwood stands are therefore of utmost importance.

Tree species with a high ability to produce straight and vigorous root suckers have a natural potential for inexpensive regeneration, which should be of interest to forest owners. Grey alder (*Alnus incana* [L.] Moench.), a native species in Sweden, readily regenerates by root suckers. It also has a fast initial growth (Rytter et al., 1989; Granhall and Verwijst, 1994), and is not subject to browsing damage (Rytter, 1996). In addition, the species has the ability to fix around 100 kg ha^{-1} yr^{-1} of atmospheric nitrogen in symbiosis with the actinomycete *Frankia* (Rytter, 1996). The aims of our study were to assess the initial production level and growth dynamics of the natural sprouting after harvest and to study the sprouting pattern of root suckers and stump sprouts.

MATERIALS AND METHODS

Site Description

Two stands of almost pure grey alder in the district of Dalecarlia in Sweden were used for the study. Stand A, Askön (~15 ha), located on a slight slope along the river Dalälven (lat. 60° 10′ N; long. 16° 7′ E; alt. 105 m), consisted of a 6-7 m high heavy naturally sprouted thicket which was about 10 years old. It contained about 5000 stems and 150 m^3 of wood ha^{-1}. Old stumps indicated that a mature grey alder stand preceded the thicket but otherwise the history is unknown. Measurements of stumps and trees were undertaken in the sample plots and in the late autumn 1991, the trees were felled.

Stand B (~0.3 ha), close to Lake Flinssjön and only a kilometre from the river Dalälven (lat. 60° 23′ N; long. 16° 6′ E; alt. 85 m), was about 35 years old and consisted of naturally regenerated seedlings established after flooding of the lake in 1957. The area is a former agricultural land surrounded by grey alder. Stand B was measured in the autumn 1991 and harvested in the winter 1991/92. It contained around 1600 living stems ha^{-1} and the standing stem volume before harvest was estimated to around 360 m^3 ha^{-1} from measurements in 16 circular plots of 50 m^2 each. Stump height was 5-10 cm in both stands.

At both sites, the soil mineral material consists of water-transported sediments. The soil in stand B has a large clay fraction (Table 1), and is more nutrient rich than in stand A. The soil in stand B also contains calcium carbonate-rich clay at deeper soil horizons.

Layout and Sampling

In both stands, four plots of 10 m × 10 m each were randomly placed in selected areas of the stands. In each plot, 9 subplots of 4 m^2 each, with a one meter border, were systematically placed. The study was performed during 3 years, and 3 subplots per plot were used each year for destructive sampling during winter. The sampling order of the subplots was randomized. The shoots in each subplot were cut and separated into living or dead stump shoots and root suckers. For each shoot, stem base diameter, length and fresh weight were recorded. In the first year all shoots were dried at 85°C to constant weight and weighed, whereas thereafter every tenth shoot was dried to confirm fresh:dry weight ratios.

An additional study was carried out with the objective to improve the information of stump shoot development after harvest. In connection to, but outside the plots, 8-10 stumps were selected annually. All shoots from each stump were removed in the autumn at the same time as the plots were harvested, and measured in the same way as the shoots of the plots. The diameter of the selected stumps was recorded in 1992 and 1994.

TABLE 1. Basic soil data from the two stands. Soil samples were taken from two different soil pits in each stand. Composite samples representing each soil layer were used for chemical analyses. All analyses were performed on air dry soil samples. CEC = Cation exchange capacity; BS = Base saturation.

Site	Soil horizon	Depth cm	Clay %	Silt %	Sand %	C %	N %	pH_{H_2O}	CEC $cmol_+ kg^{-1}$	BS %
	A	0-15/20	10	62	28	2.5	0.18	4.8	4.4	67
Stand A	B	25-35	7	81	12	0.8	0.04	5.7	2.4	72
	A	0-25	36	55	9	2.1	0.19	5.2	13.6	93
Stand B	B	30-40	43	55	2	0.3	0.03	7.0	12.9	99

Statistics

The SAS procedures ANOVA and GLM with a nested random model (SAS Institute, 1990) were used to test for differences in standing biomass, shoot number, height, and diameter, between stands, years and plots in the main study. The GLM procedure was used for the same parameters in the stump shoot study, motivated by a slight unbalance in the stump number. The confidence level was 95%, and pair-wise comparisons were done with the LSD method.

RESULTS

The total leafless aboveground biomass production reached about 1 Mg dry matter ha^{-1} during the first year in both stands (Table 2). The second and third year growth increased significantly and reached nearly 5 Mg ha^{-1} yr^{-1} in stand A and over 6 Mg in stand B. After three years there were no significant differences between the stands in standing biomass and total shoot number. At this time the shoot diameter at the base was 1.5-2.0 cm, with stump shoots being slightly larger. The arithmetic mean height of living shoots had reached 203 cm and 223 cm in stand A and B, respectively. However, shoots of 4 m height or more occurred. There were no significant differences in shoot diameter and shoot height between the stands. Results from the additional study, showed more shoots per stump in stand B, where resprouting occurred from harvested mature trees, as compared to the resprouting in stand A from a young thicket (Table 3). However, after three years, the difference was no longer significant.

Root suckers dominated the resprouting pattern in both stands and constituted 58% of the total biomass in stand A, and 81% in stand B, after three years. The shoot mortality, mainly in root suckers, was pronounced in stand B in year two, when the total number of shoots decreased with approximately 35%. In both stands, the number of root suckers remained higher than the number of stump shoots. During the third year, in stand A, stump shoot growth dominated over root sucker growth. In stand B the growth of root suckers remained superior to that of stump shoots.

DISCUSSION

Productivity studies in grey alder stands show a high growth rate when conditions are favourable (Rytter, 1996). The highest figure so

TABLE 2. Resprouting characteristics after harvest in the two grey alder stands. Mean values are arithmetic averages. Different letters in the row indicate significant differences between stands and/or years at the 95% confidence level.

Site	Stand A			Stand B		
Original stand						
Generation		2			1	
Age at harvest (yrs)		10			35	
New generation						
Age (yrs)	1	2	3	1	2	3
All shoots						
Number of shoots (no. m^{-2})	18.7^{bc}	14.8^{ab}	10.7^{a}	18.8^{bc}	23.0^{c}	11.5^{ab}
Height (cm)	52.3^{a}	119.0^{b}	187.3^{c}	45.5^{a}	115.1^{b}	202.2^{c}
Diameter (mm at base)	5.9^{a}	10.6^{b}	14.8^{c}	5.6^{a}	10.1^{b}	16.2^{c}
Dry mass (g shoot^{-1})	6.1^{a}	36.9^{b}	92.3^{c}	4.7^{a}	34.5^{b}	120.5^{d}
Dry mass (g m^{-2})	113^{a}	613^{b}	1085^{c}	94^{a}	729^{b}	1370^{c}
Production (g m^{-2} yr^{-1})	113	500	472	94	636	641
Living shoots						
Number of shoots (no. mm^{-2})	18.7^{b}	12.4^{a}	8.8^{a}	18.8^{b}	14.9^{ab}	9.5^{a}
Height (cm)	52.3^{a}	131.5^{b}	202.7^{c}	45.5^{a}	142.5^{b}	223.0^{c}
Diameter (mm at base)	5.9^{a}	11.5^{b}	16.3^{c}	5.6^{a}	12.0^{b}	17.9^{c}
Dry mass (g shoot^{-1})	6.1^{a}	43.2^{b}	108.7^{c}	4.7^{a}	47.1^{b}	142.9^{d}
Dry mass (g m^{-2})	113^{a}	609^{b}	1047^{c}	94^{a}	703^{b}	1323^{c}
Stump shoots						
Number of shoots (no. mm^{-2})	5.7^{a}	3.6^{a}	3.2^{a}	5.7^{a}	4.2^{a}	2.1^{a}
Height (cm)	60.0^{a}	127.0^{b}	203.8^{c}	50.9^{a}	98.1^{b}	215.7^{c}
Diameter (mm at base)	6.3^{a}	11.2^{b}	17.1^{c}	6.5^{a}	9.3^{b}	17.2^{c}
Dry mass (g shoot^{-1})	8.2^{ab}	41.4^{b}	125.7^{c}	6.5^{a}	26.2^{ab}	137.3^{c}
Dry mass (g m^{-2})	38^{a}	177^{ab}	459^{c}	33^{a}	112^{ab}	258^{b}
Production (g DM m^{-2} yr^{-1})	38	139	282	33	79	146
Living stump shoots						
Number of shoots (no. mm^{-2})	5.7^{b}	3.1^{ab}	2.5^{ab}	5.7^{b}	2.5^{ab}	1.6^{a}
Height (cm)	60.0^{a}	142.2^{b}	223.5^{c}	50.9^{a}	140.2^{b}	244.6^{c}
Diameter (mm at base)	6.3^{a}	12.5^{b}	19.0^{c}	6.5^{a}	13.3^{b}	19.7^{c}
Dry mass (g shoot^{-1})	8.2^{ab}	48.5^{b}	149.2^{c}	6.5^{a}	52.1^{b}	171.8^{c}
Dry mass (g m^{-2})	38^{a}	176^{ab}	445^{c}	33^{a}	107^{ab}	251^{b}
Root suckers						
Number of shoots (no. mm^{-2})	13.0^{ab}	11.2^{a}	7.5^{a}	13.1^{ab}	18.9^{b}	9.4^{a}
Height (cm)	49.3^{a}	114.8^{b}	179.5^{c}	43.0^{a}	115.0^{b}	200.4^{c}
Diameter (mm at base)	5.6^{a}	10.2^{b}	14.1^{c}	5.3^{a}	10.0^{b}	16.1^{d}
Dry mass (g shoot^{-1})	5.2^{a}	33.5^{b}	77.2^{c}	4.1^{a}	34.0^{b}	118.2^{d}
Dry mass (g m^{-2})	75^{a}	436^{b}	626^{b}	60^{a}	617^{b}	1111^{c}
Production (g DM m^{-2} yr^{-1})	75	361	190	60	557	495
Living root suckers						
Number of shoots (no. mm^{-2})	13.0^{b}	9.3^{ab}	6.2^{a}	13.1^{b}	12.4^{b}	7.9^{ab}
Height (cm)	49.3^{a}	127.6^{b}	193.2^{c}	43.0^{a}	143.0^{b}	220.2^{d}
Diameter (mm at base)	5.6^{a}	11.2^{b}	15.1^{c}	5.3^{a}	11.8^{b}	17.6^{d}
Dry mass (g shoot^{-1})	5.2^{a}	39.9^{b}	88.1^{c}	4.1^{a}	46.6^{b}	138.5^{d}
Dry mass (g m^{-2})	75^{a}	432^{b}	602^{b}	60^{a}	596^{b}	1072^{c}

TABLE 3. Development, expressed as arithmetic averages, of stump shoots from the separate stump shoot study in the two grey alder stands. Different letters in the row indicate significant differences between stands and/or years at the 95% confidence level.

Site	Stand A			Stand B		
Original stand						
Generation		2			1	
Age at harvest (yrs)		10			35	
Stump basal area (cm^2)	65.1^a	-	61.9^a	428^b	-	340^b
New generation						
Age (yrs)	1	2	3	1	2	3
Stumps sampled (no.)	8	10	10	10	10	10
All shoots						
Shoots per stump (no.)	12.2^{ab}	4.5^a	6.1^{ab}	33.1^c	13.6^b	9.1^{ab}
Shoot height (cm)	55.2^a	130.8^b	199.0^c	65.2^a	119.3^b	208.4^c
Shoot diameter (mm at base)	5.3^a	12.8^b	15.8^c	6.5^a	11.1^b	16.2^c
Shoot dry mass (g stump^{-1})	63.8^a	279.9^{ab}	654.2^{cd}	274.9^{ab}	611.5^{bc}	1018.1^d
Shoot dry mass (g shoot^{-1})	5.2^a	62.2^b	107.3^c	8.3^a	45.0^b	111.9^c
Living shoots						
Shoots per stump (no.)	12.2^{ab}	3.7^a	5.1^{ab}	33.1^c	7.7^{ab}	6.9^{ab}
Shoot height (cm)	55.2^a	147.6^b	221.4^c	65.2^a	171.0^b	240.7^c
Shoot diameter (mm at base)	5.3^a	14.6^b	17.6^{bc}	6.5^a	14.9^b	19.0^c
Shoot dry mass (g stump^{-1})	63.8^a	278.5^{ab}	649.2^{bc}	274.9^{ab}	592.8^{bc}	1002.6^c
Shoot dry mass (g shoot^{-1})	5.2^a	75.3^b	127.3^c	8.3^a	77.0^b	145.3^c

far, a mean annual increment of 8.2 Mg aboveground woody dry matter ha^{-1} yr^{-1}, originates from an irrigated and fertilized plantation on former agricultural land (Granhall and Verwijst, 1994). Also in initially poor soils the growth rate can be high provided that nutrients are supplied (Rytter et al., 1989; Rytter, 1995). However, studies in the secondary sprouting generation are rare. Given that hardwood forestry in Sweden is hampered by expensive stand establishment and problems with browsing damage, the possibility of using a species which exhibits cost-effective natural regeneration after harvest and which is not severely damaged by animals (e.g., Wentz, 1982; Danell et al., 1991) warrants evaluation. In this study, a production level of 5-6 Mg dry matter ha^{-1} yr^{-1} was reached the second year after harvest (Table 2). This level is not as high as has been recorded in willow stands on fertilized plots with genetically improved plant material (Christersson, 1986; Rytter and Ericsson, 1993; Willebrand et al., 1993), where a total of 20 Mg of dry matter ha^{-1} can be recorded two years after harvest. If compared with other sprouting species the growth level in this study can compete well. For example, Ferm (1990) estimated an average annual coppice production of 4-5 Mg woody dry matter ha^{-1} in pubescent birch (*Betula pubescens* Ehrh.) during 10 years.

There were few significant differences in growth levels and growth dynamics between the two stands (Tables 2, 3), irrespective of the different ages of the preceding stands and different site conditions (Table 1). The increased number of shoots in stand B the second year, compared with the first year, indicate that sprouting of root suckers continued also the second year. As a consequence of the high shoot density attained, a large number of dead shoots were found at the end of the season (Table 2). The somewhat decreased growth in stand A the third year may have been a combined effect of dry soil conditions and a precipitation deficit. Saarsalmi (1995) suggested that old age may be a reason for low production of naturally regenerated grey alder. The overall impression from our study, though, is that grey alder does not seem to be sensitive to stand age or soil conditions for successful stand regeneration. The older stand (B) showed more sprouts per stump than the younger (A) (Table 3), although on an area basis, there was no significant difference between the stands (Table 2). Paukkonen et al. (1992) found that repeated coppicing enhanced sprouting ability of grey alder stumps, which thus emphasizes the possibilities of successful future regeneration.

Different tree species show different strategies in resprouting. Among the most prominent coppicing species are willows and poplars, which mainly resprout through stump shoots (Sennerby-Forsse et al., 1992). In grey alder, root suckers are common and dominated in the two stands studied here, although stump sprouting showed an important magnitude. Root sucker regeneration leads to a spatially more even stand and the stem number per unit area of the preceding stand is less important than in stands regenerated by stump sprouts. Furthermore, our investigation supports the opinion that regenerations with root suckers often result in trees with straight stems and good quality.

CONCLUSIONS

Results of this study, and other available information, suggest that grey alder has the potential to be an economic competitive alternative in hardwood forestry, especially when regeneration ability and initial growth development is considered. Grey alder has also been shown to improve soil conditions, to tolerate acid soil conditions (cf. Rytter, 1996), and logs are currently well paid. Furniture manufacturers have recently increased their interest in the species. All these are factors that should be of interest for the forest owner. Furthermore, the common observation that root suckers in general grow straighter than stump sprouts and that grey alder does not attract browsing animals points towards a possible development of high quality stems in these stands.

REFERENCES

Christersson, L. 1986. High technology biomass production by *Salix* clones on a sandy soil in southern Sweden. Tree Physiology 2: 261-272.

Danell, K., L. Edenius and P. Lundberg. 1991. Herbivory and tree stand composition: moose patch use in winter. Ecology 72: 1350-1357.

Ferm, A. 1990. Coppicing, aboveground woody biomass production and nutritional aspects of birch with specific reference to *Betula pubescens*. Metsäntutkimuslaitoksen Tiedonantoja 348, 35 pp., Kannus, Finland.

Granhall, U. and T. Verwijst. 1994. Grey alder (*Alnus incana*)–an N_2-fixing tree suitable for energy forestry. In: pp. 409-413. D.O. Hall, G. Grass and H. Schemer (eds.). Biomass for Energy and Industry. Ponte Press, Bochum, Germany.

Paukkonen, K., A. Kauppi and A. Ferm. 1992. Root and stump buds as structural faculties for reinvigoration in *Alnus incana* (L.) Moench. Flora 187: 353-367.

Rytter, L. 1995. Effects of thinning on the obtainable biomass, stand density, and tree diameters of intensively grown grey alder plantations. For. Ecol. Manage. 73: 135-143.

Rytter, L. 1996. Grey alder in forestry: a review. Norwegian J. Agric. Sci. Suppl. 24: 61-78.

Rytter, L. and T. Ericsson. 1993. Leaf nutrient analysis in *Salix viminalis* (L.) energy forest stands growing on agricultural land. Z. Pflanzenernähr. Bodenk. 156: 349-356.

Rytter, L., T. Slapokas and U. Granhall. 1989. Woody biomass and litter production of fertilized grey alder plantations on a low-humidified peat bog. For. Ecol. Manage. 28: 161-176.

Saarsalmi, A. 1995. Nutrition of deciduous tree species grown in short rotation stands. Diss., University of Joensuu, Faculty of Forestry, Joensuu, Finland, 60 pp.

SAS Institute. 1990. SAS/STAT User's Guide Version 6. 4th ed., SAS Institute, Cary, N.C., USA, 1686 pp.

Sennerby-Forsse, L., A. Ferm and A. Kauppi. 1992. Coppicing ability and sustainability. In: pp. 146-184. C.P. Mitchell, J.B. Ford-Robertson, T. Hinckley and L. Sennerby-Forsse (eds.). Ecophysiology of Short Rotation Forest Crops. Elsevier Applied Science, London, Great Britain.

Wentz, F. 1982. Erfahrungen mit der Dreierpflanzung in wildverbissenen Forstkulturen auf Rutschhängen. Allgemeine Forstzeitschrift 37: 1124-1126.

Willebrand, E., S. Ledin and T. Verwijst. 1993. Willow coppice systems in short rotation forestry: effects of plant spacing, rotation length and clonal composition on biomass production. Biomass and Bioenergy 4: 323-331.

Seasonal Fine Root Carbohydrate Relations of Plantation Loblolly Pine After Thinning

Mary A. Sword
Eric A. Kuehler
Zhenmin Tang

INTRODUCTION

Loblolly pine (*Pinus taeda* L.) occurs naturally on soils that are frequently low in fertility and water availability (Allen et al., 1990; Schultz 1997). Despite these limitations, this species maintains a high level of productivity on most sites (Schultz, 1997). Knowledge of plantation loblolly pine root system growth and physiology is needed to understand how this species is adapted to soil resource limitations, and how management can be used to favor root system function.

The fine root dynamics of mature conifers is modal with a distinct

Mary A. Sword is Research Plant Physiologist, U.S. Department of Agriculture, Forest Service, Southern Research Station, 2500 Shreveport Highway, Pineville, LA 71360 USA.

Eric A. Kuehler is Plant Physiologist, U.S. Department of Agriculture, Forest Service, Southern Research Station, Pineville, LA 71360 USA.

Zhenmin Tang is Post Doctoral Research Associate, School of Forestry, Wildlife and Fisheries, Louisiana State University, Baton Rouge, LA 70803 USA.

Address correspondence to Mary A. Sword at the above address (E-mail: MSWORD/r8_kisatchie@FS.FED.US).

[Haworth co-indexing entry note]: "Seasonal Fine Root Carbohydrate Relations of Plantation Loblolly Pine After Thinning." Sword, May A., Eric A. Kuehler, and Zhenmin Tang. Co-published simultaneously in *Journal of Sustainable Forestry* (Food Products Press, an imprint of The Haworth Press, Inc.) Vol. 10, No. 3/4, 2000, pp. 295-305; and: *Frontiers of Forest Biology: Proceedings of the 1998 Joint Meeting of the North American Forest Biology Workshop and the Western Forest Genetics Association* (ed: Alan K. Mitchell et al.) Food Products Press, an imprint of The Haworth Press, Inc., 2000, pp. 295-305. Single or multiple copies of this article are available for a fee from The Haworth Document Delivery Service [1-800-342-9678, 9:00 a.m. - 5:00 p.m. (EST). E-mail address: getinfo@haworthpressinc.com].

surge of growth in spring, and the variable occurrence of a second flux of growth in fall (Gholz et al., 1986; Santantonio and Santantonio, 1987). We found that the new root growth of plantation loblolly pine peaked in May through July, and continued at a reduced rate in fall without a second surge of growth (Sword et al., 1998a, b).

We are conducting intensive research in a loblolly pine plantation to identify the physiological mechanisms that control root growth, and determine how silvicultural treatments affect these mechanisms. As a component of this research, the objectives of the present study were to simultaneously document the seasonal root growth and carbohydrate concentrations of thinned and non-thinned loblolly pine, and use these results together with canopy information to propose the sources of energy for root growth.

METHODS

This study was conducted in a 15-year-old loblolly pine plantation located on the Palustris Experimental Forest in Rapides Parish, LA. The soil is a Bearegard silt loam that is inherently low in fertility. In 1981, container-grown loblolly pine from a genetically unimproved source were planted (1.8 m $\times$ 1.8 m). In 1988, 6 - 0.06 ha plots, 13 $\times$ 13 trees each, were established. Two levels of thinning (not thinned: 2,732 trees ha^{-1}, 27.3 m^2 ha^{-1}; thinned: 721 trees ha^{-1}, 7.1 m^2 ha^{-1}) were randomly applied in three replications. In March 1995, the thinned plots (25 m^2 ha^{-1}) were re-thinned from below (not thinned: 42 m^2 ha^{-1}; thinned: 15.6 m^2 ha^{-1}).

From June 1995 to January 1996, root length (mm dm^{-2}) was quantified biweekly in five Plexiglas rhizotrons per plot of two replications that were grouped as blocks based on topography (Sword et al., 1998a). At 2- to 4-week intervals from March 1995 to January 1996, ten soil cores (6.5 cm diameter $\times$ 15 cm long), were extracted from random locations in plot peripheries. Cores were pooled and live fine roots ($<$ 1 mm in diameter) were extracted from cores, washed, frozen, lyophilized and ground in a Wiley mill (40 mesh). Live roots were distinguished based on color, pliability, cohesion between the cortex and vascular cylinder, and the presence of meristematic root tips.

Concentrations of root starch, sucrose and glucose were determined enzymatically by the method of Jones et al. (1977) with modifications for loblolly pine. After extraction, starch, sucrose and hexoses (glucose and fructose) were enzymatically converted to glucose. Glucose was assayed by glycolysis with the production of one unit of reduced nicotinamide adenine dinucleotide phosphate (NADPH) per unit of glucose. The NADPH was quantified spectrophotometrically by absorption at 320 nm. Carbohydrate concentrations are expressed as mg g^{-1} ash-free dry weight of fine root tissue.

In one replication, photosynthetic photon flux density (PPFD) at three randomly chosen, south-facing locations in the upper and lower one-third of the canopy were measured hourly. Values of PPFD were quantified as an average of two opposing photodiodes (Sword et al., 1998a). Branch environmental measurements were initiated in April 1995 and terminated in July due to electrical failure. Values of PPFD are expressed as the mean of four hourly midday measurements between 10:00 a.m. and 2:00 p.m.

Cumulative root lengths were transformed to natural logarithm (Y + 1) values where Y was equal to root length to insure that the data were normally distributed. Transformed root length data collected at each measurement interval were subjected to analyses of variance using a randomized complete block design with two replications. To obtain normally distributed root carbohydrate data for the evaluation of root carbohydrate concentrations in response to measurement interval, transformed data (natural logarithm [Y + 1], where Y = carbohydrate concentration), were grouped into three time periods (March-May 1995, June-August 1995, and September 1995-January 1996). Carbohydrate data in each time period were subjected to analyses of variance using a randomized complete block, split plot in time design with two replications. Main and interaction effects were considered significant at $P < 0.05$, and significantly different means were compared by the Ryan-Einot-Gabriel-Welsch multiple range test at $P < 0.05$, unless otherwise noted (SAS Institute, 1991).

RESULTS

Fine root starch and sucrose concentrations were significantly different within the three time periods (Table 1). Fine root starch concentration was greatest in March and April (98.8 mg g^{-1}), decreased significantly between April and May, continued to decrease during June through August and reached a minimum in late July and August

TABLE 1. Probabilities of a greater F-value associated with fine root starch, sucrose and glucose concentrations during March-May 1995, June-August 1995, and September 1995-January 1996 in a 15-year-old loblolly pine plantation subjected to two levels of thinning treatment.[1]

Source	df	Carbohydrate (mg g^{-1} dry weight)		
		Starch	Sucrose	Glucose
March-May 1995				
Block (R)	1	0.6992	0.2956	0.0025
Thinning (D)	1	0.4176	0.0852*[2]	0.0077*
Time (T)	4	0.0024**	0.0200**	0.0716
T × D	4	0.8937	0.4416	0.9361
June-August 1995				
Block (R)	1	0.9291	0.9265	0.3201
Thinning (D)	1	0.9163	0.3757	0.4799
Time (T)	4	0.0508*	0.0362**	0.2507
T × D	4	0.0142**	0.6464	0.4336
September 1995-January 1996				
Block (R)	1	0.8930	0.3720	0.8940
Thinning (D)	1	0.2256	0.1840	0.3870
Time (T)	3	0.0009**	0.0035**	0.0012**
T × D	3	0.6231	0.2731	0.3822

[1] Analyses were conducted on transformed data [natural logarithm (Y + 1), where Y = starch, sucrose or glucose concentration].
[2] Significance at P < 0.05 and 0.10 are noted by "**" and "*", respectively.

(12.0 mg g^{-1}) (Figure 1). During September-January, significant increases in fine root starch concentration were observed monthly. Significantly lower fine root sucrose concentrations were observed in March (14.1 mg g^{-1}), April (10.8 mg g^{-1}), early June (12.7 mg g^{-1}), and January (12.2 mg g^{-1}). Mean fine root sucrose concentrations in May, mid-June-August, and September-December were 19.6, 23.0, and 21.0 mg g^{-1}, respectively.

Fine root glucose concentration varied significantly in March-May ($P < 0.10$), and September-January (Table 1). Significantly greater fine root glucose concentrations were observed in March (20.4 mg g^{-1}), than in April-May (15.2 mg g^{-1}), and in December-January (23.3 mg g^{-1}), than in September-November (16.4 mg g^{-1}) (Figure 1).

The concentration of sucrose in fine roots was significantly in-

FIGURE 1. Fine root starch (A) sucrose (B) and glucose (C) concentrations of 15-year-old loblolly pine during three time periods: March-May 1995 (black bars), June-August 1995 (white bars), and September 1995-January 1996 (hatched bars). Within time periods, means associated with the same letter are not significantly different by the Ryan-Einot-Gabriel-Welsch multiple range test at $P < 0.05$ (lower case) and $P < 0.10$ (upper case). Error bars represent one standard error of the mean.

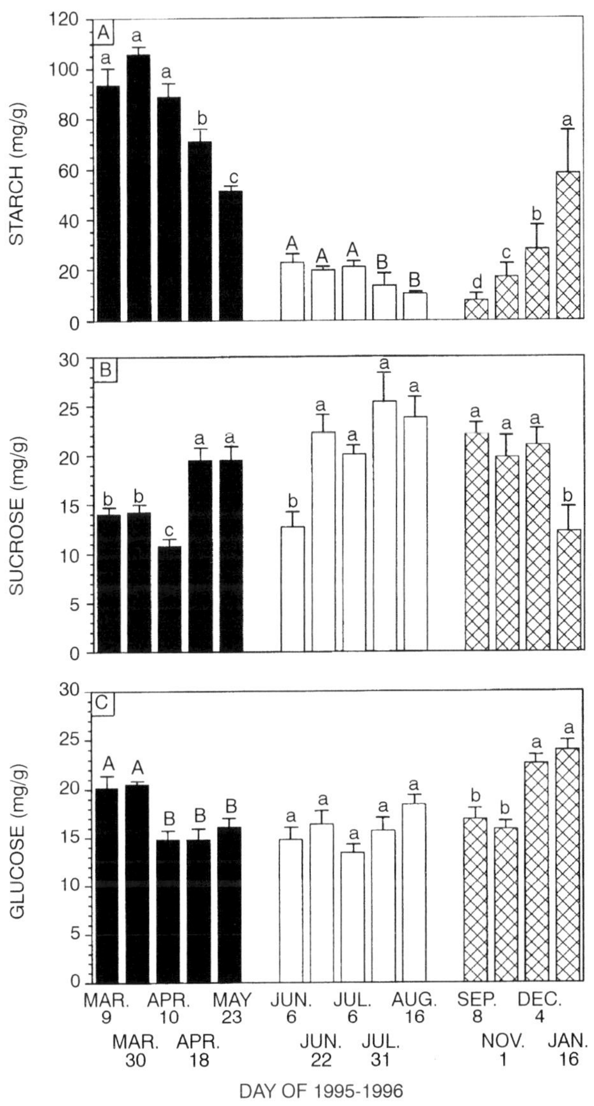

creased (21%) in response to thinning during March-May ($P < 0.10$), but was unaffected by thinning during June-August, and September-January (Table 1). Similarly, thinning was associated with a small but significant increase (2%) in fine root glucose concentration during March-May.

At each measurement interval except July 25 and August 8, the cumulative root length in rhizotrons was significantly greater on the thinned plots than on the non-thinned plots at $P < 0.05$ (Figure 2). On July 25 and August 8, cumulative root length was significantly increased in response to thinning at $P < 0.10$.

Mean midday bidirectional PPFD in the lower crown of the thinned plot (205 μmol m^{-2} sec^{-1}) was 40% greater than that on the non-thinned plot (147 μmol m^{-2} sec^{-1}). In the upper crown, bidirectional PPFD was 11% greater on the thinned plot (813 μmol m^{-2} sec^{-1}) than on the non-thinned plot (733 μmol m^{-2} sec^{-1}).

DISCUSSION

Fine root starch and glucose concentrations of loblolly pine exhibited distinct seasonal patterns. Starch concentration was greatest in March through early April, declined to a minimum in late July and August, and began accumulating in November. This pattern of root starch dynamics is similar to those reported by other investigators for southern pine species in forest stands (Adams et al., 1986; Gholz and Cropper, 1991).

The concentration of glucose in fine roots decreased between late March and mid-April, remained relatively constant between mid-April and early November, and increased in early December. This pattern may delineate the period of maximum loblolly pine root metabolism. At our study site, therefore, rhizotron measurements of loblolly pine root phenology and growth must start in March rather than May.

We have observed that the majority of loblolly pine root growth occurs in late spring and summer (Sword et al., 1998a, b). In the present study, root growth in May through June corresponded to the period of starch depletion from fine roots. This suggests that stored starch is one source of energy for spring root growth.

Fine roots consistently contained sucrose during the period of starch depletion indicating that a carbohydrate source, in addition to starch, was available for root metabolism in spring. During spring and

FIGURE 2. Cumulative length of 15-year-old loblolly pine roots in rhizotrons on thinned and non-thinned plots during May 1995-January 1996. Within all measurement intervals except July 25 and August 8, means were significantly different at $P < 0.05$. Means associated with July 25 and August 8 were significantly different at $P < 0.10$. Error bars represent one standard error of the mean.

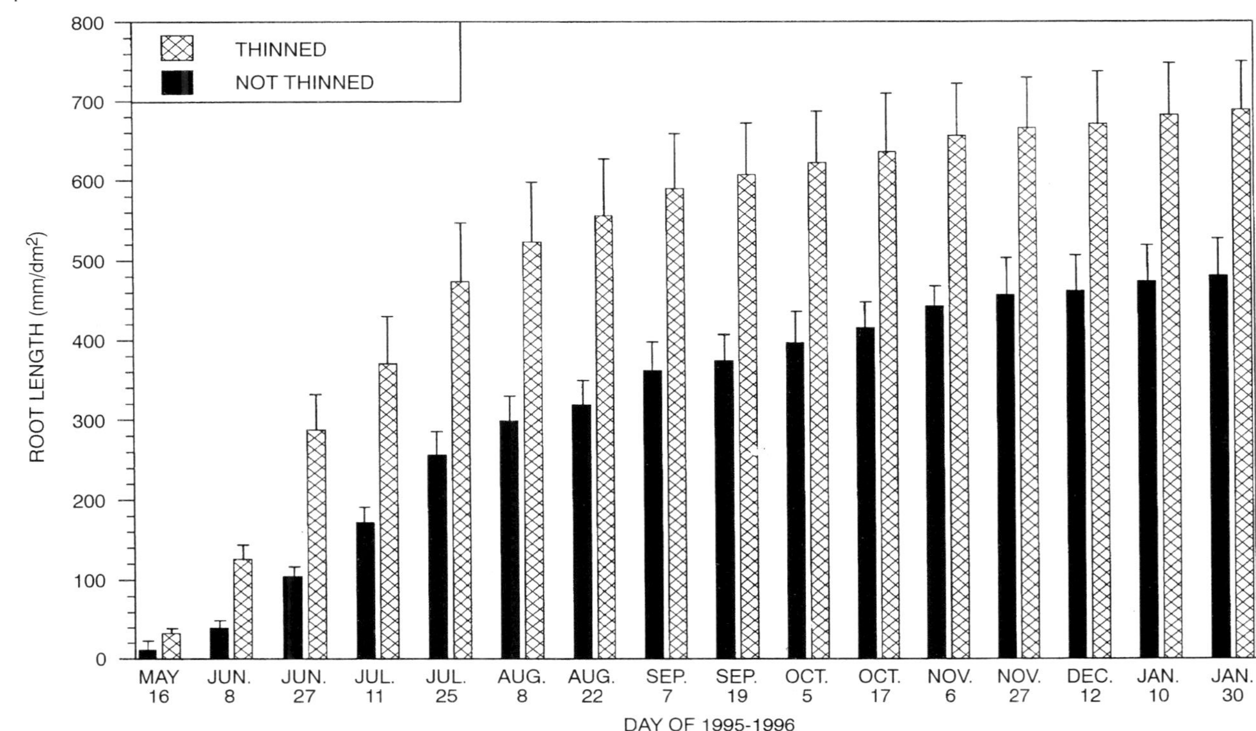

summer, the crown of loblolly pine grows vigorously with the formation of two to four successive flushes (Stenberg et al., 1994). In general, photosynthate is preferentially allocated to developing branches until new foliage becomes autonomous at approximately 50% expansion (Dickson, 1989; Dickson, 1991). In our study, the fascicles of the first flush reached 50% expansion in mid-June (Tang, personal communication). Before mid-June, therefore, it is unlikely that root growth was sustained by the age class of pre-existing fascicles that functioned as an energy source for new shoot growth.

Gordon and Larson (1970) used $^{14}CO_2$ to determine the pattern of photosynthate allocation after assimilation by one- and two-year-old needles of four-year-old red pine (*Pinus resinosa* Ait.). They found that photosynthate produced by the one-year-old needles was predominantly allocated to shoot growth, and that produced by the two-year-old needles was allocated to root growth. We hypothesize that before mid-June, sucrose produced by the foliage of the youngest pre-existing flush was allocated to new shoot growth, whereas that produced by older pre-existing foliage was allocated to root growth.

In addition to photosynthate produced by older foliage, stored carbohydrates may have been mobilized in the stem and older foliage, translocated to the root system and used as an energy source for root growth in spring (Egger et al., 1996;, Hansen et al., 1996; Schier, 1970). After labeling red pine seedlings with $^{14}CO_2$, Schier (1970) found that mobilized ^{14}C in the roots originated primarily from two-year-old needles. Egger et al. (1996) found that high concentrations of starch and sugar were stored in the pre-existing needles, sapwood and bark of Norway spruce (*Picea abies* [L.] Karst.) seedlings in winter and spring, and suggested that a portion of these reserves was mobilized for root growth.

Fine root sucrose concentration was variable during the growing season. The increase in fine root sucrose concentration observed between April and May may have been caused by a positive effect of spring warming on photosynthesis (Teskey et al., 1994). In mid-June, the concentration of sucrose in fine roots increased and remained elevated until January. This increase in fine root sucrose concentration may signify a shift in the allocation of photosynthate produced by the youngest flush of the previous year. Specifically, autonomy of the newly developed fascicles in mid-June may have resulted in basipetal allocation of photosynthate produced by the youngest flush of the previous year and a mid-June increase in the availability of sucrose for root metabolism.

On forest sites with limited mineral nutrient and water resources, silvicultural manipulation of root carbohydrate availability could be valuable if, in turn, new root growth increased. In our study, new root growth was stimulated by thinning. We did not observe an effect of thinning on fine root starch concentration suggesting that thinning did not influence the role of root starch in early root growth. Thinning, however, was associated with an increase in fine root sucrose and glucose concentrations in March through May. Elevated concentrations of fine root sucrose and glucose on the thinned plots during this period may be attributed to an increase in the allocation of photosynthate from older foliage, or greater mobilization of stored carbohydrates from the older foliage and/or stem.

Similar to Gravatt et al. (1997), we observed an increase in the availability of light in the lower crown in response to thinning. In 1995, Yu (1996) found that the leaf area per tree on the thinned plots was greater than that on the non-thinned plots. As older foliage became shaded, its carbon balance may have decreased below the light compensation point resulting in senescence (Schoettle and Fahey, 1994). Perhaps the positive effect of thinning on loblolly pine root growth was caused by several factors. First, thinning increased light availability in the lower crown and leaf area per tree. As a result, the basipetal translocation of photosynthate or mobilized carbohydrates from the older fascicles and/or stem during spring was increased and early root growth was stimulated. Maintenance of a nearly constant difference in cumulative root length between thinned and non-thinned plots after late June, and the absence of additional significant effects of thinning on root carbohydrate concentrations after May, suggest that the positive effect of thinning on root growth occurred early in the growing season and persisted through January.

CONCLUSIONS

Seasonal patterns of root growth and carbohydrate dynamics in our study suggest that both root starch and translocated sucrose were sources of energy for loblolly pine root growth. Thinning resulted in a greater concentration of sucrose in fine roots during spring, and more cumulative root length throughout the growing season. The positive effect of thinning on loblolly pine root sucrose concentration and growth may have been governed by increases in light availability and the leaf area of pre-existing foliage. Further research is warranted to investigate the influence of different age classes of foliage on loblolly pine root growth, the effect of the stand environment on fascicle abscision, and the effect of premature foliage loss on root growth.

REFERENCES

Adams, M.B., Allen, H.L. and C.B. Davey. 1986. Accumulation of starch in roots and foliage of loblolly pine (*Pinus taeda* L.): effects of season, site and fertilization. Tree Physiol. 2: 35-46.

Allen, H.L., P.M. Dougherty and R.G. Campbell. 1990. Manipulation of water and nutrients–Practice and opportunity in southern U.S. pine forests. For. Ecol. Manage. 30:437-453.

Dickson, R.E. 1989. Carbon and nitrogen allocation in trees. In: Dreyer, E., Aussenac, G., Bonnett-Masimbert, M., Dizengremel, P., Favre, J.M., Garrec, J.P., Le Tacon, F., Martin, F. eds. Forest Tree Physiology, Ann. Sci. For. 46 (suppl.). Paris: Elsevier and Institut National de la Recherche Agronomique: 631s-647s.

Dickson, R.E. 1991. Assimilate distribution and storage. In: Raghavendra, A.S., ed. Physiology of Trees. New York: John Wiley & Sons: 51-85.

Egger, B., Einig, W., Schlereth, A., Wallenda, T., Magel, E., Loewe, A. and R. Hampp. 1996. Carbohydrate metabolism in one- and two-year-old spruce needles, and stem carbohydrates from three months before until three months after bud break. Physiol. Plant. 96: 91-100.

Gholz, H.L., Hendry, L.C. and W.P. Cropper. 1986. Organic matter dynamics of fine roots in plantations of slash pine (*Pinus elliotii*) in north Florida. Can J. For. Res. 16: 529-538.

Gholz, H.L., and W.P. Cropper, Jr. 1991. Carbohydrate dynamics in mature *Pinus elliottii* var. *elliottii* trees. Can. J. For. Res. 21: 1742-1747.

Gordon, J.C. and P.R. Larson. 1970. Redistribution of ^{14}C-labeled reserve food in young red pines during shoot elongation. For. Sci. 16: 14-20.

Gravatt, D.A., Chambers, J.L. and J.P. Barnett. 1997. Temporal and spatial patterns of net photosynthesis in 12-year-old loblolly pine five growing seasons after thinning. For. Ecol. Manage. 97: 73-83.

Hansen, J., Vogg, G. and E. Beck. 1996. Assimilation, allocation and utilization of carbon by 3-year-old Scots pine (*Pinus sylvestris* L.) trees during winter and early spring. Trees 11: 83-90.

Jones, M.G.K., Outlaw, W.H. and O.L. Lowry. 1977. Enzymatic assay of 10^{-7} to 10^{-14} moles of sucrose in plant tissues. Plant Physiol. 60: 379-383.

Santantonio, D. and E. Santantonio. 1987. Effect of thinning on production and mortality of fine roots in a *Pinus radiata* plantation on a fertile site in New Zealand. Can. J. For. Res. 17: 919-928.

SAS Institute. 1991. SAS/STAT User's Guide Release 6.03. SAS Institute, Cary, NC.

Schier, G.A. 1970. Seasonal pathways of ^{14}C-photosynthate in red pine labeled in May, July, and October. For. Sci. 16: 1-13.

Schoettle, A.W. and T.J. Fahey. 1994. Foliage and fine root longevity of pines. Ecol. Bull. 43: 136-153.

Schultz, R.P. 1997. Loblolly Pine, The Ecology and Culture of Loblolly Pine (*Pinus taeda* L.). Agricultural Handbook 713. Washington, DC.: U.S. Department of Agriculture, Forest Service: 495 p.

Stenberg, P., Kuuluvainen, T., Kellomäki, S., Grace, J.C., Jokela, E.J. and H.L. Gholz. 1994. Crown structure, light interception and productivity of pine trees and stands. Ecol. Bull. 43: 20-34.

Sword, M.A., Chambers, J.C., Gravatt, D.A. and J.D. Haywood. 1998a. Ecophysiological responses of managed loblolly pine to changes in stand environment. In: Mickler, R.A., ed. The Productivity and Sustainability of sOuthern Forest Ecosystems in a Changing Environment. New York: Springer Verlag: 185-206.

Sword, M.A., Haywood, J.D. and C.D. Andries. 1998b. Seasonal lateral root growth of juvenile loblolly pine after thinning and fertilization on a Gulf Coastal Plain site. In: Waldrop, T.A., ed. Proceedings of the Ninth Biennial Southern Silvicultural Research Conference, February 25-27, 1997, Clemson, SC. Gen. Tech. Rep. SRS-20. USDA Forest Service, Southern Research Station, Asheville, NC: 194-201.

Sword, M.A., Gravatt, D.A., Faulkner, P.L. and J.L. Chambers. 1996. Seasonal root and branch growth of 13-year-old loblolly pine five years after fertilization. Tree Physiol. 16: 899-904.

Teskey, R.O., Whitehead, D. and S. Linder. 1994. Photosynthesis and carbon gain by pines. Ecol. Bull. 43: 35-49.

Yu, S. 1996. Foliage and crown characteristics of loblolly pine (*Pinus taeda* L.) six years after thinning and fertilization. Baton Rouge, LA: Louisiana State University. 77 p. M.S. thesis.

Root Architectural Plasticity to Nutrient Stress in Two Contrasting Ecotypes of Loblolly Pine

R. L. Wu
J. E. Grissom
D. M. O'Malley
S. E. McKeand

R. L. Wu, J. E. Grissom, D. M. O'Malley, and S. E. McKeand are affiliated with the Department of Forestry, North Carolina State University, Raleigh, NC 27695-8008, USA.

Address correspondence to: R. L. Wu, Forest Biotechnology Group, Department of Forestry, Box 8008, North Carolina State University, Raleigh, NC 27695-8008 (E-mail: rwu@unity.ncsu.edu).

The authors thank Wen Zeng, Zhigang Lian, Anthony McKeand, Yi Li, Hongxiu Liu, Helen Chen, Paula Zanker and Jun Lu for technical assistance, Scot Surles for tipmoth control, and Mary Topa, Bill Retzlaff and Bailian Li for helpful discussion regarding this study. The authors also thank Mary Topa, Bill Retzlaff and two anonymous referees for constructive comments on an earlier version of this manuscript.

This research was supported by the Department of Energy (Grant #DE-FC07-97IO13527), the Department of Forestry, NCSU, Agricultural Research Service, NCSU, Tree Improvement Program, NCSU, and Forest Biotechnology Program, NCSU.

[Haworth co-indexing entry note]: "Root Architectural Plasticity to Nutrient Stress in Two Contrasting Ecotypes of Loblolly Pine." Wu, R. L. et al. Co-published simultaneously in *Journal of Sustainable Forestry* (Food Products Press, an imprint of The Haworth Press, Inc.) Vol. 10, No. 3/4, 2000, pp. 307-317; and: *Frontiers of Forest Biology: Proceedings of the 1998 Joint Meeting of the North American Forest Biology Workshop and the Western Forest Genetics Association* (ed: Alan K. Mitchell et al.) Food Products Press, an imprint of The Haworth Press, Inc., 2000, pp. 307-317. Single or multiple copies of this article are available for a fee from The Haworth Document Delivery Service [1-800-342-9678, 9:00 a.m. - 5:00 p.m. (EST). E-mail address: getinfo@haworthpressinc.com].

INTRODUCTION

Plant species are frequently composed of genetically differentiated populations (ecotypes), each of which is adapted to a different set of environmental conditions. These adjustments to environmental conditions can operate through genetic variation and phenotypic plasticity of the organisms (Schlichting, 1986, 1989). Phenotypic plasticity is the degree to which the phenotypic expression of a genotype varies under different environmental conditions (Sultan, 1987). Phenotypic plasticity may be important for plant survival in a heterogeneous environment that varies spatially and temporally (Bradshaw, 1965). Plasticity in morphological, physiological, reproductive and life history traits has been reported for many plant species (reviewed in Scheiner, 1993; Wu and Stettler, 1998).

Although root systems play an important role in plant growth and development, most studies of phenotypic plasticity have focused on the aboveground parts of plants because roots are difficult to measure. Physiological studies of tree roots have shown that high amounts of assimilates (approximately 60%) are consumed by roots under infertile conditions, whereas this proportion is decreased when trees receive fertilization (Cannell, 1989). It is not clear why decreased carbon allocation to roots under fertilization can fulfill water and nutrient requirements of rapid aboveground growth. Alterations in root architecture have been suggested to play a role in regulating water and nutrient uptakes (Lynch, 1995).

The term "root architecture" refers to the geometric deployment of root axes in a three-dimensional space (Lynch, 1995). The architecture of a root system can be defined based on the allometric relationships among root number, length, and thickness, or based on the allometric topology among the taproot, first-, second- and third-order lateral roots. Root systems show considerable architectural variation among species and populations of a given species (Lynch, 1995). We investigated patterns of ecotype variation in root architecture and response to nutrient supply in two ecologically contrasting populations of loblolly pine (*Pinus taeda* L.). We hypothesized that phenotypic plasticity of root architecture to soil nutrition would respond differently in each ecotype.

MATERIALS AND METHODS

Plant material used in this experiment was chosen from two contrasting ecotypes of loblolly pine from environments that differ in resource availability. One of the ecotypes, "Lost Pines," Texas, is adapted to droughty conditions and low soil fertility and termed "xeric," whereas the other, Atlantic Coastal Plain, is adapted to more moderate conditions and termed "mesic" (van Buijtenen, 1978). In May 1997, the seeds from five open-pollinated families of each ecotype were germinated in vermiculite. Following germination, seedlings were transplanted to 40 cm deep and 20 cm diameter plastic pots filled with pure sand, and placed in an open site at the Horticulture Field Laboratory at North Carolina State University, Raleigh. The experiment was laid out in a complete randomized design receiving two different nutrient levels and with eight seedlings per family in each level.

Nutrient treatments were initiated in late July 10 weeks after planting. The seedlings in the high nutrient regime were fertilized at 50 ppm N solution (Peters 15-16-17) every morning, and those in the low nutrient level at 10 ppm N every other morning. These solutions of nutrient were distributed separately to drip emitters placed at each pot, using a RainBird™ irrigation system. Tipmoths (*Rhyacionia* sp.) were controlled by spraying pesticide on trees once every two weeks.

Seedlings were harvested in early November, after 14 weeks of treatment. Plants were separated into above- and belowground components for biomass measurements. Root system volume, taproot length and basal diameter and the number, length, and basal diameter of lateral roots, separately into different orders, were all determined. Root system volume was measured by displacement of water in a graduated cylinder. Taproots were severed into 5 cm regions. In each region, the number of first-order laterals was counted and a representative first-order lateral root was chosen. The length and basal diameter were measured on each chosen first-order lateral, and its constituent second-order laterals were counted. A similar sampling strategy was used for measurement of the second-order laterals, though only in the top two regions (10 cm) of the taproot. For the third-order laterals, only their numbers were counted. Total number, total length and total basal area (calculated by diameter$^2 \times \pi/4$) of the second-order lateral roots were estimated for a representative first-order lateral root in the top two regions of the taproot. The taproot length, ranging from 30 to 40 cm, was excluded from all analyses, since the pots used were not deep enough to allow free growth of the taproot.

A two-way analysis of variance was used to detect effects due to differences between ecotypes, nutrient treatments, and ecotype × treatment interaction by assuming that all these effects were fixed due to their non-random sampling. Data were log transformed when traits were not normally distributed. Allometric relationships were characterized using exponential functions of the form $y = ax^b$, where x and y are either total plant biomass or the number, length, or area of different plant parts, and a and b represent the coefficient and exponent of the allometric equation, respectively (Niklas, 1994). Using this approach, we studied patterns of architectural deployment (total plant biomass vs. root architectural parameters) within a plant affected by the ecotype, treatment and ecotype × treatment interaction effects. Allometric analysis was also employed to examine the treatment effect for each ecotype and the ecotypic effect in each treatment.

RESULTS

Total plant biomass, root biomass, and root volume were not significantly different between the trees from "Lost Pines" Texas (xeric) and the Atlantic Coastal Plain (mesic) sources. However, significant differences between the two ecotypes were observed for root architectural traits (Table 1). All root traits, except for the first-order lateral number and length, were highly responsive to nutrient availability. The two ecotypes diffe ed in the response of root system volume to nutrient stress, as indicated by significant ecotype × treatment interaction. This interaction did not occur in any of the other root traits, indicating that two ecotypes responded to nutrient treatments in the same way.

The analysis of variance based on allometric relationships (using total plant biomass as a co-variable) revealed significant ecotype and treatment effects for root biomass (Table 1). The proportion of root system volume to total plant biomass was affected by treatment, with the proportion decreasing with heavy fertilization. Significant variation in architectural traits were explained by ecotype and treatment effects when total plant biomass was used as a co-variable.

For each ecotype, seedlings displayed greater total plant biomass,

TABLE 1. Mean squares due to the ecotype, treatment and ecotype × treatment interaction effects for loblolly pine seedling root traits. The transformation (trans) used for each dependent variable are indicated.

Trait	Trans	Ecotype	Treatment	Eco. × T	Res.
Degree of freedom		1	1	1	146
ANOVA without a co-variable					
Total plant biomass	none	45^{ns}	20195***	0^{ns}	211
Root biomass	none	109^{ns}	971**	7^{ns}	39
Root volume	none	2.2^{ns}	50.3***	14.6*	3.4
Taproot basal area	none	1.35**	26.94***	0.11^{ns}	0.17
Root number First-order	none	146*	36^{ns}	11^{ns}	23
Second-order	none	76*	89*	0^{ns}	14
Third-order	log	0.09^{ns}	4.03**	0.56^{ns}	0.35
Root length First-order	none	25.2**	0.0^{ns}	2.7^{ns}	2.7
Second-order	log	2.9*	3.5**	0.2^{ns}	0.5
Root basal area First-order	log	4.11***	8.07***	0.18^{ns}	0.28
Second-order	log	4.58*	14.97***	0.00^{ns}	1.27
ANOVA with total plant biomass as co-variable					
Root biomass	log	0.154**	1.176***	0.045^{ns}	0.015
Root volume	log	0.003^{ns}	1.329***	0.510***	0.031
Taproot basal area	log	0.919***	1.578***	0.021^{ns}	0.034
Root number First-order	log	0.378**	0.098^{ns}	0.032^{ns}	0.050
Second-order	log	2.467**	5.240***	0.001^{ns}	0.251
Third-order	log	0.581^{ns}	5.669***	0.809^{ns}	0.273
Root length First-order	log	0.782**	0.259^{ns}	0.115^{ns}	0.082
Second-order	log	2.89**	11.58***	0.14^{ns}	0.40
Root basal area First-order	log	4.091***	2.579***	0.120^{ns}	0.151
Second-order	log	4.597*	0.119^{ns}	0.008^{ns}	1.046

* $P < 0.05$; ** $P < 0.01$; *** $P < 0.001$; and ^{ns} Nonsignificant.

root biomass and root system volume in high compared to low fertility, but this responsiveness was stronger for total plant biomass than for the two root traits (Figures 1a, b and c). Root system volume of "mesic" was much more sensitive to nutrient availability than "xeric" (Figure 1c). Although the taproots were similar in length in all

FIGURE 1. Phenotypic means of total plant biomass (a), root biomass (b), root system volume (c), and the taproot basal area (d) for the Lost Pines Texas ("xeric," open) and Atlantic Coastal Plain ("mesic," solid) ecotypes in low and high fertility treatments

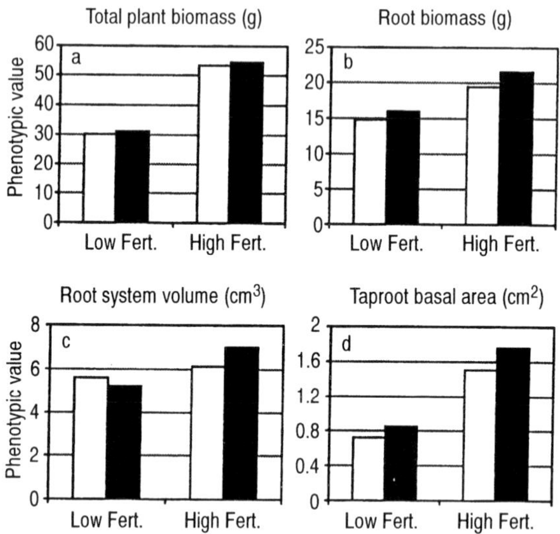

cases possibly due to the small pots used (data not shown), they were significantly thicker in high fertility than low fertility, as well as for "mesic" than "xeric," especially in high fertility ($P = 0.0001$) (Figure 1d). Different patterns and degrees of response to nutrient treatment were observed among the lateral numbers for the three branching orders of roots. The number of first-order lateral roots was similar between the two nutrient regimes for both ecotypes, but it was significantly higher for "mesic" than "xeric" in low fertility (Figure 2a). The number of second-order lateral roots generated from a given first-order lateral decreased significantly with increased nutrient availability (Figure 2b) and more so for third-order laterals (Figure 2c). In each nutritional level, there were more second-order lateral roots for "xeric" than "mesic" ($P < 0.05$). Similar trends held for the lateral root length (Figures 2d, e), except for no difference in the second-order length between the two ecotypes in the low nutrient treatment. The basal area of lateral roots was strikingly larger in high than low fertility, but the pattern of ecotypic variation differed between first and second orders (Figures 2f, g). In both nutritional levels, first-order

FIGURE 2. Phenotypic means of the number, length and basal area of lateral roots separated for first, second and third orders for the Lost Pines Texas ("xeric," open) and Atlantic Coastal Plain ("mesic," solid) ecotypes in low and high fertility treatments. Variables in the second- and third-order laterals were measured for a representative first-order lateral root.

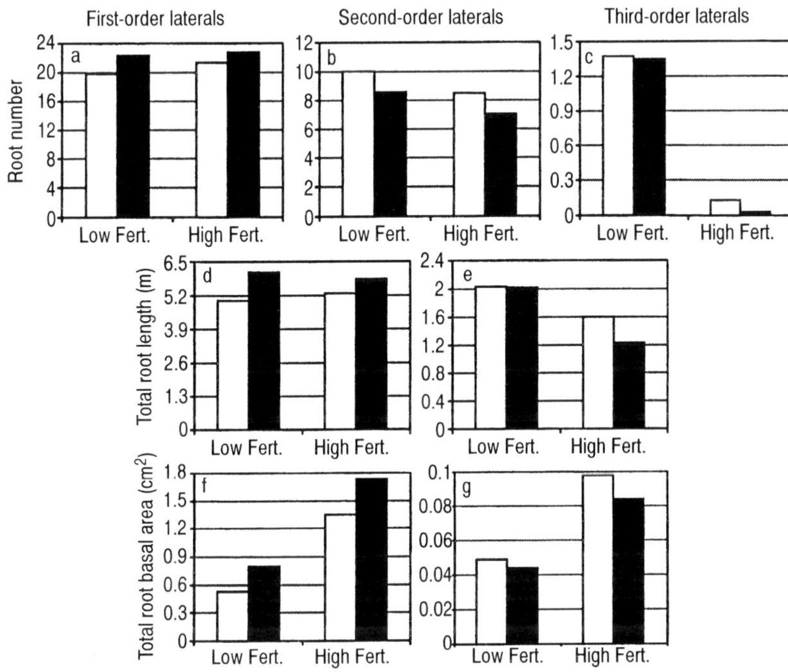

laterals of "mesic" were thicker than "xeric," whereas the ecotype response was reversed for the second-order laterals.

DISCUSSION

Many physiological studies have been carried out to investigate the effect of nutritional levels on biomass allocation of forest trees (Ledig et al., 1970; Cromer and Jarvis, 1990; Li et al., 1991; Thomas et al., 1994; Gebauer et al., 1996). How a root system responds to nutrient availability by altering its morphology has received less attention. As important determinants of the capacity for resource acquisition, root architectural traits may vary independently of biomass allocation

(Cody, 1986; Körner and Renhardt 1987; Berntson et al., 1995; Lynch, 1995). We present strong evidence that pine root architectural traits are highly plastic in contrasting nutrient regimes. Under high nutrient conditions, first-order roots of both ecotypes tend to be less branched, and there are fewer, shorter, but thicker high-order laterals in the high compared to the low fertility condition. A similar phenomenon has also been found in other plants. For example, beans developed vigorous, highly branched root systems with a larger number of apices when nutrients were limited (Lynch and van Beem, 1993).

Total plant biomass increased more than root biomass or root system volume at high nutrient level for loblolly pine seedlings. High levels of nutrients can decrease the partitioning of assimilates to the root system (reviewed in Cannell, 1989). Our results suggested that this response may be manifested through decreased root numbers and lengths in higher-order laterals (Figure 2b, c, and e). However, the total basal area of roots of all orders was dramatically increased under higher nutrient levels. Since root basal area determines the surface area of root exposed to the soil, increased basal area may indicate that the function of roots in the uptake of water and nutrients is increased by fertilization for the young pine seedlings. With a higher level of nutrition, a proportionately smaller root system could acquire sufficient resources to support a larger shoot.

Although the two ecotypes did not display significant differences in absolute root biomass, they differed in two kinds of root traits, root morphology and branching pattern, regardless of whether allometric relationships were considered. The "mesic" ecotype appeared to invest more carbon in the taproot and in the first-order lateral roots, whereas "xeric" displayed heavier second-order lateral growth, especially in high fertility. The taproot and first-order laterals for "mesic" showed greater thickness, but the second-order laterals for "xeric" showed greater root number, length and thickness.

Findings from a concurrent field study indicate that "mesic" had much greater stem volume growth at four years of age than "xeric" in fertilized plots (McKeand et al., 2000). Thus, our seedling study results suggest that greater growth of "mesic" at four years may be due to reduced costs for higher-order root production. Our results suggest that the root surface area of lower-order, but still functioning, laterals may be increased for the four-year old trees experiencing high nutrient regimes or conditions. In an earlier study, deeper root systems and wider ranging laterals were observed for the drought-hardy ecotype compared with a drought-susceptible ecotype in eastern Texas (van Buijtenen et al., 1976). All these ecotypic differences suggest that these plants have evolved root systems for the efficient exploitation of soil resources in their native environments.

In this study, we did not find significant ecotypic variation in the plasticity of root architectural traits, although such variation occurred in root system volume. Since phenotypic plasticity is controlled by genetic factors, as observed in many other plant species (Bradshaw 1965; Schlichting, 1986; Wu, 1997), a more detailed understanding of the genetic basis for phenotypic plasticity in root traits should be intriguing to students of root physiology and genetics.

CONCLUSIONS

In this experiment we evaluated root architectural plasticity and ecotypic differentiation of root system development in loblolly pine seedlings. The main findings were:

- Root architectural traits, such as root number, length, and thickness of the taproot and lateral roots, were responsive to nutrient levels. Roots were fewer in number, less branched, and shorter, but much thicker for the trees in the higher nutrient treatment.
- In the high nutrient treatment, the partitioning of assimilates to roots was proportionately decreased, but root surface area exposed to the soil was increased, relative to the lower nutrient treatment. Thus, total plant biomass in young pine seedlings appeared to increase in the high nutrient treatment by enhancing the efficiency of roots to acquire soil resources.
- The mesic Atlantic Coastal Plain ecotype displayed greater thickness of first-order lateral roots, whereas second-order lateral growth including formation, elongation, and thickening was more prominent for xeric "Lost Pines" in Texas. Thus, greater volume growth of "mesic" in the field could be due to its balance between minimum root production and maximum root surface area for water and nutrient uptake.

REFERENCES

Berntson, G.M., E.J. Farnsworth and F.A. Bazzaz. 1995. Allocation, within and between organs, and the dynamics of root length changes in two birch species. Oecologia 101: 439-447.

Bradshaw, A.D. 1965. Evolutionary significance of phenotypic plasticity in plants. Adv. Genet. 13: 115-155.

Cannell, M.G.R. 1989. Physiological basis of wood production: a review. Scand. J. For. Res. 4: 459-490.

Cody, M.L. 1986. Roots in plant ecology. Trend. Ecol. Evol. 1: 76-78.

Cromer, R.N. and P.G. Jarvis. 1990. Growth and biomass partitioning in *Eucalyptus grandis* seedlings in response to nitrogen supply. Aust. J. Plant Physiol. 17: 503-515.

Gebauer, R.L.E., J.F. Reynolds and B.R. Strain. 1996. Allometric relations and growth in *Pinus taeda*: The effect of elevated CO_2 and changing N availability. New Phytol. 134: 85-93.

Körner, C. and U. Renhardt. 1987. Dry matter partitioning and root/leaf area ratios in herbaceous perennial plants with diverse altitudinal distribution. Oecologia 74: 411-418.

Ledig, F.T., F.H. Bormann and K.F. Wenger. 1970. The distribution of dry matter growth between shoot and roots in loblolly pine. Bot. Gaz. 131: 349-359.

Li, B., H.L. Allen and S.E. McKeand. 1991. Nitrogen and family effects on biomass allocation of loblolly pine seedlings. For. Sci. 37: 271-283.

Lynch, J. 1995. Root architecture and plant productivity. Plant Physiol. 109: 7-13.

Lynch, J. and J. van Beem. 1993. Growth and architecture of seedling roots of common bean genotypes. Crop Sci. 33: 1253-1257.

McKeand, S.E., J.E. Grissom, J.A. Handest, D.M. O'Malley and H.L. Allen. 2000. Responsiveness of diverse provenances of loblolly pine to fertilization–age 4 results. J. Sust. For. 10(1/2): 87-94.

Niklas, K.J. 1994. Plant allometry: The scaling of form and process. University of Chicago Press, Chicago, IL.

Scheiner, S.M. 1993. Genetics and evolution of phenotypic plasticity. Annu. Rev. Ecol. Syst. 24: 35-68.

Schlichting C.D. 1986. The evolution of phenotypic plasticity in plants. Annu. Rev. Ecol. Syst. 17: 667-693.

Schlichting C.D. 1989. Phenotypic integration and environmental change. Bioscience 39: 460-464.

Sultan S.E. 1987. Evolutionary implications of phenotypic plasticity in plants. Evol. Biol. 21: 127-178.

Thomas, R.B., J.D. Lewis and B.R. Strain. 1994. Effects of nutrient status on photosynthetic capacity in loblolly pine (*Pinus taeda* L.) seedlings grown in elevated atmospheric CO_2. Tree Physiol. 14: 847-960.

van Buijtenen, J.P. 1978. Response of Lost Pines' seed sources to site quality. In: pp. 228-234. Proc.5th Amer. For. Biol. Workshop. Gainesville, FL.

van Buijtenen, J.P., M.V. Bilan and R.H. Zimmerman. 1976. Morpho-physiological characteristics related to drought resistance in *Pinus taeda* L. In: pp. 349-358.

M.G.R. Cannell and F.T. Last (eds). Tree physiology and yield improvement. Academic Press, London.

Wu, R.L. 1997. Genetic control of macro- and micro-environmental sensitivity in *Populus*. Theor. Appl. Genet. 94: 104-114.

Wu, R. and R.F. Stettler. 1998. Quantitative genetics of growth and development in *Populus*. III. Phenotypic plasticity of crown structure and function. Heredity. 81:299-310.

PART FOUR: FRONTIERS OF CONSERVATION

Changes in Desiccating Seeds of Temperate and Tropical Forest Tree Species

K. F. Connor
F. T. Bonner
J. A. Vozzo
I. D. Kossmann-Ferraz

INTRODUCTION

Seeds have traditionally been divided into two storage classes (Roberts, 1973). Orthodox seeds, which include those of most temperate

K. F. Connor, F. T. Bonner (retired), and J. A. Vozzo are Plant Physiologists, U.S. Forest Service, Forestry Sciences Laboratory, P.O. Box 928, Starkville, MS 39760-0928 USA.

Isolde Kossman-Ferraz is Scientist, Instituto Nacional de Pesquisas da Amazônia, Coordenacão de Pesquisas em Silvicultura Tropical, Manaus-AM, Brasil.

[Haworth co-indexing entry note]: "Changes in Desiccating Seeds of Temperate and Tropical Forest Tree Species." Connor, K. F. et al. Co-published simultaneously in *Journal of Sustainable Forestry* (Food Products Press, an imprint of The Haworth Press, Inc.) Vol. 10, No. 3/4, 2000, pp. 319-326; and: *Frontiers of Forest Biology: Proceedings of the 1998 Joint Meeting of the North American Forest Biology Workshop and the Western Forest Genetics Association* (ed: Alan K. Mitchell et al.) Food Products Press, an imprint of The Haworth Press, Inc., 2000, pp. 319-326. Single or multiple copies of this article are available for a fee from The Haworth Document Delivery Service [1-800-342-9678, 9:00 a.m. - 5:00 p.m. (EST). E-mail address: getinfo@haworthpressinc.com].

tree species, undergo a period of desiccation before being shed from the tree; they can be dried easily to moisture contents of less than 12% and stored for long periods of time. Recalcitrant seeds, however, do not undergo a significant maturation drying phase, are sensitive to moisture loss and low temperatures and have a short storage lifespan. Most tropical and a few important genera of temperate tree species, notably *Quercus*, *Aesculus*, some *Acer* (Bonner, 1990) and *Castanea* (Pritchard and Manger, 1990) have recalcitrant seeds.

The physiological basis of seed recalcitrance is as yet unknown. Hypotheses suggesting possible causes have been proposed (Roberts, 1973; Flood and Sinclair, 1981; Finch-Savage et al., 1994; Berjak and Pammenter, 1997), but the end result is that intact recalcitrant seeds cannot be stored for long periods of time. Thus, if the seed crop of a recalcitrant species fails, nurseries will be unable to draw upon a storage reserve of seeds in order to meet the demands of growers. A series of experiments on both temperate and tropical recalcitrant-seeded tree genera were initiated with the goal of relating changes in the physiology, biochemistry, and ultrastructure of desiccating seeds to loss of viability. Two tropical species, *Carapa guianensis* Aubl. and *Guarea guidonia* (L.) Sleumer (American muskwood) and two temperate species of oak, *Quercus nigra* L. (water oak) and *Quercus alba* L. (white oak) were selected for our experiments. This paper summarizes results (Connor et al., 1996, Connor et al., 1998a, Connor et al., 1998b) obtained over a 7-year period.

MATERIALS AND METHODS

Seeds were either collected locally or express-shipped to the Starkville laboratory shortly after shedding. They were imbibed overnight in tap water at room temperature before the experiment to ensure full hydration. Desiccation of the seeds was carried out on a laboratory benchtop at a room temperature of 27 ± 2°C and an ambient RH of 40 ± 10%. Random subsamples of seeds were periodically removed to provide specimens between full imbibition/high germination and reduced viability from desiccation. Each subsample was further divided for the following tests:

1. *Germination*: At each sampling time, seeds were germinated as two replications of 17-50 seeds each, depending on the species

and seed abundance. Seeds were placed on trays lined with Kimpak®/blotter paper and germinated in a Stults® wet box set on a 20-30°C temperature cycle with 8 h of light at 30°C and 16 h of dark at 20°C. Germination was considered complete when the cotyledons emerged.

2. *Differential scanning calorimetry (DSC) and moisture content (MC) analyses*: Tissue samples were sealed into aluminum pans and placed in a Perkin Elmer® DSC-7. Samples were cooled from 30°C to −150°C at 10°C/min, held at −150°C for 5 min, then warmed back to 30°C at the same rate. The onset temperature for melting water in the seed and the enthalpy (heat content) value of the melt were determined. Moisture content (MC) of whole seeds was determined on three to six subsamples by cutting seeds into halves or quarters and drying in an oven set at 107°C. MC was expressed as a percentage of fresh weight. To determine distribution of moisture within the seeds, tissues were dissected, rapidly weighed to the nearest 0.1 mg on an electronic balance, and immersed in 20 ml of anhydrous methanol (MEOH). After dehydration in the MEOH for 48 h, MCs were measured on an aliquot of the MEOH by Karl Fisher analysis with an Aquastar® V1B automatic titrator (Association of Official Agricultural Chemists, 1965).
3. *Gas chromatography (GC)*: Seeds were chopped, the pieces immediately immersed in liquid nitrogen (LN_2), and the entire sample ground either by a LN_2-cooled Wiley mill equipped with a 20-mesh screen or with a LN_2-cooled mortar and pestle. A portion of the sample was placed in 2:1 chloroform ($CHCl_3$)/MEOH and the lipids extracted and purified as described in Connor et al. (1996).
4. *Ultrastructure* (*Carapa* and *Quercus* only): Three-mm^3 plugs of tissue were immediately placed in a 2.5% glutaraldehyde solution of pH 7.2 phosphate buffer, fixed in osmium tetroxide, dehydrated in acetone, embedded in Epon®, sliced at 0.5 to 0.75 m, and stained with uranyl acetate (Vozzo and Song, 1989). Sections were observed at 1.2 MeV using high-voltage transmission electron microscopy (HVEM).

Experiments were replicated in time; results summarized below represent two years of data for *Carapa*, *Guarea*, and *Q. alba* and three years for *Q. nigra*.

RESULTS AND DISCUSSION

Germination

Initial seed germination was high in all species (Table 1); fungal contamination in year 1 *Guarea* seeds resulted in a drop in viability to 3% by day 3 of the experiment despite a 30 second surface sterilization wash in a 10% solution of commercial bleach. In year 2, seed coats were removed prior to germination testing, and viability remained higher than that of year 1 seeds throughout the experiment. Generally, viability of these recalcitrant seeds did not fall below 50% of the original fresh value until at least 5 days of exposure to ambient laboratory conditions.

DSC and Moisture Content (MC)

Karl Fisher analyses demonstrated that the axis MC of all species was high throughout the experiment (Figure 1). Despite lower mois-

TABLE 1. Germination (%) results of temperate and tropical recalcitrant seeds as they were desiccated.

Drying Time (days)	*Q. nigra*			*Q. alba*		*Carapa*		*Cuarea*	
	1991	1992	1993	1993	1994	1992	1994	1992	1993
Fresh	90	97	94	79	90	80	69	57	74
1	96	--	--	86	--	--	--	23	54
2	97	--	--	90	92	--	--	--	34
3	93	--	--	84	--	70	42	3	38
4	89	--	--	--	68	--	--	--	--
5	83	68	71	74	68	60	15	--	--
6	46	--	--	--	32	--	--	0	20
7	52	49	48	36	--	40	5	--	--
8	48	--	--	--	14	--	--	0	24
9	21	15	19	38	--	--	--	--	--
10	23	--	--	--	0	10	--	--	--
11	13	--	--	--	--	--	0	--	--
12	--	4	5	--	0	--	--	--	--

Each number represents the average germination of two replications.
Number of seeds per replication varied from 17 to 50 depending on species and availability.

ture in the surrounding cotyledonary tissue, axis MC never fell below 20%. In fact, when water oak cotyledon tissue reached the lethal MC of 15% (Agmata 1982), the axis MC was still 27% (Connor et al., 1996). In *Carapa*, axis MC was still at least 45% when seed viability had been reduced to 5%. It appears that the axes in these recalcitrant species may act as moisture sinks; and despite differences in seed sizes and chemical composition, the MC of all tissue in these recalcitrant-seeded species remained relatively high even when viability dropped below 50% of the original value (Table 2). Predictably, statistics re-

FIGURE 1. Axis moisture dynamics for desiccating temperate and tropical recalcitrant seeds.

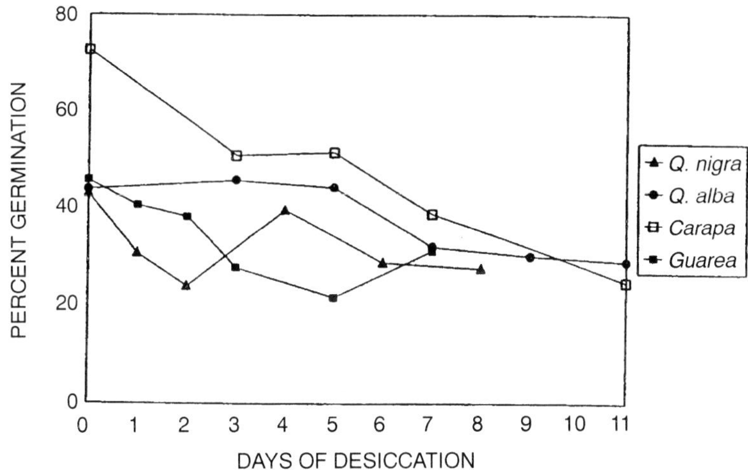

TABLE 2. Moisture contents of seeds after a 50% reduction in viability.

Species	Year	% Germination	% Moisture (fresh weight)			DCOT[1]
			Intact	Axis	PCOT[1]	
Carapa	1994	15	28.6	51.4	59.4	35.4
Guarea	1993	20	20.5	21.6		21.3
Q. alba	1993	36	28.0	34.1	37.7	33.7
Q. nigra	1992	49	16.0	29.3	21.7	13.0

[1] PCOT = cotyledon tissue proximal to the axis; DCOT = cotyledon tissue distal to the axis.

vealed that the relationship between declining MCs of the various seed tissues and declining viability was high (r^2 = 0.46-0.99).

DSC thermographs also showed a strong but variable relationship between moisture enthalpy (heat content) values of the seed tissues and seed viability (r^2 = 0.25-0.99). Generally, the axes and intact seeds appeared most sensitive to moisture loss. However, while useful for showing general declines in seed viability, DSC analyses could not pinpoint drops in viability when germination remained above 50%. Enthalpy values only gradually declined in the axes, while viability dropped more dramatically. High enthalpy values in the axes might seem good indicators of high seed viability, but they are mostly a reflection of high MC. Since the axes may act as moisture sinks, high enthalpy values are maintained despite sharply declining seed viability. Therefore, such analyses cannot be used as predictors of seed deterioration.

Lipid Analyses

Petroleum ether soxhlet extractions on fresh seeds yielded 0.1901 g lipid/g dry wt of seed tissue in *Guarea*, 0.284 g/g in *Q. nigra*, 0.057 g/g in *Q. alba*, and 0.486 g/g in *Carapa*. Analyses determined that there were no significant patterns of gain or loss of individual fatty acids as the seeds desiccated and viability was lost (Connor et al., 1996; Connor et al., 1998a; Connor et al., 1998b). Also, results could vary from year to year. While 1994 *Carapa* seeds revealed a strong relationship between viability and a decline in palmitic and stearic acids and an increase in oleic acid (r^2 = 0.60, 0.84. and 0.98, respectively), the pattern was not repeated in 1992, where r^2 values dropped as low as 0.05. At this time, the lipid analyses have provided little insight into the cause or nature of tree seed recalcitrance.

Electron Microscopy

The cotyledonary tissues of both *Carapa* and *Q. nigra* exhibited cell wall trauma after only 3 days of drying. Axis tissue in these 2 species, however, showed greater resistance to desiccation-imposed damage. This strengthens the significance of the moisture analysis results, indicating that the high MC of the axes and the possibility that they might be moisture sinks may protect them from desiccation damage.

Like the cells of *Q. nigra*, there were enlarged spherosomes and highly vacuolated cytoplasm present in the *Q. alba* samples. However, there was no evidence of cell wall trauma in either the cotyledon or axis tissue even after 7 days of drying. Thus, whereas in *Carapa* and *Q. nigra*, high in lipid, cell wall damage was evident, no damage was visible in low-lipid *Q. alba*. Perhaps in these seeds, changes in lipid composition may not be an important indicator of seed deterioration; but the overall hydrophobic nature of lipids in general may be an important factor in cell wall trauma.

CONCLUSIONS

- Viability for all species tested did not drop below 50% of the original level until seeds had dried at least 5 days.
- Relationships between axis and intact seed MCs and seed viability were generally strong.
- Axis MC rarely dropped below 20%; this suggests axes may act as moisture sinks.
- DSC analyses cannot be used as predictors of seed deterioration.
- Lipid analyses provided little insight into the nature of seed recalcitrance but the hydrophobic nature of lipids may be a factor in cell trauma, since seeds high in lipid showed damage after only three days of drying.
- None of the tests provided clear clues to explain the physiological basis of recalcitrance.

REFERENCES

Association of Official Agricultural Chemists. 1965. Official methods of analysis of the association of official agricultural chemists. Association of Official Agricultural Chemists, Washington, DC.

Agmata, A. 1982. Physiological Effects of Drying *Quercus Nigra* L. Acorns. MS Thesis, Mississippi State Univ. 59 pp.

Berjak P., and N.W. Pammenter. 1997. Progress in the understanding and manipulation of desiccation-sensitive (recalcitrant) seeds. In: Ellis R.H., Black M., Murdoch A.J., Hong T.D., eds. Basic and Applied Aspects of Seed Biology. Proceedings of the Fifth International Workshop on Seeds, Reading: Kluwer Academic Publishers: 689-703.

Bonner, F.T. 1990. Storage of seeds: potential and limitations for germplasm conservation. For. Ecol. Manag. 35: 35-43.

Connor, K.F., F.T. Bonner and J.A. Vozzo. 1996. Effects of desiccation on temperate recalcitrant seeds: differential scanning calorimetry, gas chromatography, electron microscopy, and moisture studies on *Quercus nigra* and *Quercus alba*. Can. J. For. Res. 26: 1813-20.

Connor, K.F., F.T. Bonner and J.A. Vozzo. 1998a. Effects of desiccation on the recalcitrant seeds of *Carapa guianensis* Aubl. and *Carapa procera* DC. Seed Tech. 20(1): 71-82.

Connor, K.F., F.T. Bonner and J.A. Vozzo. 1998b. Effects of desiccation on recalcitrant seeds: results from lipid and moisture studies on *Guarea guidonia* (L.) Sleumer. Seed Tech. 20(1): 32-42.

Finch-Savage, W.E., G.A.F. Hendry and N.M. Atherton. 1994. Free radical activity and loss of viability during drying of desiccation-sensitive seeds. Proc. Royal Soc. Edinburgh 102B: 257-60.

Flood, R.G., and A. Sinclair. 1981. Fatty acid analysis of aged permeable and impermeable seeds of *Trifolium subterraneum* (subterranean clover). Seed Sci. Tech. 9: 475-477.

Pritchard, H.W., and K.R. Manger. 1990. Quantal response of fruit and seed germination rate in *Quercus robur* L. and *Castanea sativa* Mill. to constant temperatures and photon dose. J. Exp. Bot. 41: 1549-57.

Roberts, E.H. 1973. Predicting the storage life of seeds. Seed Sci. Tech. 1: 599-514.

Vozzo, J.A., and M.J. Song. 1989. High-voltage electron microscopy of cell walls in *Pinus taeda* seeds. In: Turner, J.W., ed. Tropical Seed Research. ACIAR Proc. No. 28: 78-80.

Red and Black Spruce Introgression in Montane Ecosystems in New England and New York

G. J. Hawley
D. H. DeHayes
G. J. Badger

INTRODUCTION

In the northern portion of the red spruce (*Picea rubens* Sarg.) range, red spruce are often sympatric with black spruce (*P. mariana* [Mill.] BSP) and hybridization and introgression are suspected. Morphological similarity between the species coupled with the decline of montane populations of red spruce and no apparent decline of black spruce have provided renewed interest on the genetic relationship between the two species, especially in regions of range overlap. Determination of the extent of the contribution of black spruce to the germ plasm of

G. J. Hawley is Senior Researcher and D. H. DeHayes is Professor of Forest Biology, School of Natural Resources, 303 Aiken Center, School of Natural Resources, University of Vermont, Burlington, VT 05405 USA.

G. J. Badger is Biostatistician, Biometry Facility, Hills Agricultural Sciences, University of Vermont, Burlington, VT 05405 USA.

Address correspondence to G. J. Hawley at the above address (E-mail: ghawley@nature.snr.uvm.edu).

This research was supported by the USDA Forest Service, Northeastern Forest Experiment Station, Spruce-Fir Research Cooperative and the McIntire-Stennis research program.

[Haworth co-indexing entry note]: "Red and Black Spruce Introgression in Montane Ecosystems in New England and New York." Hawley, G. J., D. H. DeHayes, and G. J. Badger. Co-published simultaneously in *Journal of Sustainable Forestry* (Food Products Press, an imprint of The Haworth Press, Inc.) Vol. 10, No. 3/4, 2000, pp. 327-333; and: *Frontiers of Forest Biology: Proceedings of the 1998 Joint Meeting of the North American Forest Biology Workshop and the Western Forest Genetics Association* (ed: Alan K. Mitchell et al.) Food Products Press, an imprint of The Haworth Press, Inc., 2000, pp. 327-333. Single or multiple copies of this article are available for a fee from The Haworth Document Delivery Service [1-800-342-9678, 9:00 a.m. - 5:00 p.m. (EST). E-mail address: getinfo@haworthpressinc.com].

high elevation declining spruce forests of the northern Appalachians is critical to accurate and meaningful interpretation of results from studies addressing spatial variation in the health of spruce forests.

Based on morphological assessments, estimates of the degree of introgression between red and black spruce vary considerably (Gordon, 1976, Morgenstern and Farrar, 1964; Manley, 1971; Morgenstern et al., 1981). More recently, molecular and DNA content studies have also led to differing views of the level of introgression among these species. Using nuclear DNA content, Berlyn et al. (1990) reported a west-to-east gradient with a higher proportion of black spruce towards the east, and increases in the degree of introgression along an elevational gradient up Mt. Washington, NH with only black spruce found above 1220 m. Using nuclear and organelle DNA analysis, Bobola et al. (1996) found that hybridization and introgression were not a major factor in the pattern of variation on the elevational gradient used by Berlyn et al. (1990) on Mt. Washington or on Mt. Lafayette, NH. They did, however, detect extensive introgression between red and black spruce on Isle au Haut, ME. Using species-specific RAPD fingerprints, Perron and Bousquet (1996) found more hybrids in sympatric than allopatric spruce populations in Quebec. Eckert (1988) used isozyme data and found that in 16 plots located in VT, NH, and NY that 22% of trees sampled as red spruce were actually black spruce, while only 2% were hybrids. In this study, we used isozyme analysis to examine the genetic relationship between allopatric red and black spruce and to investigate genetic relationships between red and black spruce in eleven montane spruce populations in NY, VT, NH and ME.

MATERIALS AND METHODS

To establish baseline genetic differences between non-introgressed red and black spruce, we examined the genetics of trees from three red spruce populations (Clingman's Dome, TN; Roan Mt., TN; and Whitetop Mt., VA) located outside the range of black spruce, and three black spruce populations (Upsala, ON; Minchin Lake, ON; and Red Lake, ON) located outside the range of red spruce. The red-black spruce relationship was examined in eleven montane populations throughout NY and New England (Table 1).

Using electrophoretic procedures similar to those described by Jech and Wheeler (1984), genetic variability estimates were generated us-

TABLE 1. Genetic distances and discriminant analysis results of allopatric red and black spruce and northern montane spruce populations in NY, VT, NH and ME.

Population			Genetic distance from non-introgressed:		Discriminant classification of individual trees in population		
Origin	Sample size	Elevation	Red spruce	Black spruce	Red spruce	Intermediate	Black spruce
					(% of population/#trees)		
Allopatric red spruce							
Whitetop Mt., VA	30	1660	0.010	0.103			
Roan Mt., TN	30	1860	0.006	0.094			
Clingmans's Dome, TN	30	1830	0.006	0.097			
Mean			**0.007**	**0.098**	**98.9/89**		**1.1/1**
Allopatric black spruce							
Upsala, ON	15	480	0.089	0.006			
Minchin Lake, ON	14	360	0.099	0.004			
Red Lake, ON	Bulk	350	0.107	0.006			
Mean			**0.098**	**0.005**	**0/0**		**100/29**
Northern Montane							
Whiteface Mt., NY	31	1100	0.008	0.063	90.3/28	6.5/2	3.2/1
	28	1220	0.009	0.061	85.7/24	7.2/2	7.1/2
Camel's Hump, VT	17	1000	0.013	0.068	76.5/13	0/0	23.5/4
Mt. Mansfield, VT	28	800	0.014	0.059	71.4/20	3.6/1	25.0/7
	5	1150	0.039	0.031	60.0/3	0/0	40.0/2
Mt. Washington, NH	27	1100	0.008	0.067	85.2/23	11.1/3	3.7/1
Mt. Washington, NH	26	875	0.015	0.058	65.4/17	7.7/2	26.9/7
	29	1220	0.026	0.036	58.6/17	3.4/1	37.2/11
	7	1520	0.069	0.035	28.6/2	0/0	71.4/5
Greenville, ME	36	520	0.007	0.067	91.7/33	2.8/1	5.6/2
Beech Mt., Acadia National Park, ME	20	240	0.011	0.068	100/20	0/0	0/0

ing seed tissue for 29 gene loci encoding 17 enzyme systems. For most populations, individual tree genotypes were inferred by examining 7 haploid megagametophytes per tree for 25 to 30 trees from each population. For the Red Lake, Ontario black spruce population (represented by a bulk seed collection of approximately 100 trees), electrophoretic assessments from 120 megagametopyhtes were used to obtain allele frequency estimates. Nei's genetic distance (Nei, 1978) was used as a measure of genetic similarities and differences among species/populations. Genetic distance measures range from 0.0 for genetically identical populations to 1.0 for populations with no genes in common. In addition, allele frequencies for 8 diagnostic loci derived from the allopatric red and black spruce populations was used to generate an optimal discriminant function. Genotypes from each tree in the eleven test populations were then classified as red spruce, black spruce or intermediate based on congruence with the discriminant function.

RESULTS AND DISCUSSION

Genetic Relationships of Montane Spruce with Allopatric Red and Black Spruce

An examination of allele frequencies at 29 loci indicates that red and black spruce are distinctive, although they are not fixed for different alleles at any locus examined. The population mean genetic distance between southern Appalachian red spruce and Ontario black spruce populations ranged from 0.089 to 0.107 with a overall mean of 0.098, well outside the range among non-introgressed populations within each species (Table 1). Interspecific genetic distances are in the range of those reported for other closely related and potentially introgressed species such as lodgepole pine (*Pinus contorta* Dougl.) and jack pine *(P. banksiana* Lamb.) (Wheeler and Guries, 1987).

With the exception of populations above 1100 m on Mt. Mansfield and Mt. Washington, all northern montane populations exhibited genetic distances from allopatric red spruce (range from 0.007 to 0.015, Table 1) well within the reported range described for 19 geographically distant red spruce populations (Hawley and DeHayes, 1994). Interestingly, despite their genetic similarity to southern Appalachian red spruce populations, these 8 northern montane red spruce populations had an average genetic distance with Ontario black spruce of only about 0.06 (range of 0.058 to 0.068). This indicates a much closer genetic relationship between northern montane red spruce populations and Ontario black spruce than between southern Appalachian red spruce and Ontario black spruce. This substantial difference in genetic relationship between regional red spruce populations most likely reflects their varied biogeographical histories, geographic isolation, and differential selection pressures. There may have been introgression between co-existing northern montane red spruce and black spruce in northeastern North America glacial refugia that was not possible in the isolated southern Appalachian red spruce refugia.

High elevation populations on Mt. Mansfield and Mt. Washington exhibited genetic distance measures outside the range of those for multiple populations of both red (Hawley and DeHayes, 1994) and black spruce (Boyle and Morgenstern, 1987) indicating genetic intermediacy between the two species. In fact, the highest elevation population on Mt. Washington is more genetically similar to black than red spruce (Table 1). Although it is tempting to interpret genetic intermediacy as evidence of hybridization and introgression between species, an alternative explanation is that these populations merely consist of a mixture of both black and red spruce trees. Discriminant analysis provided a vehicle to examine the genetic relationship of individual trees.

Discrimination of Individual Trees

Discriminant function scores of individual trees within each population provided some surprising insights. For example, the genetically intermediate high elevation populations on Mt. Mansfield and Mt. Washington were classified by the discriminant function as containing a mixture of red and black spruce trees with very few intermediate trees (Table 1). In fact, these populations contained the highest proportion of black spruce of any of the northern montane populations. Surprisingly, lower elevation populations on Mt. Mansfield, Mt. Washington, and Camels Hump, which exhibited genetic distances well within the range reported for red spruce, also contained 24 to 27% black spruce trees and there appears to be an increasing proportion of black spruce with increasing elevation (Table 1). Despite this elevational trend, the proportion of trees classified as intermediate in these populations is greatest in the low elevation populations that contained the least black and the most red spruce. This provides evidence for some hybridization between red and black spruce in these populations, but suggests hybrids are not competitive at the highest elevation of the spruce-fir zone where both red and black spruce intermingle. Our results contrast with those of Berlyn et al. (1990) who, based on nuclear DNA content analysis, concluded that all trees on Mt. Washington above 1220 m were black spruce and that hybrids predominate between 1000 and 1220 m.

Based on discriminant function scores of individual trees, many of the remaining northern montane populations, which exhibited genetic distances expected for pure red spruce, also contained a low proportion of both black spruce and trees intermediate between black and red spruce. In fact, with the exception of the montane population on Beech Mt., these montane spruce populations throughout New England and New York consist of about 85 to 90% red spruce with low and approximately equally proportions of black spruce (3-7%) and trees intermediate (3-11%) between red and black spruce (Table 1). These data contrast with Berlyn et al. (1990) who reported no black spruce on Whiteface Mt. and all intermediate spruce on Camels Hump. Differences between our results and those of Berlyn et al. (1990) could be explained by the fact that quantity of nuclear DNA, which they used as a measure of introgression, does not necessarily reflect genetic differences and, therefore, may not be a viable test for introgression among species. In fact, Bobola et al. (1996), whose conclusions were also considerably different than those of Berlyn et al. (1990), concluded that the elevational differences in nuclear DNA found by Berlyn et al. (1990) were not due to introgression between red and black spruce. Our data also suggest that fewer black spruce exist in these montane populations than the 22% reported by Eckert (1988). We detected only red spruce in our montane population from Beech Mt., while Bobola et al. (1996) reported a high proportion of hybrid and introgressed spruce on nearby Isle au Haut, ME. This difference is likely explained by the fact that our spruce were collected near the top of Beech Mt. (240 m) whereas Bobola et al. (1996) made their collection along the coast at low elevation (15-30 m) either in, or adjacent to, a bog where black spruce is likely to grow. Finally, our data do indicate some small degree of gene exchange between red and black spruce throughout most montane populations in New England and New York and suggest a hybrid habitat for introgressed spruce in low elevation populations dominated by red spruce.

REFERENCES

Berlyn, G.P., J.L. Royte, and A.O. Anoruo. 1990. Cytophotometric differentiation of high elevation spruces: physiological and ecological implications. Stain Technol. 65:1-14.

Bobola, M.S. R.T. Eckert, A.S. Klein, K. Stapelfeldt, K.A. Hillenberg, and S.B. Gendreau. 1996. Hybridization between *Picea rubens* and *Picea mariana*; differences observed between montane and coastal island populations. Can. J. For. Res. 26: 444-452.

Boyle, J.B., and E.K. Morgenstern. 1987. Some aspects of the population structure of black spruce in Central New Brunswick. Silvae Gen. 36:53-60.

Eckert, R.T. 1988. Genetic variation in red spruce and its relation to forest decline in the northeastern United States. IUFRO, Air Poll. and Forest Decline. Interlaken, Switzerland. 6 pp.

Gordon, A.G. 1976. The taxonomy and genetics of *Picea rubens* and its relationship to *Picea mariana*. Can. J. For. Res. 54: 781-813.

Hawley, G.J., and D.H. DeHayes. 1994. Genetic diversity and population structure of red spruce *(Picea rubens)*. Can. J. Bot. 72 (12):1778-1786.

Jech, K.S., and N.C. Wheeler. 1984. Laboratory manual for horizontal starch gel electrophoresis. Weyerhaeuser Research Report 050-3210: 61 pp.

Manley, S.A.M., 1971. Identification of red, black and hybrid spruces. Dep. Environ., Can. Forest. Serv., Inform. Rep. N-X-200, 43 pp.

Morgenstern, E.K, A.G. Corriveau, and D.P. Fowler. 1981. A provenance test of red spruce in nine environments in eastern Canada. Can. J. For. Res. 11:124-131.

Morgenstern, E.K, and J.L. Farrar. 1964. Introgressive hybridization in red spruce and black spruce. Faculty of Forestry. University of Toronto. Tech. Rep. No. 4.

Nei, M. 1978. Estimation of average heterozygosity and genetic distance from small number of individuals. Genetics 89: 583-590.

Perron, M., and J. Bousquet. 1996. Natural hybridization and genetic variation between black spruce and red spruce. Proc. 14th North Am. For. Bio. Work. June 16-20, Quebec City, Can. Abst. p. 88.

Wheeler, N.C., and R.P Guries. 1987. A quantitative measure of introgression between lodgepole and jack pine. Can. J. Bot. 65:1876-1885.

Sea Buckthorn: A Near-Ideal Plant for Agroforestry on the Canadian Prairies

S. Jana
W. R. Schroeder

INTRODUCTION

Sea buckthorn, *Hippophae rhamnoides* L., is a member of the family, Elaeagnaceae, is a Eurasian shrub, with wide adaptation and diverse use. The plant bears attractive and edible berries, and is often used as an ornamental. Sea buckthorn is an extremely winter-hardy, drought resistant shrub. It interacts synergistically with other trees and shrubs in mixed stands. It is an excellent source of food and shelter for wildlife. It fixes atmospheric nitrogen, prevents soil erosion and helps to conserve soil moisture. The Prairie Farm Rehabilitation Authority (PFRA), Agriculture and Agri-Food Canada introduced sea buckthorn from only one Siberian population to the Canadian prairies in the early 1930s as a shelterbelt plant. Since all sea buckthorn populations in the

---

S. Jana is affiliated with the Department of Plant Sciences, University of Saskatchewan, Saskatoon, Saskatchewan, Canada S7N 5A8.

W. R. Schroeder is affiliated with the Prairie Farm Rehabilitation Authority (PFRA), Agriculture and Agri-Food Canada, Indian Head, Saskatchewan, Canada.

The authors acknowledge financial support from the Saskatchewan Agricultural Development Fund and the Tri-Council Prairie Ecosystem Research Fund, and the technical help provided by M.A. Chowdhury and V. Nikolic.

[Haworth co-indexing entry note]: "Sea Buckthorn: A Near-Ideal Plant for Agroforestry on the Canadian Prairies." Jana, S., and W. R. Schroeder. Co-published simultaneously in *Journal of Sustainable Forestry* (Food Products Press, an imprint of The Haworth Press, Inc.) Vol. 10, No. 3/4, 2000, pp. 335-339; and: *Frontiers of Forest Biology: Proceedings of the 1998 Joint Meeting of the North American Forest Biology Workshop and the Western Forest Genetics Association* (ed: Alan K. Mitchell et al.) Food Products Press, an imprint of The Haworth Press, Inc., 2000, pp. 335-339. Single or multiple copies of this article are available for a fee from The Haworth Document Delivery Service [1-800-342-9678, 9:00 a.m. - 5:00 p.m. (EST). E-mail address: getinfo@haworthpressinc.com].

Canadian prairies are derived from the progeny of a single Siberian sample, the prairie populations of this introduced species are expected to have limited genetic diversity due to the founder effect. One of the objectives of this research was to determine the amount of genetic diversity in local populations of sea buckthorn in comparison with that of native species of the family Elaeagnaceae in the prairies. The second objective was to determine the utility of sea buckthorn as an agroforestry plant in this region.

MATERIALS AND METHODS

Four species of the family Elaeagnaceae were used in our molecular diversity study, including sea buckthorn and another introduced species, Russian olive (*Elaeagnus angustifolia* L.). Buffaloberry (*Shepherdia argentea* Nutt.) and silverberry (*Elaeagnus commutata* Bernh.) were the two native species used for the study. Leaf samples were collected from geographically separated populations of the four species in Saskatchewan: four populations of sea buckthorn, four populations of Russian olive, eight populations of buffaloberry, and eight populations of silverberry. Samples, varying in size from 9 to 35 individuals, were stored at $-40°C$. DNA extracted from the leaf samples were used for the random amplified polymorphic DNA (RAPD) analysis using the procedure of Williams et al. (1990). The RAPD protocol was optimized for each of the four species. One hundred primers obtained from the University of British Columbia were screened and 7-8 primers that produced several scorable bands were selected. RAPD bands were scored in '0' and '1,' where '0' indicated absence and '1' indicated presence of a band. Only well-resolved and repeatable bands were scored. Seven primers were used to produce 43 scorable bands in sea buckthorn, 86% of which were polymorphic. In Russian olive, eight primers produced 74 scorable bands with 36.5% polymorphic bands. Eight primers produced 75 scorable bands in buffaloberry with 80% polymorphic bands and 67 scorable bands in silverberry with 55.2% polymorphic bands. The RAPD data were used to perform an analysis of molecular variance (AMOVA) using the Euclidean distance matrix (Excoffier et al., 1992) to partition the total molecular variance into within and between populations components. Shannonian diversity index (Bowman et al., 1971) was calculated as an additional measure of diversity.

Plant surveys were conducted in 1996 and 1997 to gather information on the distribution and adaptation of sea buckthorn to different ecological zones in Saskatchewan.

RESULTS

The banding pattern generated by a particular primer varied with the species. Several primers that generated scorable bands in one species did not produce any band in other species. Moreover, when DNA fragments generated by a primer were compared over different species, very few sharing bands were observed. For this reason, we analyzed the RAPD data separately for each species. Then, we compared genetic diversity for RAPD markers in sea buckthorn samples with that of Russian olive and the two native shrubs, silverberry and buffaloberry. Despite the fact that a small number of plants from only one Siberian location were introduced on the prairies, genetic diversity in sea buckthorn was similar to that of buffaloberry, and higher than silverberry and Russian olive.

In all four species, most of the total molecular diversity was attributable to the within-population component, ranging from 98.3% in Russian olive to 71.7% in silverberry (Table 1). Assuming that RAPD fragments constitute a random representative of the genome, a comparison of the diversity was made among the species regardless of the differences in primer set used in respective species. Buffaloberry had the highest total molecular variance (4.64), followed by sea buckthorn (4.38), whereas Russian olive had the lowest molecular variance (3.05). Shannonian information index was the highest for sea buckthorn (0.31), followed by buffaloberry (0.21) (Table 1), while Russian olive had the smallest (0.13) diversity index. In all four species, the major part of the total genetic diversity was due to the within-populations component. Sea buckthorn and Russian olive had the highest within-populations diversity (87.6%) and silverberry had the lowest within-populations diversity (60%).

We conducted a two-year (1996-97) population survey of sea buckthorn in Saskatchewan, which revealed that the species has wide adaptation. It is an excellent colonizer, which grows well under both moisture-limiting and waterlogged conditions, and it is extremely winter-hardy. It was found to grow successfully and bear fruit on poor marginal soils, where no other shrubs would grow. It is an efficient

TABLE 1. Analysis of molecular variance (AMOVA) and Shannonian index for four species in the Elaegnaceae family in Saskatchewan.

	Variance Components (AMOVA)[a]			Shannonian index		
Species	Total	Among Sites	Within Sites	Total	Among Sites	Within Sites
Sea Buckthorn	4.38	0.66 (15.1)	3.72** (84.9)	0.31	0.04 (12.4)	0.27 (87.6)
Russian Olive	3.05	0.06 (1.7)	2.99** (98.3)	0.13	0.02 (12.4)	0.11 (87.6)
Buffalo Berry	4.64	0.74 (16.0)	3.90** (84.0)	0.21	0.05 (26.5)	0.16 (73.5)
Silverberry	3.51	0.99 (28.3)	2.52** (71.7)	0.16	0.06 (40.0)	0.10 (60.0)

[a] Figures in parentheses are in percent.
**Significant at the 1% level of significance.

nitrogen-fixer that interacts synergistically with other trees and shrubs, such as poplars and pines, in mixed stands. It is an excellent source of food and shelter for wildlife in both pure and mixed stands.

We conducted a market survey that established that numerous commercial products are made from sea buckthorn fruit and seed, and that there is an increasing demand for sea buckthorn in the cosmetic and pharmaceutical industries in North America. Thus it combines favorable ecological and medicinal properties with excellent commercial potential. In addition, sufficient genetic diversity allows sea buckthorn to meet adaptive challenges in the harsh prairie environments, both natural and agricultural.

DISCUSSION AND CONCLUSIONS

We compared variation for RAPD markers in sea buckthorn populations of Saskatchewan with that of two native shrubs, silverberry and buffaloberry, and another introduced shrub, Russian olive. Despite the fact that a limited number of plants from only one Siberian population were introduced in the province, molecular diversity as determined by RAPD analysis in sea buckthorn was similar to that of buffaloberry, and higher than the diversity in silverberry and Russian olive. Although sea buckthorn in Saskatchewan had a restricted origin, diversity discerned by the RAPD markers was found to be comparable to that of the two native prairie species. This result was contrary to our expectation that inbreeding and random genetic drift would lead to the loss of genetic diversity in the prairie population, and thus limit the colonizing ability of the species in the harsh prairie environments.

Its numerous ecologically beneficial properties, along with its excellent commercial prospects, make sea buckthorn a prime candidate for sustainable agroforestry development on the forest-poor Canadian prairies. The abundance of genetic variation in local and exotic populations (Yao, 1994; Yao and Tigerstedt, 1993, 1994) of sea buckthorn may be utilized toward this end.

REFERENCES

Bowman, K. D., K. Hutchenson, E. P. Odum and L. R. Shenton. 1971. Comments on the distribution of indices of diversity. International Symposium on Statistical Ecology, Vol. 3, 315-359. University Park: Pennsylvania State University Press.

Excoffier, L., P. E. Smouse and J. M. Quatro. 1992. Analysis of molecular variance inferred from metric distances among DNA haplotypes: Application to humans mitochondrial DNA restriction data. *Genetics* 131: 479-491.

Williams, J. G. K., A. R. Kubelik, K. J. Livak, J. A. Rafalski and S. V. Tingey. 1990. DNA polymorphisms amplified by arbitrary primers are useful as genetic markers. *Nucleic Acids Research* 18: 6531-6535.

Yao, Y. and P. M. A. Tigerstedt. 1993. Isozyme studies of genetic diversity and evolution in *Hippophae*. *Genetic Resources and Crop Evolution* 42: 153-164.

Yao, Y. 1994. Genetic diversity, evaluation and domestication in sea buckthorn (*Hippophae rhamnoides* L.). Academic Dissertation, University of Helsinki.

Yao, Y. and P. M. A. Tigerstedt. 1994. Genetic diversity in *Hippophae* L. and its use in plant breeding. *Euphytica* 77: 165-169.

Some Aspects of the Impact and Management of the Exotic Weed, Scotch Broom (*Cytisus scoparius* [L.] Link) in British Columbia, Canada

Raj Prasad

INTRODUCTION

Scotch broom (*Cytisus scoparius* [L.] Link) and gorse (*Ulex europeus* L.) are two related exotic weeds (Peterson and Prasad, 1998) which pose a serious threat to native plant species in forested and other landscapes in southwestern British Columbia. Scotch broom was introduced to Vancouver Island in 1850 by Captain Walter Grant who had seeds planted near Sooke, B.C. (Zielke et al., 1992). After nearly one and a half centuries, Scotch broom has expanded its range, occupying many roadsides, hydro right-of-ways, and other disturbed areas along coastal B.C., and an isolated area along Kootenay Lake in the interior of the province. Scotch broom has several characteristics

---

Raj Prasad is affiliated with the Pacific Forestry Centre, Natural Resources Canada, 506 West Burnside Road, Victoria, B.C. V8Z 1M5, Canada (E-mail: Rprasad@PFC.Forestry.Ca).

The author wishes to thank David Peterson and Randy Lauzon for technical assistance.

[Haworth co-indexing entry note]: "Some Aspects of the Impact and Management of the Exotic Weed, Scotch Broom (*Cytisus scoparius* [L.] Link) in British Columbia, Canada." Prasad, Raj. Co-published simultaneously in *Journal of Sustainable Forestry* (Food Products Press, an imprint of The Haworth Press, Inc.) Vol. 10, No. 3/4, 2000, pp. 341-347; and: *Frontiers of Forest Biology: Proceedings of the 1998 Joint Meeting of the North American Forest Biology Workshop and the Western Forest Genetics Association* (ed: Alan K. Mitchell et al.) Food Products Press, an imprint of The Haworth Press, Inc., 2000, pp. 341-347. Single or multiple copies of this article are available for a fee from The Haworth Document Delivery Service [1-800-342-9678, 9:00 a.m. - 5:00 p.m. (EST). E-mail address: getinfo@haworthpressinc.com].

which promote its invasiveness, suppression or displacement of native plant species:

a. reduced leaves and active stem photosynthesis during unfavourable periods (Nilsen et al., 1993),
b. nitrogen fixation capability,
c. profuse seed production and longevity of seed banks,
d. rapid vertical growth and intense spatial competition,
e. adaptability to various ecological niches, and
f. lack of natural enemies (parasite-predator complex).

Generally, Scotch broom is managed by:

a. manual cutting or pulling which is labor-intensive and costly,
b. mechanical cutting and pulling, which is very expensive,
c. burning, which not only stimulates seed germination from seedbanks but also may be undesirable in certain areas due to smoke pollution,
d. using herbicides such as 2,4-D and Garlon, but they are not wholly effective and are ecologically controversial, and
e. biologically, by sheep grazing or use of seed-eating insects, none of which are completely successful.

An innovative approach using bioherbicides whereby indigenous plant pathogens are exploited and employed to control weeds in forestry (Johnston and Parkes, 1994; Prasad, 1996) is gaining momentum and shows great promise. Therefore, we evaluated several fungi (*Chondrostereum purpureum*, Pouzar, *Pleiochaeta seteosa* L. and *Fusarium tumidum* Sherb.) against Scotch broom and gorse, tested their potential under the greenhouse conditions and found that one of them (*Fusarium tumidum*) demonstrated a high degree of control. The response to the other two fungi was either erratic or low and they were, therefore, dropped from further screening.

The objectives of the present study were to determine the nature and extent of invasiveness of the Scotch broom in forested areas and its impacts on growth and development of Douglas fir seedlings (*Pseudotsuga menziesii* Mirabel/Franco) and to evaluate the potential of a fungus (*Fusarium tumidum*) for control of Scotch broom under the greenhouse conditions.

METHODS

Field Experiments to Measure the Impact of Broom on Douglas Fir Seedlings

A site near Maple Mountain, North Cowichan Municipality, Duncan, B.C. was set up. The site is a recent cutover area (5.1 ha) planted with (2+1) Douglas fir seedlings in 1994, Scotch broom has rapidly invaded this site originally starting from the logging roadside, forming a dense canopy and over-topping the conifer seedlings on flat, open terrain. Plots with Douglas fir seedlings were established all across this site where broom was over topping the seedlings. Ten plots (each 20 m × 20 m) were selected randomly to include 100 conifer seedlings competing with a dense canopy of broom. From 5 of these plots, broom was manually cut and completely removed, while in the other 5 plots, the broom canopy was left intact to compete against the conifer seedlings. The impact of competition was measured by recording the changes in height and root collar diameter (r.c.d.) of the seedlings by the standard procedure of using a ruler and caliper in the following two years. Data were also collected during August on the photosynthetically-active radiation (PAR) falling at the canopy and ground levels with the aid of a Sunfleck ceptometer (Decagon Inc., Washington).

Evaluation of the Bioherbicide Agents on Growth of Scotch Broom Seedlings Under Greenhouse Conditions

Three fungal pathogens (*Fusarium tumidum*, *Pleiochaeta setosa* and *Chondrostereum purpureum*) were cultured under laboratory conditions on puffed wheat cereal and when enough mycelia and spores were produced, a formulation of these pathogens using a slurry of 5% vegetable oil and a sticker (Bond[1] @ 0.1%) were prepared and sprayed on the broom seedlings grown under greenhouse conditions. There were ten replicates of check (control) and ten replicates of each treatment and the experiment was randomized.

Scotch broom seedlings were grown from seeds obtained from the experimental site at Maple Mountain and cultivated in pots containing a mixture of soil and peat moss. Three stages of seedlings (one, three and six months old) were raised in the greenhouse under a constant condition of temperature (76°F), light (μ md m^{-2} s^{-1}) and nutrition

(1/4 Hoagland solution). These seedlings were then sprayed to the point of run-off with a suspension (1×10^6 spores/ml) and maintained for one month under the same conditions for monitoring the effects of the bioherbicidal treatment.

It was found that only *Fusarium tumidum* was effective and results, therefore, are reported from these experiments only. Procedures for obtaining the culture, multiplication and preparation of the spray were carried out in a standard manner and are briefly described by Prasad (1997). The effectiveness of the bioherbicide in controlling the broom seedlings under the greenhouse conditions was determined by measuring the height of the surviving individuals. Data were analyzed by the analysis of variance (least significant difference LSD) method.

RESULTS

Impact of Scotch Broom on Conifer Seedlings Under Field Conditions

The data presented in Table 1 demonstrate clearly that Scotch broom is very effective in blocking the infiltration of light–a 71% reduction in PAR. These data also reveal that both the height and volume (root collar diameter) growth of Douglas fir are considerably depressed (46-45%) by the Scotch broom.

TABLE 1. Impact of Scotch broom on light infiltration and growth of Douglas fir seedlings two years after cutting*

	Photosynthetically-active radiation		Growth of Douglas fir			
			Height		Root collar diameter	
Treatment	(μ md $m^{-2}s^{-1}$)	(percent)	(mm)	(percent)	(mm)	(percent)
A. Scotch broom (not removed–uncut)	282.75	29	46.02	54	6.75	55
B. Scotch broom (removed–cut)	969.03	100	86.75	100	12.25	100
LSD (.05)	107.41		12.24		1.84	

*Treatments statistically analyzed by the least significant difference (LSD) method and are highly significant.

Effects of Bioherbicides on Scotch Broom Seedlings Under Greenhouse Conditions

Fusarium tumidum has a marked effect on the suppression of height growth of broom seedlings (Table 2). All three stages of growth are affected, maximum efficacy resulting at the 6 month stage when over 70% of the seedlings are killed or retarded in growth.

DISCUSSION

Scotch broom has been reported to be inhibitory to growth of conifer (*Pinus radiata*) seedlings in New Zealand (Williams, 1981) and Australia (Smith and Hosking, 1994) but this is the first report to demonstrate that Scotch broom and probably its related ally, gorse, both compete and reduce the growth of the crop species, Douglas fir in B.C. The reduction in height (46%) and in root collar diameter growth (45%) is significant, and thus, the establishment and subsequent growth of these seedlings have economic repercussions since Douglas fir is the preferred forest species in coastal B.C. Any retardation in growth in early stages is likely to prolong the rotation cycle and harvesting costs of this economic species. Because the incident radiation (PAR) falling on the seedlings is reduced by significant magnitude (71%), it is tempting to suggest that broom inhibits growth by curtailing the photosynthesis of these seedlings. However, once the

TABLE 2. Effects of spraying a formulation of *Fusarium tumidum* on growth and survival of Scotch broom at three stages of growth*

Treatment	Height growth (cm)	Percentage of control
1. One month old seedlings		
Check	4.41a	100.0
Treated	2.20b	49.8
2. Three month old seedling		
Check	15.33a	100.0
Treated	8.31b	54.3
3. Six month old seedlings		
Check	26.20a	100.0
Treated	6.90b	26.4

*Data analyzed by Duncan Multiple Range Test ($P < 0.05$). Treatments at each stage are significantly different from the corresponding checks as denoted by different letters.

conifer seedlings grow over the canopy of the broom, this limitation may decline and probably other factors such as moisture deficit, root competition and allelopathy may come into play but our data provide a strong evidence to show that Scotch broom limits the PAR in early stages of growth.

Control of broom by various means, probably using chemical herbicides, might be most cost-effective but sensitivity of early flushes of Douglas fir seedlings to glyphosate, coupled with opposition by the public to using chemicals on crown lands of B.C., must also be considered. Therefore, biological control such as use of eco-friendly bioherbicide (*Fusarium tumidum*) may provide an acceptable alternative. Certainly, this bioagent is found naturally in the forest and can be developed for large-scale treatment (Johnston and Parks, 1994). Another feature of this bioherbicide is that its chlamydospore can persist in the soil for a long time and may suppress the germination of seedlings from the seed banks. Production of a large number of seeds and survival of these seeds in the soils for 15-30 years is a cause for concern. However, this natural control with chlamydospores may eventually check the population explosion of Scotch broom (Peterson and Prasad, 1998) and arrest its further spread in forestry situations.

NOTE

1. Bond is a synthetic latex (45%) plus aliphatic oxyalkylated alcohol (10%) manufactured by Loveland Industries Inc., Colorado, USA.

REFERENCES

Johnston, P.R. and S.L. Parkes. 1994. Evaluation of the mycoherbicide potential of fungi found on Broom and Gorse in New Zealand. Proc. 47th N.Z. Plant Protection Conf. 121-124. Canterbury, NZ.

Nilsen, E., D. Karpa, H. Mooney, and C. Field. 1993. Patterns of stem photosynthesis in two invasive legumes (*Cytisus scoparius*, *Spartium junceum*) of the California coastal region. Am. J. Bot. 80: 1126-1136.

Peterson, D. and R. Prasad. (1998). The biology of Canadian weeds, *Cytisus scoparius* (L.) Link. Can. J. Pl. Sci. 78(3):479-504.

Prasad, R. 1997. Evaluation of some fungi for bioherbicidal potential against Scotch broom under greenhouse conditions. Proc. Weed Science Society of America, Chicago, Feb. 8-12. 38:46

Prasad, R. 1996. Development of bioherbicides for weed management in forestry. Proc. 2nd Intl. Weed Congress, Copenhagen. Edit. H. Brown: 1197-1203.

Smith, J. and J. Hosking. 1994. Broom in Australia. Oreg. Dept. Agric. Weed Contr. Prog., Broom/Gorse Q. 3: 1-4.

Williams, P.A. 1981. Aspects of ecology of broom (*Cytisus scoparius*) in Canterbury, NZ. N.Z.J. Bot. 32: 373-383.

Zielke, K., J. Boateng, N. Caldicott and H. Williams. 1992. Broom and Gorse: A forestry perspective problem analysis. B.C. Ministry of Forests, Queens's printer Victoria, B.C.: 20.

Pleistocene Refugia for Longleaf and Loblolly Pines

R. C. Schmidtling
V. Hipkins
E. Carroll

INTRODUCTION

Longleaf pine (*P. palustris* Mill.) and loblolly pine (*P. taeda* L.) are two species that are common to the coastal plain of the southeastern United States. The current natural range of the two species is largely overlapping. Loblolly pine occurs in 13 southeastern states (Figure 1). Longleaf pine is the more austral of the two species, occurring further south into peninsular Florida, but not occurring naturally in Oklahoma, Arkansas, Tennessee, Maryland and New Jersey (Critchfield and Little, 1966). The two species are closely related. They sometimes hybridize naturally (Chapman, 1922), and creating the artificial hybrid is not difficult if longleaf pine is the female parent (Snyder and Squillace, 1966).

Very little is known about the location of the southern pines during the Wisconsin glaciation because of the lack of macrofossils. Palyno-

R. C. Schmidtling is affiliated with the Southern Institute of Forest Genetics, 23332 Highway 67, Saucier, MS 39574 and the USDA Forest Service (E-mail: schmidtl@datasync.com).

V. Hipkins and E. Carroll are affiliated with the National Forest Genetic Electrophoresis Laboratory, Placerville, CA and the USDA Forest Service.

[Haworth co-indexing entry note]: "Pleistocene Refugia for Longleaf and Loblolly Pines." Schmidtling, R. C., V. Hipkins, and E. Carroll. Co-published simultaneously in *Journal of Sustainable Forestry* (Food Products Press, an imprint of The Haworth Press, Inc.) Vol. 10, No. 3/4, 2000, pp. 349-354; and: *Frontiers of Forest Biology: Proceedings of the 1998 Joint Meeting of the North American Forest Biology Workshop and the Western Forest Genetics Association* (ed: Alan K. Mitchell et al.) Food Products Press, an imprint of The Haworth Press, Inc., 2000, pp. 349-354. Single or multiple copies of this article are available for a fee from The Haworth Document Delivery Service [1-800-342-9678, 9:00 a.m. - 5:00 p.m. (EST). E-mail address: getinfo@haworthpressinc.com].

FIGURE 1. Natural range of loblolly pine showing proportion of trees with high concentrations of cortical limonene.

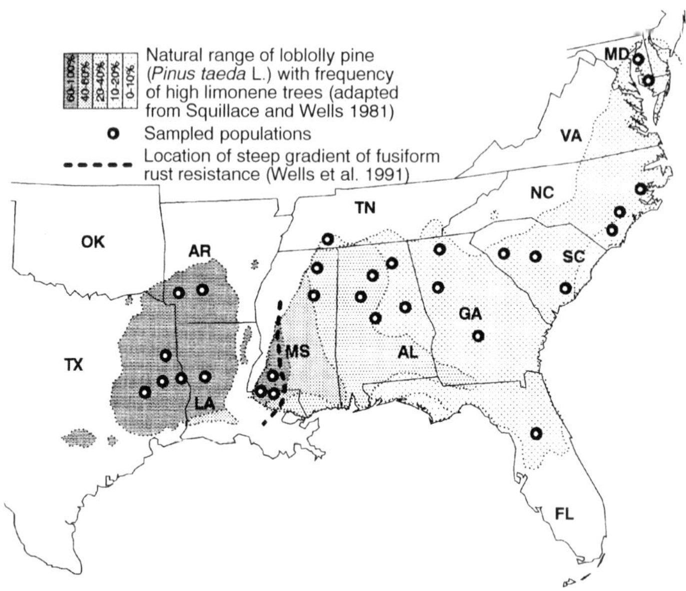

logical records are not conclusive, due to the difficulty in identifying pine pollen to the species level. Macrofossils of spruce (*Picea* sp.) dating from the Pleistocene have been found within the current range of both longleaf and loblolly pine in southern Louisiana, central Georgia and southern North Carolina, indicating that the climate was considerably colder at that time (Watts, 1983). It is reasonable to assume that longleaf and loblolly pine were situated south of their present range during the Pleistocene. The exact location of these refugia is a matter of some speculation (Wells et al., 1991), but considering the relatively uniform land form of the lower Coastal Plain of the southeastern United States, there appears to be only two possibilities; south Florida/Caribbean or south Texas/northeast Mexico.

Genetic data can be used to infer location of glacial refugia and migration patterns (Wheeler and Guries, 1982). The present study examines geographic patterns of allozyme variation in longleaf and loblolly pines to provide evidence for the location of Pleistocene refugia for the two species.

MATERIALS AND METHODS

Allozyme frequencies were studied in range-wide collections from 23 populations of longleaf pine (Schmidtling and Hipkins, 1998) and 33 populations of loblolly pine (Figure 1). The longleaf pine data were from megagametophytes of approximately 30 individual trees per population, and included three seed orchard sources and an old-growth stand. In loblolly pine, three different sets of range-wide collections were used; nine seed orchard populations averaging 46 clones per source, 14 populations from the Southwide Southern Pine Seed Source Study (Wells and Wakeley, 1966) averaging 46 megagametophytes per source, and 10 bulk seed collections, averaging 66 embryos per source.

In both species, gel electrophoresis was used to resolve enzyme systems phosphoglucose isomerase (PGI), fluorescent esterase (FEST), malic enzyme (ME), aconitase (ACO), phosphoglucomutase (PGM). 6-phosphogluconate dehydrogenase (6PGD), glutamic oxaloacetate transaminase (GOT), leucine aminopeptidase (LAP), isocitrate dehydrogenase (IDH), and malate dehydrogenase (MDH), for a total of 10 enzyme systems and 16 loci. In longleaf, an additional 4 enzyme systems and 6 loci were assayed; alcohol dehydrogenase (ADH), triosephosphate isomerase (TPI), glycerate-2-dehydrogenase (GLYDH), and glucose-6-phosphate dehydrogenase (G6PD).

Allozyme data were used to provide several estimates of genetic variation using BIOSYS I (Swofford and Selander, 1989), including mean number of alleles per polymorphic loci (N_a), percent loci polymorphic (P_l, 95% criterion), observed heterozygosity (H_o), and expected heterozygosity (H_e), as well as the standard measures of genetic differentiation, the F statistics FIS, F_{IT} and F_{ST}.

Additional measures of diversity computed were (N_r) number of rare alleles per tree (one that occurs at a frequency of 0.05 or less in the overall population) and the frequency of the most common allele (A_c) averaged over all loci. A "diversity index" was computed by taking the mean of the standardized scores for N_a, P_l, H_e, N_r, and $A_c{}^{-1}$ for each population. Each measure of genetic diversity is first standardized by subtracting the mean and dividing by the standard deviation, and the "index" is the unweighted mean of the five scores.

RESULTS AND DISCUSSION

F statistics for both species indicated very little inbreeding with F_{IS} and F_{IT} near zero or slightly negative. F_{ST} was between 0.03 and 0.04, indicating that differences among populations accounted for between 3 and 4 percent of the total variation in both species. There was little difference in allele frequencies due to origin of the populations, i.e., orchard versus wild populations, for loblolly (Schmidtling et al., 1994) or longleaf (Schmidtling and Hipkins, 1998).

In longleaf pine, there was a linear decrease in allozyme variation from west to east. Correlations of longitude of the seed source with N_a, H_o, P_l, N_r, and A_c^{-1} were r = 0.604, 0.787, 0.718, 0.549 and 0.728, respectively, which are all significant at the 0.01 level. The best correlation was between longitude and the composite function, the allozyme variation index (Figure 2A).

In loblolly pine, there was no east-west trend in allozyme variation, and there appeared to be a tendency for more variation in the central part of the natural range (Figure 2B).

Provenance tests have shown that substantial geographic variation in growth, survival, and disease susceptibility exists in loblolly pine as well as in longleaf pine (Wells and Wakeley, 1966, 1970). Growth is generally related to latitude or temperature at the seed source (Schmidtling, 1997). Geographic variation in both species parallels that of other forest tree species; seed sources from warmer climates grow faster than those from colder climates, if these sources are not transferred to very different climates.

East-west variation in adaptive traits such as growth, disease re-

FIGURE 2. Plot of allozyme variation index vs. longitude of the seed source for range-wide collections of (A) longleaf pine and (B) loblolly pine.

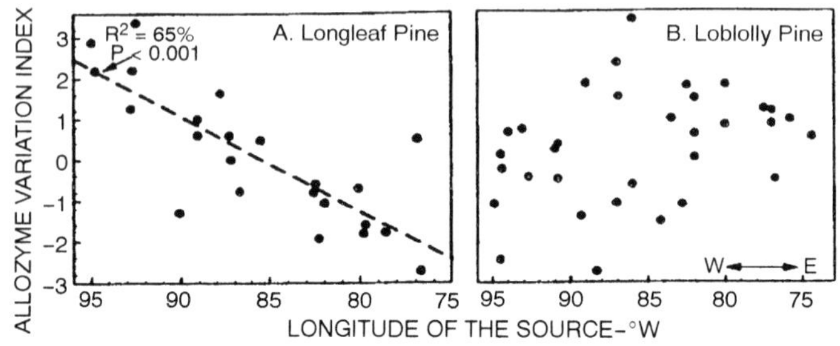

sistance and survival is minimal in longleaf pine, but is quite important in loblolly pine (Wells and Wakeley, 1966, 1970). In loblolly pine, western sources are slower growing, survive better, have greater resistance to fusiform rust (*Cronartium quercuum* [Berk.] Miyabe ex Shirai f.sp. *fusiforme*), and have greater concentrations of limonene in cortical gum than eastern sources (Squillace and Wells 1981, Figure 1). The isolating effect of the pineless Mississippi River Valley has been proposed as the reason for these differences (Wells and Wakeley, 1966), but if geographic variation in limonene concentration and fusiform rust resistance are considered (Figure 1) the division between western and eastern sources appears to be east of the Mississippi River.

CONCLUSIONS

It is proposed that the continuous linear decrease in allozyme variation in longleaf pine from west to east is a result of migration from a single refugium in the west (south Texas or northeast Mexico) after the Pleistocene, with a loss of variability due to stochastic events during migration. The lack of such a trend in allozymes in loblolly pine, coupled with the distinct east versus west variation in fusiform rust resistance and other adaptive traits suggest that loblolly pine was located in two refugia during the Pleistocene: in Texas/Mexico and Florida/Caribbean, as proposed by Wells et al. (1991).

The dashed line east of the Mississippi, in southeast Louisiana and west Mississippi (Figure 1) shows the location of a very steep gradient in fusiform rust resistance which can best be explained by assuming the confluence of two populations (Wells et al., 1991). This, as well as the steep gradient in terpene concentration in the same location suggests the merging and mixing of the two populations after the retreat of the Wisconsin glaciation.

Seed movement guidelines should take into account the differences among the two species. It does not appear that east-west movement of longleaf seed sources need be restricted, especially if seed of the more diverse western populations are moved eastward. More caution should be used in east-west movement of loblolly pine, since there appear to be two different populations.

REFERENCES

Chapman, H.H. 1922. A new hybrid pine (*Pinus palustris* × *Pinus taeda*). J. Forestry 20:729-734.

Critchfield, W.B. and E.L. Little, Jr. 1966. Geographic distribution of the pines of the world. Misc. Pub. 991. USDA Forest Service, Washington, DC.

Schmidtling, R.C. 1997. Using provenance tests to predict response to climatic change. Chapter 27 In: Ecological Issues and Environmental Impact Assessment, pp. 621-642. Gulf Publishing Co., Houston, TX.

Schmidtling, R.C. and V. Hipkins. 1998. Genetic diversity in longleaf pine (*Pinus palustris* Mill.): Influence of historical and prehistorical events. Can. J. For. Res. 28:1135-1145.

Schmidtling, R.C., B. Carroll and T. LaFarge. 1994. Genetic diversity of selected loblolly pine populations versus natural populations. In: Proc. 13th N. Amer. For. Biol. Workshop, Baton Rouge, LA. June 1994 (abs.) p. 66.

Snyder, E.B. and A.E. Squillace. 1966. Cone and seed yields from controlled breeding of southern pines. Research Paper SO-22. USDA Forest Service, Southern For. Exp. Sta., New Orleans, LA, 7 pp.

Squillace, A.E. and O.O. Wells. 1981. Geographic variation of monoterpenes in cortical oleoresin of loblolly pine. Silvae Genetica 30: 127-135.

Swofford, D.L. and Selander, R.B. 1989. BIOSYS-1, A computer program for the analysis of allelic variation in population genetics and biochemical systematics. Release 1.7. Illinois Natural History Survey, Champaign, IL.

Watts, W.A. 1983. A vegetational history of the eastern United States 25,000 to 10,000 years ago. In: S.C. Porter (ed.) The Late Pleistocene. Late-Quaternary Environments of the United States, pp. 294-310. University of Minnesota Press, Minneapolis.

Wells, O.O., G.L. Switzer and R.C. Schmidtling. 1991. Geographic variation in Mississippi loblolly pine and sweetgum. Silvae Genetica 40: 105-118.

Wells, O.O. and P.C. Wakeley. 1966. Geographic variation in survival, growth, and fusiform rust infection of planted loblolly pine. Forest Sci. Monograph 11. 40 p.

Wells, O.O. and P.C. Wakeley. 1970. Variation in longleaf pine from several geographic sources. Forest Sci. 16: 28-45.

Wheeler, N.C. and R.P. Guries. 1982. Biogeography of lodgepole pine. Can. J. Bot. 60: 1805-1814.

Predator Guild Structure of Canopy Arthropods from a Montane Forest on Vancouver Island, British Columbia

Neville N. Winchester
Laura L. Fagan

INTRODUCTION

Conservation of Biodiversity in Northern Temperate Rainforests

In temperate zones some of the last remaining tracts of intact ancient coniferous forests occur in the Pacific Northwest of North America (Franklin, 1988) and the "coastal temperate rainforest" of British Columbia represent approximately 25% of the worldwide coastal temperate rainforests (Kellog, 1992). The ongoing fragmentation of these landscapes has heightened the awareness for a need to understand/determine the endemic fauna and flora (Scudder, 1994) and apply sys-

Neville N. Winchester and Laura L. Fagan are affiliated with the Biology Department, University of Victoria, P.O. Box 3020, Victoria, B.C. V8W 3N5, Canada.

The authors thank K. Zolbord, K. Jordan and K. Nelson for assistance in several aspects of this project.

Funding for this project was made possible with financial support from Forest Renewal B.C. grants to N.N. Winchester from S. McNay, Research Branch, Ministry of Forests.

[Haworth co-indexing entry note]: "Predator Guild Structure of Canopy Arthropods from a Montane Forest on Vancouver Island, British Columbia." Winchester, Neville N., and Laura L. Fagan. Co-published simultaneously in *Journal of Sustainable Forestry* (Food Products Press, an imprint of The Haworth Press, Inc.) Vol. 10, No. 3/4, 2000, pp. 355-361; and: *Frontiers of Forest Biology: Proceedings of the 1998 Joint Meeting of the North American Forest Biology Workshop and the Western Forest Genetics Association* (ed: Alan K. Mitchell et al.) Food Products Press, an imprint of The Haworth Press, Inc., 2000, pp. 355-361. Single or multiple copies of this article are available for a fee from The Haworth Document Delivery Service [1-800-342-9678, 9:00 a.m. - 5:00 p.m. (EST). E-mail address: getinfo@haworthpressinc.com].

tem-based conservation approaches across a wide range of forest types.

Arthropods in Northern Temperate Rainforests

Arthropods, primarily insects, are an integral part of most ancient forests and may comprise 80-90% of the total species in these systems (Asquith et al., 1990). The study of forest canopies in determining the structure of arthropod assemblages and the systematics of canopy arthropods has increased rapidly during the last 20 years (Stork and Best, 1994; Stork et al., 1997). Canopies of natural forests in temperate regions contain largely undescribed and little understood assemblages of arthropods that have expanded estimates of the total number of insect-arthropod species (Schowalter, 1989; Winchester and Ring, 1996a, b; Behan Pelletier and Winchester, 1998; Winchester 1997a, b). In this note we explore, for the first time, the arthropod predator guild from a montane forest and ask the following questions:

1. Does the predator guild dominate the arthropod guild structure in *Abies amabilis* and *Tsuga heteropylla*?
2. Is there a significant effect of sampling date, sampling site (elevation) and crown height on mean predator population intensities?

MATERIALS AND METHODS

Study Area

The study area is located in the Mt. Cain area (50°13′N, 126°18′W), near the community of Woss, on the northern part of Vancouver Island, British Columbia, Canada. The montane study area ranges in elevation from 700 to 1,100 metres and encompasses an area of approximately 400 square km, spanning between the Cain and Maquilla peaks. Samples sites include 5 transects that range in elevation between 730 and 1,059 metres within the Montane Very Wet Maritime Coastal Western Hemlock variant (CWHvm2) and the Windward Moist Maritime Mountain Hemlock variant (MHmm1). Both variants are characterized by short, cool, moist summers and long, wet, cold winters with a deep snowpack (e.g., mean annual precipitation in CWHvm2 is 2,787 mm; mean annual snowpack in MHmm1 is 820 cm; Meidinger and Pojar, 1991).

Dominant tree species in the CWHvm2 are western hemlock (*Tsuga heterophylla*), mountain hemlock (*T. mertensiana*) and western red cedar (*Thuja plicate*), interspersed with varying numbers of amabilis fir (*Abies amabilis*). Overstory trees range in age from 200 to 1,000 years. Drier sites at lower elevations support Douglas fir (*Pseudotsuga menziesii*), while yellow-cedar (*Chamaecyparis nootkatensis*) is present in higher, moister elevations.

Sampling

A branch clipping program was conducted in single *Abies amabilis* and *Tsuga heteropylla* on 5 transects (ranging in elevation between 700 and 1,200 m). The sampling procedure was modified after Winchester (1997a). In each tree, 3 samples were taken at random from each of 2 heights (low and high crown). A total of 60 branch samples were collected for each of 3 samples periods, early June, late July, and late August. All insects were remove from each sample and prepared for identification. The total number of arthropods sorted from the branch clipping samples was 1,662.

Statistical Analysis

For the purposes of this note, we examined trends in the predator guild (excluding all microarthropods, such as mites) using total number of individuals. We used analysis of variance (ANOVA) to examine the effects of transect, sampling date and sampling height on mean number of predators per kilogram of dry plant material (population intensities) within each tree species. Because only 1 tree of each species was sampled on each transect, date, height, and the interaction of date and height were nested within transect in our ANOVA model. Data were square root transformed (square root of number of predators per kg + 0.5) prior to analysis.

RESULTS

Canopy Guild Composition

Table 1 summarizes mean population intensities for arthropods in each guild by tree species and with all factors pooled. The canopy

TABLE 1. Mean number of individuals per kg dry plant material (S.D. = standard deviation) in each guild for each tree species (n = 90) and with all factors pooled (Total, n = 180). Percent is calculated by dividing the mean number of individuals/kg in each guild by the total of the mean number of individuals/kg in all guilds (*100).

	Abies amabilis	*Tsuga heteropylla*	Total	
Guild	Mean no./kg (S.D.)	Mean no./kg (S.D.)	Mean no./kg (S.D.)	%
Predators	15.84 (17.75)	14.98 (26.68)	15.41 (22.60)	36.8
Phytophages	3.26 (8.44)	8.56 (16.78)	5.91 (13.51)	14.1
Parasitoids	2.05 (8.52)	0.23 (1.25)	1.14 (6.14)	2.7
Epiphytes	0	0	0	0
Scavengers	4.77 (11.25)	4.38 (11.29)	4.58 (11.24)	10.9
Tourists	19.77 (149.68)	10.03 (58.35)	14.90 (113.38)	35.5

arthropod fauna in this study is dominated by the predator (36.8%) guild, and there appears to be remarkable consistency in the population intensities of the predator guild among the two tree species.

Within hemlock (*Tsuga heteropylla*) trees, we found no significant effects of sampling date, height, or the interaction of date and height on mean predator population intensities. The effect of transect on mean predator population intensities in hemlock trees approached significance [P ($F_{4,60}$(2.44) = 0.056].

In fir (*Abies amabilis*) trees, there was a significant interaction between sampling date and height [P ($F_{10,60}$(2.38) = 0.019], and a significant effect of height [P ($F_{5,60}$(2.45) = 0.044] on mean predator population intensities. There were also significant differences in mean predator population intensities between transects [P ($F_{4,60}$(3.39) = 0.014].

DISCUSSION

The resident canopy arthropod fauna in this study is dominated by individuals in the predator guild, supporting previous studies in coniferous forests (see Winchester, 1997a). Predator proportions are higher in our study than those reported by Winchester (1997a) but are in agreement with the general observation of maintenance of a high predator loading in a structurally and functionally diverse forest ecosystem such as that found at Mt. Cain. The predator guild is composed primarily of 30 arachnid species from 15 families and is similar to numbers reported from Vancouver Island canopy sites by Winchester (1997a, 1998) and supports reports of the numerical dominance of spiders in forest ecosystems (Nielson, 1975; Bassett, 1991).

Mean population intensities of spiders exhibit uniformity between both tree species although some arachnid species do not seem able to utilize the entire elevational range of the montane forest. This distribution may be the result of a myriad of factors that are coupled with the physical characteristics of the sample site. Temporal sequencing was not apparent and supports observations by Winchester (1997a, 1998) where canopy spiders have been shown to be poorly synchronized with herbivore accumulations and may be able to wait or switch prey items based on availability. Input from the forest floor and adjacent riparian zones may provide a 'pulse' source to canopy spiders during times of low numbers of resident herbivores, thus providing opportunities to maintain a high proportion of predators throughout the growing season.

Where differences in mean population intensities of predators in *Abies amabilis* are evident, crown height was an important factor shaping predator guilds. We suggest that habitat characteristics of *A. amabilis* provide structural features that play a key role in determining mean predator population intensities. Individuals in the predator guild exhibit non-uniformity between crown height, time and elevation, which may be a result of a myriad of factors that are coupled with the physical characteristics of the tree. Factors may include plant chemistry, plant architecture and plant health (see Winchester, 1998).

CONCLUSION

Patterns of community structure on coniferous trees in the montane, examined at the guild level, indicate that in terms of number of individuals, the predator guild is numerically dominant. The high proportion of predators indicates that herbivory in these mature, structurally complex forests may be relatively insignificant. Predator guild population intensities are not affected by time, crown height, and elevation in *Tsuga heteropylla*. This is not the case, however, in *Abies amabilis*. The summarizing of these key patterns and documentation of changes across the elevational gradients presented in this montane forest should identify ecological roles of arthropods. Knowledge of the predators in this forest ecosystem will help provide an understanding of the diversity, habitat requirements and system processes that occur in these forests. It is possible that this information will provide a framework for the long-term management and retention of biodiversity in montane forests on Vancouver Island.

REFERENCES

Asquith, A., J.D. Lattin and A.R. Moldenke. 1990. Arthropods, the invisible diversity. Northwest Environ. J. 6: 404-405.

Basset Y. 1991. The seasonality of arboreal arthropods foraging within an Australian rainforest tree. Ecol. Entomol. 16: 265-278.

Behan-Pelletier, V.M. and N.N. Winchester 1998. Arboreal oribatid mite diversity: colonizing the canopy. Applied Soil Ecology. 9:45-51.

Franklin, J.F. 1988. Structural and functional diversity in temperate forests. *In* pp. 166-175. E.O. Wilson (ed.). Biodiversity. National Academy Press, Washington, D.C.

Kellog, E. (ed.) 1992. Coastal Temperate Rainforests: Ecological Characteristics, Status and Distribution Worldwide. Occasional Paper Series No. 1, Ecotrust and Conservation International, Portland, OR.

Meidenger, D. and J. Pojar. 1991. Ecosystems of British Columbia. B.C. Ministry of Forests, Victoria, BC.

Nielsen, B.O. 1975. The species composition and community structure of the beech canopy fauna in Denmark. Videnskabelige Medelelser fra Dansk Naturhistorisk Forening 138: 137-170.

Schowalter, T.D. 1989. Canopy arthropod community structure and herbivory in old-growth and regenerating forests in western Oregon. Can. J. For. Res. 19: 318-322.

Scudder, G.G.E. 1994. An annotated systematic list of the potentially rare and endangered freshwater and terrestrial invertebrates in British Columbia, J. Entomol. Soc. Brit. Columbia, Occasional paper 2.

Stork, N.E. and V. Best. 1994. European Science Foundation–results of a survey of European canopy research in the tropics.

Stork, N.E., J. Adis and R.K. Didhan (eds.) 1997. Canopy Arthropods. Chapman and Hall, London.

Winchester, N.N. 1997a. Canopy arthropods of coastal Sitka spruce forests on Vancouver Island, British Columbia, Canada. *In* pp. 151-168. N.E. Stork, J.A. Adis, and R.K. Didham (eds.). Canopy Arthropods. Chapman and Hall, London.

Winchester, N.N. 1997b. Arthropods of coastal old-growth Sitka spruce forests: Conservation of biodiversity with special reference to the Staphylinidae. *In* pp. 363-376. A.D. Watt, N.E. Stork, M.D. Hunter (eds.). Forests and Insects. Chapman and Hall, London.

Winchester, N.N. 1998. Conservation of Biodiversity: Guilds, Microhabitat Use and Dispersal of Canopy Arthropods in the Ancient Sitka Spruce Forests of the Carmanah Valley, Vancouver Island, British Columbia. Ph.D. Thesis. University Victoria.

Winchester, N.N., and R.A. Ring 1996a. Centinelan extinctiona: extirpation of Northern temperate old-growth rainforest arthropod communities. Selbyana 17 (1): 50-57.

Winchester, N.N. and R.A. Ring 1996b. Arthropod diversity of a coastal Sitka spruce forest on Vancouver Island, British Columbia, Canada. Northwest Science 70 (special issue): 94-103.

Index